ARMAGEDDON

STONES

DAVID SOLOMON

The Dead Saints Chronicles
Book III

Publisher Information:
Library of Congress Cataloging-in-Publication Data
Solomon, David
The Armageddon Stones/The Dead Saints Chronicles III/David Solomon
p. cm.
Includes bibliographical references and index.
ISBN: 978-0-9972454-4-8
1. Spiritual formation 2. Prophecy. Title ISBN identifies
Printed in the United States of America
First Edition

Armageddon Stones

The Dead Saints Chronicles
Volume III

Planetary Phenomena, Prophecies, Myth, and the Future

Author
David Solomon
Publisher/Contributor/Cover Design by
Delynn Solomon

Other books by David Ben Solomon

Dead Saints Chronicles: A Zen Journey Through the Christian Afterlife – w John Anthony West

Training Wires of the Soul – with Delynn Solomon

www.dsmediapub.com

CONTENTS

THIS BOOK IS DEDICATED TO THE AUTHOR, DAVID BEN SOLOMON
MY TEACHER, MY LOVER, MY FRIEND

DAVID BEN SOLOMON
January 23, 1959 – 3:10 a.m. April 18, 2016

David began his journey with this book 25 years ago. He had hoped, potentially with a co-author, it would be published shortly after his passing. Yet the only files under Armageddon Stones were brief summaries of his work. I was sure he had written more. Five years later, while working on Training Wires of the Soul *I opened a file with an obscure name, and it was his draft manuscript. It has taken another two years to bring it to you.*

Seven years after his passing, I send it off today, July 13, 2023, to prepare for print. I know the work is less complete than David would like yet I would not doubt that he himself would still be editing it today. Every day new (or ancient) information is brought forth that sheds light on his theories, connects dots, and even could have sent his research in a different direction. Yet what he does share is a plethora of research, theories, ideas and information worthy of consideration for those who are on their quest.

Without knowledge of publishing, editing, money, or the topic itself I had committed to David that I would bring his baby to light. Though he started this project long before he would be on an actual "deadline" he felt leaving his books would be a legacy to his children and a guide to anyone who may be searching.

David knew how to dream big, think big, live large and make things happen. I knew I could not replace him. Trying to get into his brain, pick up where he left off or even begin to understand all the concepts, it has taken me this time to get it done. As you journey through his book, please have grace as I have edited this book to the best of my knowledge.

I have been and continue to be shaped, cultivated, and changed by my journey with David. Receiving God nods and breadcrumbs, encouragement of friends, family and even strangers along the way has made it possible for me to bring this to you, from David with love.

Delynn Solomon

Introduction

March 3, 2016

On Jim Sinclair's website, the top of the page posted this quote on January 30, 2015, at 10:34 am. "There are two mistakes one can make along the path of truth –not going all the way and not starting. ~**Buddha** ...Two hours later, by chance, I downloaded a book on Kindle: The Fall of a Thousand Suns: Comets, Meteors in History. I opened the first page, and on the top was the quote. "There are two mistakes one can make along the path of truth—not going all the way and not starting. ~**Buddha**

I had never seen this quote before, much less twice within a few hours on the same day. So what are the odds of reading Buddha's quote twice, two hours apart, from two very different sources? God's same-day coincidences often happen in 2's. Another author, Michael Flipp, wrote an entire book about the coincidence of **2's** called '2! Signs and Coincidences from God.'[1] The Divine message? Use the comet impact book and economic collapse prophecies of Jim Sinclair in Vol. III of the Dead Saints Chronicles, the Armageddon Stones ~**Chronicle 596**

Such are a few of the strange beginnings in the writing and completion of the *Armageddon Stones*.

My initial epiphany with the book was inspired by Luis Alvarez's discovery of the iridium layer and his theory that dinosaurs' mass extinction was caused by an asteroid or comet colliding with Earth. His theory was confirmed two years after he died.

It was the impact of a six-mile-wide asteroid leaving the massive crater at Chicxulub off the Yucatan Peninsula.

An x marks the spot smoking gun that creates enough heat, ash, and darkness to kill off the dinosaurs. I just knew from the statement that innumerable meteors in the comet would cause the end of the world spoken of by John in the Book of Revelation. The comet's impact destroyed civilizations the world over. The poles shifted and caused the Great Flood. The Bible stories were true. We had geological proof.

It was 1988. I was about to marry my first wife, Emiko. I began to learn all I could about Comet Lore. I read Zacharia Sitchen's *Earth Chronicles* about Planet Nibiru, a rogue planet that triggered the Pole Shift. By 1991, I had met John Anthony West and wound-up part of the exciting announcement that hit front-page headlines worldwide. I inspired Tom Kay to write the book *When the Comet Runs* and later

---

1 Michael Flipp 2012. 2! Signs and Coincidences from God.

walked beside Dr. Schoch, associate professor of Natural Sciences at Boston University, having an opportunity to talk about my Armageddon Stone theories. He and John Anthony West worked together on the Mystery of the Sphinx documentary. I had videos dating back to 1992 discussing my theories. All laid out nicely. I was looking forward to the recurring period of the Armageddon Stones' return. The period of the eternal return spoken of by all religions, by the prophets, and in the Book of Revelation. I believed the keys to deciphering the future were all there.

It was a Galileo astronomy solution. There was a need to learn Greek and Hebrew, astronomy, and paleontology. In 1992 I visited Victor Clube during a stay in England, which allowed me to talk to him about his ideas about comet Encke and the 1908 Tunguska explosion and his belief that we are headed for a future of many of these types of bombardments. I had a chance to visit historian Daniel Warner, who studied depopulation anomalies since 11,000 B.C. and how fossils are created, which require silt and water to cover decaying plant and animal life.

Shortly after Paul Solomon died in 1994, I dreamt about being in heaven with him and he directed me to finish writing the *Last Judgment*, a title I later changed to the *Armageddon Stones*. I had no idea it would evolve into the last of the Dead Saints Chronicles Trilogy (changed name of the trilogy, which would be completed only at the very last moments of my life. Or at least that is what I thought.

In 1994, I also received a $40,000 grant to do more research. By then, I was in my second marriage. I spent my time, before the internet days, running to libraries near VB and the Library of Congress. It is frustrating to write a book requiring so much research, before the internet age. After a year, I ran out of funds. The elusive recurring impacts theory I never found. At first, I thought it was 3600 years, but while we see impacts that could be grouped in periods of 1500 years, but still there was no hard evidence. I didn't find the key: Bellamy, Velikovsky, numerous white papers. Venus could not be the answer. It was not a comet. Its orbit has been stable and circular for millions of years. Bellamy thought it was the Moon. Is it a meteor stream? I narrowed it down to four meteor showers. The Leonids. Beta Taurids. Orionids. And Draconids. All based on mythology, numbers and historical impact evidence. And numbers. Before the comet chronicles, I became a scholar on Prophecy. That's why I was led to Nostradamus. Mythology, Jewish, Cayce, Solomon, Malachi, Mother Shipton. If you compare their prophecies side by side, common denominators jump out. Where was the master key? I studied Mythology as a common thread. It did not have the filters of religion. There was no doubt, the ancient mythologies passed down to religion, believed the clock to Doomsday was counting down to zero. I didn't want to be another date setter. And then

I had to put the book away. The birth of my kids. I was building a company...for 15 years. Sell the company. Begin Near-death research in in July 2011. The Near-death lighting experience. The Dead Saints Epiphany, then on June13, 2013.

Cancer.

Then the Race. The Armageddon Stones were still the last thing on my list. The Zen Journey was of most importance. JAW's writing assistance and commitment in 2014.

The guidance on October 29th, 2014, from my decease Bonsai sensei to split the Chronicles in two. The November 13th dream of Jesus spreading grass seed. And his appearance in a dream 50 days later on January 1, 2015, a Jubilee year, a Holy Year, beginning the Schmeda; a year dedicated to writing and forgiveness of debt.

Then the dream of Jesus January 2, 2016, a day after the Schmeda year ended. I still wasn't sure if I would have time to complete Armageddon Stones. At one point in January as we prepared Zen Journey for print, I removed the mention of the Armageddon Stones from both Zen and Training wires. I would not have the time. I had 100 files I had moved from 1.44 floppies I had transferred to to DVD in 2003, Thank God.

The Research was still there. It was the book Akito and everyone wanted. I always said, it would be the most interesting, but not the most important. What was unique in 1991 were now movies like *Deep Impact* and *Armageddon. Many* books have come out about comet impacts and likely comet impacts that would happen in our future.

Then I had the coincidence of ordering *The Fall of a 1000 Suns*, about Halley's Comet. And the Buddha quote on the first page, and on Jim Sinclair's JS Mineset website mentioned above. Sometime in mid-January, I told Delynn I am finished with Training Wires. All we need to do is add Chronicle notes and your assignment to write chapters about your walk with my cancer after I am gone.

I need to begin an attempt to complete the Armageddon Stones. I have to try. Not yet David. We still have to bring Zen Journey to print. There's a lot still to do. Delynn was right. More than I thought. There is a lot to do. But I felt differently. I could do both. It seemed impossible, but I felt like the camel of the impossible. I could make the desert journey. God still had the waters life in me.

It was early February 2016. I woke up at 4:00 am and assessed my old files. There were 20 years of comet impact research to update. The Chelyabinsk February 2013 explosion over Russia was curiously dismissed. Then several near-Earth orbit near-misses reported. Then the 2004 discovery of Apophis. A statement by Victor Cube that we were entering 400 years of meteor and comet bombardment.

I was running out of time. I knew it.

How could I possibly complete the Armageddon Stones in 60 days? I was physically declining weekly, now almost daily. The first week of February 2016 passed. The second week passed. March came, and we sent the final proof to the printer. I finished the back cover finally. Media articles were written. But finally, I was done with *A Zen Journey through the Christian Afterlife*. Could I now - please get to work on the Armageddon Stones?

Could I be the camel of the impossible and complete the work I was commissioned to do in 1994? I promised Akito that I would. Emiko would find him in Japan. We would find a translator and get him a copy.

Then a string of extraordinary miracles began to occur. I felt like Samuel, who was being fed by Ravens in his cave. I discovered the Prayer which ended Training Wires. I wrestled with an Angel. Yes. Lord, I will do it. And if necessary, I will do it on my own. So, the race was on to finish the Trilogy. I began 18-hour days of typing, beginning on February 25, 2016. I created a new chapter outline; created titles, and then began dropping rough word files into them. Some were completely written, but many were an absolute mess, with a dozen confusing iterations. It was maddening. I didn't have files properly identified. Many quotes need paraphrasing or referencing. But by March 5, 2016, I had 26 chapters defined, titled, and thought through than ever before. I had a book.

My commitment to move forward to finish the Armageddon Stones invoked miracles. God sending ravens and doves to feed me information and understanding. I put the dots together quickly. I didn't have to work at it. A friend shows up unannounced on Thursday, February 2016 after 39-years years, which helped clarify a dream my daughter Angela had about the future in 2028, adding facts she didn't think to tell me about nearly three years before, 34 days after my cancer was diagnosed. It helped connect dots, just as Paul Solomon told me I could and would do. Integrating many Dead Saints prophecies and other ancient prophecies were the key. Then James Yax shows up unannounced three days later on Sunday morning and shared a dream he had on his birthday February 8, 1972, about the Apostle Paul, Saint Peter and Jesus, a story I will tell you about later in the book.

I had *the Armageddon Stones* open to the Pole Shift chapter, when he told me another dream he had in 1977 about a Future Pole shift and how many would be saved from the resulting Flood of the Apocalypse, again a dream that put dots together in a remarkable web of intrigue. Why now?

God sent the previous flood on the 17th day. I was his 17th man. This warning or preparation was my job to complete. If so, I had very little time do it, At least I could make sure the important message was given. Whatever I have time to finish, I will finish.

It's almost like a scene from *Close Encounters of the Third Kind*. Instead of throwing mud into my living room and building the Devil's Tower UFO landing site, I was writing the same thing, but not from any exploratory mission or saving a few evolved children from a Solar Flare, but the rescue of many millions or more from the floods of a Pole Shift. A rapture? A rescue?

Previously, I had never paid attention to Islamic prophecies. It was only in 2014, that a friend lent to me the book, *The First Muslim*, where I read the story of Mohammed's life. I was quite surprised that the man I read about was quite different than the man and the Prophet portrayed by Christianity and Judaism. The Holy Imam's, priests of Islam, passed down prophecies that focused the ending of a 1500 year-period that started with the death of Mohammed in 622 and a period between 1979 and 2069 when it was believed the 12th Madhi would appear on Judgment Day. The Christian parallels were striking. Were both religions describing the same Day and the same event?

I did not find this out on my own. By accident, I found the book, *From Adam to Apophis*, by Nicola Costa, who did research in End Time Prophecies of Mohammed, prophecies I would never have been privy to. I discovered the Qur'an was written over a period of 22 years. I discovered today was the 22nd year since Paul Solomon died. Today, March 5 was 22 days until my dream of March 27, 2016. Easter. There are 22 paths on the Tree of Life.

Another mystery of the 82 was solved yesterday. It was the average days the Comet Halley shines in the heavens when it appears every 76 years. It is also sura 82 describing the Day of Judgment.

Few know that Mohammed, born in 570 A.D., was born under the Year of the Elephant, when flaming fire fell from the skies, destroying hundreds of them. A myth? (described) The day with the 12th Madhi descends from the sky to cleave the atmosphere. It mirrors the Return of One like the son of Man who cleaves the sky. Madhi is not a Man. The Son of Man is not a Man. It is a blinding light, a serpent of fire, the Ancient of Days that will cause the killing of 75 to 90 percent of the world in a single day. It is the central story of the Revelation of Saint John the Divine.

I went back to my files and reviewed all the math. All the Galileo codes. Measure the Temple within and without. Everything became clear. It was there in front of me all the time. The original flood was recorded on the 17th day of the month. That day equals the first week of November of the Gregorian calendar the week of All Hollows Day, Halloween. Knight Jazdnick discovered it was a VERY ancient holiday that remembered the mass extermination of life on Earth. It is still celebrated today. We were on the precipice of that day happening again. How do I know? Twenty-five prophecies of the Dead Saints. Their prophecies connect ancient dots, and point to a period soon to come, events that will unfold rapidly over the next 20 years or less. A convergence of time. A dangerous time. It was a time of Apocalypse and the Great War of Armageddon, but also of rapture and rescue, of Light, conquering Darkness, not only on Earth, but in Heaven.

Thank you to Flem and Rose Flem-Ath, authors of *When the Sky Fell*, *In Search of Atlantis*, who by chance had their Afterword written by John Anthony West. Why did I pull his book off my library shelf? He had mythology research I needed. Thank you, Dick Dingus, for my Hospice Pastor, who will be on watch and my assistant to find files I cannot get to or lift. He is the on-call minister for Heartland Hospice.

I have hundreds of details to wrap up. Dick will help ensure I find everything I need. Any final edits or paraphrasing will be done later, but the Armageddon Story is complete. So, with that I leave you to read 40 years of research, made possible to be put to the electronic word during the last 40 days of my life. May it be a blessing and a guide and bring you faith in the difficult years of tribulation ahead, the greatest in all of history. We will see our Lord Jesus Christ come again. Every eye shall see him by men and women and children of every Faith. When all is done, after years of darkness and pain, we will see the dawning of a new world without borders and a new people who have turned their swords in to plowshares led by a child, the Christ.

Part I

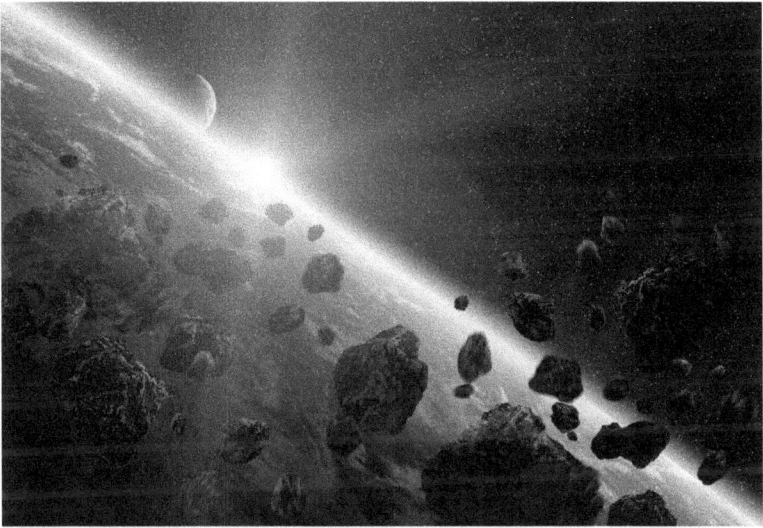

Armageddon Genesis

1

The Armageddon Stones

Then will the Earth no longer stand unshaken, and the sea will bear no ships; heaven will not support the stars in their orbits, nor will the stars pursue their constant courses in heaven. The fruits of the Earth will rot, the soil will turn barren, and the very air will sicken in sullen stagnation. [Then God] will cleanse the world from evil, now washing it away with water floods, burning with fiercest fire. Hermes

High above Mecca, two constellations, Orion and Leo, shine menacingly over the night sky of the Arabian Nation of Saudi Arabia. In the dry desert sands below, in the middle of Mecca's Grand Mosque, is the *Kaaba*, the 'House of God,' a cubed-shaped place of worship, sacred to Islam, placed within the Grand Mosque in the Plains of Arafat. In one of the greatest religious observances of Islam, up to two million Muslims per day circumambulate (Tawaaf) the *Kaaba* seven times, before having an opportunity to touch or kiss the black stone.

It is a tradition that may go back to Joshua's siege of Jericho (Joshua 6:13- 16): "So on the second day they marched around the city once and returned to camp. They did this for six days. On the seventh day, they got up at daybreak and marched around the city seven times in the same manner, except that on that day they circled the city seven times. The seventh time around, when the priests sounded the trumpet blast, Joshua commanded the army, "Shout! For the Lord has given you the city!"

It may also likely be an astronomical reference to the movement of the seven known planets around Earth (Sun, Moon, Mercury, Venus, Mars, Jupiter, Saturn). The Islamic faith believes *the black stones are angels that fell from the sky during the time of Adam and Eve (an important tradition discussed in later chapters)* but were passed down through the Abrahamic generations to Mohammed in 600 A.D. and placed in the Kaaba.

The Kaaba is a cubical building in Mecca toward which Muslims pray five times daily. The Black Stone, set in the northeastern outside corner of the Kaaba, is considered to be the most sacred treasure of Islam. The Kaaba also served as a center of worship for pre-Islamic Arabs and was reputed to contain 360 idols encircling it, an Egyptian astronomical model of the number of days in a year. In 630 A.D., the triumphant prophet Mohammed returned to Mecca and cleansed the temple of the idols after honoring the Black Stone. What is less commonly known is that eight objects placed in small, 8 x 6-inch vulva-like recession in the cornerstone, are most likely meteorites or meteoric glass from a meteorite impact.

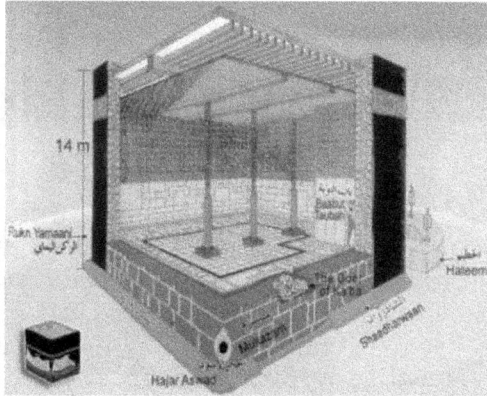

Elsebeth Thomsen of the University of Copenhagen proposed that the Black Stone may be a glass fragment or impactite from the impact of a fragmented meteorite that fell some 6,400 (± 2,500 years ago) at Wabar, a site in the Rub' al Khali desert 1,100 km east of Mecca. The craters at Wabar are notable for the presence of blocks of silica glass, fused by the heat of the impact and impregnated with beads of a nickel-iron alloy from the meteorite (most of which was destroyed in the impact). Some of the glass blocks are made of shiny black glass, with a white or yellow interior and gas-filled hollows, which allow them to float on water--one of the attributes of the Sacred Black Stone.[2] While several measurements suggest the Wabar Crater is only a few centuries old, the Black Stone still fits well with desert impactite and meteor impact traditions that go back to the days of Abraham, when "fire and brimstone" destroyed Sodom and Gomorrah. (See chapter Biblical Firefall)

Strange, celestial circumstances surrounded Mohammed's birth. A large number of birds flew over an army of elephants. As the birds flew they dropped particles of dust and stones onto them. Then a plague broke

---

2 Wikipedia, also Sky and Telescope, 1997. p. 44.

out coupled with a huge flood in which many drowned. Mohammed was born 9 or 12 Rabi-ul-Awwal 52 or 53 BH April 570 or 571 A.D. Most texts tell us Mohammed was born on the 12 of Rabi al-alwal of the Year of the Elephant, *a date that corresponds to June 8. Legends also debate that* the incident with the elephants happened some 50 days before, which would date it approximately to April 20, 570. Significantly, some sources say he was born April 20. *"When the Prophet Mohammed was born, 50 nights before the start of the year known as the Elephant Year, which was Monday the 2nd day of Rabi I... the planets were shooting and the devils were struck by them When the Quraish saw it, they did not recognize it that it was the planets shooting and instead it was the last day"* [3]

Forty-one years before his birth, Comet Halley appeared. Keep this number in mind.

Another Celestial fire legend associated with the Birth of Mohammed is described in Islamic Hadiths. "A great increase in falling stars, a light that lit up the east and the west. That night, most of the idols in the Ka'ba toppled over and fell to the ground."

The above paragraphs describing stones and falling stars surrounding Mohammed's birth, perhaps gives some insight why Mohammed prominently describes prophecies about Judgment Day in the Qur'an. Meteorites that fell from Heaven were considered angels and thus were revered in the Kaaba. For this reason alone, we should carefully consider Mohammed's prophecies, because they mirror Christian, Jewish, Mayan, Aztec, and Dead Saint predictions of our very near future.

Why Revere Meteorites?

It is important to note that the Christian, Jewish, and ancient Islamic tradition placed great importance on the stars and planets travelling through the night sky. Comets and stones that fell from heaven were both signs from God, angels, or even the devil. In some cases, the stones, mostly iron meteorites, were worshipped.

Legends record stones of fire and ice that fall from the night sky— Armageddon Stone traditions that reach far into pre-history, backed by thousands of years of mythology, told and retold in all the world's great religions.

It is not a matter of if the Armageddon Stones will strike, it is a matter of when.

As introduced at the beginning of this chapter, the prophecies of the Armageddon Stones hint that two stellar locations, Leo and Orion, are

---

3 Al-Ya'qubi, Tarikh al-Ya'qubi, vol. II p. 4.

4 | ARMAGEDDON STONES

the entry points for the army of the Armageddon Stones, which bring a *recurring* world-destroying rain of ice, fire and stone.

The Armageddon Stones are like an intricately devised jigsaw puzzle...or rather collections of such puzzles, in which after the main component pieces have been put together to make a whole picture, yet there remain elements unaccounted for which baffle the most ingenious attempts to fit them into a coherent whole.

What is fascinating is the ancient belief believes the recurring fire from Heaven occurs in predictable cycles. Ancient catastrophes— *History— becomes future catastrophe—prophecy.*

Stories of world consuming stones of fire and ice are inexorably intertwined with stories of world ending wars; point to the final chapter of modern prophecies made by the Dead Saints, ancient prophecies of Apocalypse—*the Armageddon Stones*—legends that are found in sacred literature throughout this habitable Earth. Legends describe an unusual, destructive band of comet debris periodically pummels our planet. If we examine the historical validity of the world - destroying, we will find in nearly every case, direct observation of an unusual variety of astronomical phenomenon ranging the gamut from the explosion of comets and the subsequent rain of fiery stones falling from the sky, to the deviation of the sun, moon, and planets from their courses in the heavens, indicating a sudden shift in the geographic axis of the Earth followed by a world-wide deluge.

The periodic fall of cosmic debris must have been an event so overwhelming in extent, so disastrous in its effect on human civilization, it must have left a terrible and indelible mark on those who survived. They told their descendants of their experience; what they said was handed down from one generation to the next through fireside storytelling and became preserved in the imagery of myth known as the *Apocalypse*. The ancient story is hammered into cave walls, painstakingly laid out on papyrus, built into stone temples, and even immortalized in stellar mythology.

They describe what must have been truly Earth-shattering cataclysms, not inundations caused merely by localized prolonged rain (although this is a regular feature of one type of deluge myth) which set the low-lying parts of some country under water or by earthquakes (another striking trait of many deluge tales) which caused considerable waves to sweep over some island or surge far in land on some seacoast. The events must have been tremendous, universal, and rather sudden convulsions, to impress themselves so deeply upon man's mind and memory.

Now perhaps we can understand why God commanded Adam and Eve; "Be fruitful and multiply, and replenish the earth, and subdue it." Herein the term *replenish* is insight into a past catastrophe that appears to have extinguished all life forms from a prior age (see Jeremiah 4:23-28) & (see Ezekiel 28:13-16) If this is true, then the first verse in Genesis may be taken as a statement of fact followed by the second statement that the earth "became without form and void" as relating to a prior fall and judgment, with the following being in effect, re-creation. Recreation mythology unveils the etymology of the word "apocalypse," which comes down to us from the Greek word *apokulupsis*, meaning to "*uncover what is hidden.*"

A Planetary Near-Death Experience is Coming

Earth's geological record bears witness to a cyclic bombardment by interplanetary debris, and by almost all accounts, the Earth is due for another major cosmic event.

With over forty-years of mythological and scientific research, I have come to the disquieting conclusion that ancient and modern prophecies point like hands of a giant clock. This conclusion, rooted in both the distant and the recent past, ties in the prophecies of hundreds and even thousands of years ago with the cosmic theories and scientific realities of the present. Physical evidence for past catastrophes is so strong and so worldwide in its distribution that the possibility of cometary or asteroid impact cannot be ignored.

Prophecies of the world's end by fire, ice, water, or explosion, although made in different ages and by different cultures during the past 13,000 years, seem to agree that the celestial fire will soon return, renewing the world. If the proposed cycle of destruction proves to be true, it will revolutionize the way we think about the development of life on our planet. Instead of the slow, steady replacement of one species by another, as Darwin envisioned, we would have instead a situation in which the Darwinian progression goes on for an uninterrupted period of time, —and then a catastrophic collision occurs, wiping out a good fraction of the species that have evolved during that time, and the whole process starts all over again.

In our search for answers to the eternal question of good and evil, God and Devil have their roots among the stars, and why they are the ultimate paradox of life and death. The primary cause remains a dilemma stands today unsolved. The apocalypse stands waiting to be solved as the prophecy it truly is the periodic return of the celestial-serpent-dragon-comet—

The Armageddon

Roots of Armageddon

Armageddon conjures mystical images of prophecy, war, and fear. Almost everyone has heard of Armageddon as it is one of the most recognizable place names in all of history, but it is not a city. There is no metropolis there. It is not even a town. The place is called Megiddo. It sits on a plain on a hill in north, central Israel overlooking the plains of Jezreel. During the time of King Solomon, he kept his stables for horses there. Later, King Ahab built a mighty fortress with incredible feats of engineering that still rather stagger the imagination. The plains of Jezreel and this little hill, Megiddo, were extremely important in the ancient world because it was a crossroads.

Egypt and Israel and all of the countries around the Mediterranean depended for commerce upon caravans that came from the East, which had to pass through the plains of Jezreel. Whoever controlled those plains, controlled the commerce of the world. So the plains of Jezreel experienced more battles than any other place in the world at that time. And it is why Megiddo was best *known as a battleground, the site of ultimate destruction, where the last battle between the forces of Good and Evil will occur, that will usher in the last days of planet Earth.*

Scripture describes this future cataclysmic war fought in an arena set both in Heaven and on Earth:

Book of Judges Chap 5:19-20:

The Kings came and fought; they fought the Kings of Canaan into Taanach, by the waters of Megiddo. They fought from heaven; the stars in their courses fought against Sisera."

Revelation 16:6:

And he gathered them together in a place called in the Hebrew tongue, Armageddon.

Revelation 19:19:

And I saw the beast, the kings of the Earth, and their armies, gathered together to make war against Him who sat on the horse and against His army.

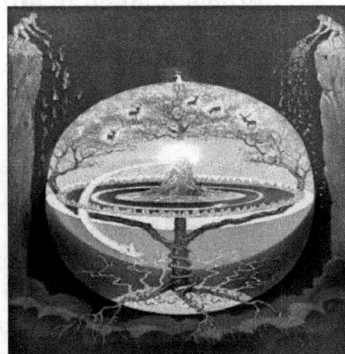

Zechariah 12:11:

In that day there shall be a great mourning in Jerusalem, like the mourning at Hadad Rimmon in the plain of Megiddo. In Hebrew, Megiddon.

Paul Solomon, a Seer who died in March 1994, wrote in *"Love and Fear,"* an interesting commentary about the Battle of Armageddon:

Everyone considers the Battle of Armageddon to be a prophecy about the future. Perhaps it is. However, more importantly as an opposing tactician, I would keep everyone intrigued with this future Battle. I would encourage the prophets and the preachers to speak of it in the future tense. That way they won't be looking for it here and now.

Religion is still in vogue, despite the efforts of intellectuals and humanists to promote "reason" over "superstition." People universally identify with good and agree that evil should be defeated. The more "religious" people might even be persuaded to ensure that good wins out by whatever "means" necessary, including killing. People are so universally agreed that good and love and light should prevail, that surely darkness doesn't really stand a chance. It seems obvious that the powers of darkness must use confusion, sabotage, espionage, disguise, and subversive activity to do serious battle.

If the two sides of Armageddon were forced to wear uniforms, carry battle standards, and show colors, the forces of evil would surely be stamped out. Good people are ready to slay, maim, and utterly annihilate the army of Darkness. Countries and religions stand ready with weapons of war and nuclear arms who are prepared to wipe darkness off the face of the Earth to win the War.

The Sages and Dead Saints declare that every living being is fighting for one side or another in the Great Battle. They also say that many do not know which side they are really on. Which side are you really on? Who are you fighting for?

Sons of Light / Sons of Darkness

The Great War has been described long before Christ by a little-known prophet of little noticed religion based in Northern Persia, a people of a religion ISIS recently attempted to exterminate.

These little-known people carry the burden of a prophecy about an ancient story of a War in Heaven between the Sons of Light and the Sons of Darkness that go back at least 3,500 years to ancient Iran, to the prophet *Zoroaster, founder of one of the world's oldest monotheistic religions.*

One thousand eight hundred fifty years later, in 1947-1948, Bedouin shepherds while exploring Caves on hillsides surrounding Qumran, discovered a collection of hundreds of scrolls and texts fragments at Qumran near the Dead Sea. The scrolls were compiled by an apocalyptic Jewish sect known as the Essenes (or Yahad), which lived in Qumran from the second century B.C. until around A.D. 70. Portions of every Hebrew Bible book except Esther and Nehemiah were found—and several complete copies of the Book of Isaiah. The scrolls also include 14 apocrypha and the Essene sect's own rule books.

The scrolls included the *War Scroll,* a story about a future war between the Sons of Light and the Sons of Darkness, found in Qumran Cave 1. *The Shrine of the Book,* a wing of the Israel Museum in Jerusalem, is built to symbolize the War Scroll. It is a white dome symbolizing the Sons of Light, and a black basalt wall—symbolizing the Sons of Darkness. The Scrolls speak about the End Times of planet Earth, when Darkness is to be destroyed and Light will live in peace for all eternity.[4]

Both Isaiah (750 A.D. – 645 A.D.) Zoroaster, echo the philosophy in the War Scroll that describes both a Great War and Global Apocalypse that had a set timeline.[5] The War Scroll is believed to be the SOURCE of information Jesus passed on to his apostles who integrated Zoroaster's prophecies into their writings and the description of the War of Armageddon in the Book of Revelation that describes a 40-year war as "an appointed time" divided into "seven lots," the first six lots occurring in 33 years. The seventh lot of seven years will happen at the HIS appointed time *"when "the great hand of God shall overcome Belial (sons of darkness) and all the angels of his dominion, and al- the men of his forces shall be destroyed forever.*[6]

The School of Prophets

"Afterward you will come to the hill of God where the Philistine garrison is; and it shall be as soon as you have come there to the city, that you will meet a group of prophets coming down from the high place with harp, tambourine, flute, and a lyre before them, and they will be prophesying.

*"Then the Spirit of the LORD will come upon you mightily, and you shall prophesy with them and be changed into another man." **1 Samuel 10:5-6***

4 http://www.britannica.com/EBchecked/topic/635670/The-War-of-the-Sons-of-Light-Against-the-Sons-of-Darkness.
5 Edition by Cyril Glasse (Author)
6 War Scroll, (Col. 11: 6-16)

Samuel & Saul- 1050 B.C.

It's Biblical history, but to hear it today may seem strange. Why the need to develop prophets? Men of renown who could see the future, heal and perform wonders for God. Who was Elijah?

Elijah & Elisha-853 B.C.

Elijah figures predominantly in Judaism, Christianity, and Islam. When his Mission was over, God took him up to Heaven. All three religions believe Elijah will return as a Harbinger at the End Times. Elijah's name means, 'God is God.'

Jonah-786 B.C.

Go to Adam to Apophis. In the belly of the whale. The Book of Jonah is one of the Minor Prophets in the Bible. It tells of a Hebrew prophet named Jonah son of Amittai who is sent by God to prophesy the destruction of Nineveh but tries to escape the divine mission.

Isaiah Ben Amoz- 750-645 B.C.

An itinerant prophet who roamed ancient Persia. In the 8th century, he composed the Book of Isaiah, which became one of the most popular works among Jews in the Second Temple period (c. 515 B.C.E - 70 CE). Later in Christian circles it was held in such high regard as to be called "the Fifth Gospel." Isaiah became prophet of One God and Monotheism. Isaiah 44:6 contains the first clear statement of monotheism: "I am the first and I am the last; besides me there is no god,"[7] In Isaiah 44:09–20 this is developed into a satire on the making and worship of idols, mocking the foolishness of the carpenter who worships the idol that he himself has carved.

While Yahweh had shown his superiority to other gods before, in Second Isaiah he becomes the sole God of the world.

This model of monotheism became the defining characteristic of post-Exilic Judaism and became the basis for Christianity and Islam.[8]

---

7 Robert Karl (1997). No Other Gods: Emergent Monotheism in Israel. Continuum.

8 Coogan, Michael D. (2009). A Brief Introduction to the Old Testament. Oxford University Press.

Zoroaster- 750 B.C.

Zoroaster became the founder of one of the world's oldest monotheistic religions in Iran, the Zoroastrians. There is some contention whether Isaiah influenced Zoroaster (also called Zarathustra), but it is clear their teachings were similar. Both Zoroaster and Isaiah carried the burden of a prophecy about an ancient story of a War in Heaven between the Sons of Light and the Sons of Darkness; stories written about a future war between Light and Darkness described in *The War Scroll.*

Ezekiel – 622 B.C. -570 B.C.E. (?)

The author of the Book of Ezekiel presents himself as Ezekiel, the son of Buzzi,[Ezekiel 1:3] born into a priesthood (Kohen) lineage[1] of the patrilineal line of Ithamar, and resident of Anathoth. Apart from identifying himself, the author gives a chronology for the first divine encounter which he will present. He states that it happened "in the thirtieth year",[Ezekiel 1:1, 2] which may be a reference to his age at the time. In such a case, the approximate year of birth is 622 B.C. He also dates the event 5 years after the exile of King of Judah Jehoiachin by the Babylonians, a recurring dating pattern throughout the book. Josephus claims that under at the request of Nebuchadnezzar II, Babylonian armies exiled three thousand Jews[2] from Judah, after deposing King Jehoiachin in 598 B.C. The book of Ezekiel was written by the Great Assembly (Knesset Hagedolah) because a prophet was not allowed to write down the prophecies while being outside of Israel. Ezekiel was 50 years old when he began to have visions of a new Temple. He served as a prophet for at least 22 years. Ezekiel last experienced an encounter with God in April 570 B.C.[Ezekiel 29:16].9

Daniel & Babylonian Captivity

During the Battle of Carchemish in 605 B.C.E, Nebuchadnezzar, the king of Babylon, besieged Jerusalem, resulting in numbers of deportees sent from Jerusalem to Babylon in 597 B.C.E for the first, with others dated at 587/586 B.C.E, and 582/581 B.C.E respectively, and were held captive there. There the Prophet Daniel interpreted Nebucadnezzar's dream of four future kingdoms.

---

9 Wikipedia: Ezekiel

Vision of the beasts from the sea

In the first year of Belshazzar Daniel (Chapter 7) has a dream of four monstrous beasts arising from the sea. The fourth, a beast with ten horns, devours the whole Earth, treading it down and crushing it, and a further small horn appears and uproots three of the earlier horns. The Ancient of Days, judges and destroys the beast, and "one like a son of man" is given everlasting kingship over the entire world. A divine being explains that the four beasts represent four kings, but that "the holy ones of the Most High" would receive the everlasting kingdom. The fourth beast would be a fourth kingdom with ten kings, and another king who would pull down three kings and make war on the "holy ones" for "a time, two times and a half," after which the heavenly judgment will be made against him and the "holy ones" will receive the everlasting kingdom.[18]

Vision of the ram and goat
Main article: Daniel 8

In the third year of Belshazzar Daniel (Chapter 8) has vision of a ram and goat. The ram has two mighty horns, one longer than the other, and it charges east and west and north, overpowering all other beasts. A goat with a single horn appears from the west and destroys the ram. The goat becomes very powerful until the horn breaks off and is replaced by four lesser horns. A small horn that grows very large, it stops the daily temple sacrifices and desecrates the sanctuary for two thousand three hundred evening and mornings (1150 days) until the temple is cleansed. The angel Gabriel informs him that the ram represents the Medes and Persians, the goat is Greece, and the "little horn" is a wicked king.[19]

Vision of the Seventy Weeks
Main article: Prophecy of Seventy Weeks

In the first year of Darius the Mede, Daniel (Chapter 9) meditates on the word of Jeremiah that the desolation of Jerusalem would last seventy years; he confesses the sin of Israel and pleads for God to restore Israel and the "desolated sanctuary" of the Temple. The angel Gabriel explains that the seventy years stand for seventy "weeks" of years (490 years), during which the Temple will first be restored, then later defiled by a "prince who is to come," "until the decreed end is poured out."[20]

Vision of the kings of north and south
Main article: Daniel 11

Daniel 10: In the third year of Cyrus Daniel sees in his vision an angel (called "a man", but clearly a supernatural being) who explains that he is in the midst of a war with the "prince of Persia", assisted only by Michael, "your prince." The "prince of Greece" will shortly come, but first he will reveal what will happen to Daniel's people.

Daniel 11: A future king of Persia will make war on the king of Greece, a "mighty king" will arise and wield power until his empire is broken up and given to others, and finally the king of the south (identified in verse 8 as Egypt) will go to war with the "king of the north." After many battles (described in great detail) a "contemptible person" will become king of the north; this king will invade the south two times, the first time with success, but on his second he will be stopped by "ships of Kittim." He will turn back to his own country, and on the way his soldiers will desecrate the Temple, abolish the daily sacrifice, and set up the abomination of desolation. He will defeat and subjugate Libya and Egypt, but "reports from the east and north will alarm him," and he will meet his end "between the sea and the holy mountain."

Daniel 12: At this time Michael will come. It will be a time of great distress, but all those whose names are written will be delivered. "Multitudes who sleep in the dust of the earth will awake, some to everlasting life, others to shame and everlasting contempt; those who are wise will shine like the brightness of the heavens, and those who lead many to righteousness, like the stars for ever and ever." In the final verses the remaining time to the end is revealed: "a time, times and half a time" (three years and a half).

Daniel fails to understand and asks again what will happen and is told: "From the time that the daily sacrifice is abolished and the abomination that causes desolation is set up, there will be 1,290 days. Blessed is the one who waits for and reaches the end of the 1,335 days."

After the fall of Babylon to the Persian king Cyrus the Great in 539 B.C.E, exiled Jews began to return to the land of Judah. According to the biblical book of Ezra, construction of a second temple in Jerusalem began at this time.

87 B.C. Comet Halley

In 87 B.C., Comet Halley appeared over Jerusalem and the Dead Sea, sparking Messianic fervor throughout the Holy Land. Was the Holy War about to happen?

It was only 15 years later in 69 B.C., when the *Teacher of Righteousness* (in Hebrew: מורה הצדק *Moreh ha-Tzedek*) appeared, a figure found in some of the Dead Sea Scrolls at Qumran, most prominently in the *Damascus Document*.10 This document speaks briefly of the origins of the sect, the Essenes, (or Yahad) who after 'groping' blindly for the way. "God... raised for them a Teacher of Righteousness to guide them in the way of His heart".[2] *The Teacher* is extolled as having proper understanding of the Torah, qualified in its accurate instruction,[3] and being the one through whom God would reveal to the community "the hidden things in which Israel had gone astray."[4]

Who were the Essenes?

The Teacher of Righteous led them. They became Scripture copyists. The remains of Qumran have been excavated, showing a small, dense settlement, with ritual baths and communal dining room, and, most poignantly, a scriptorium where the scrolls may have been composed. They surrendered all their possessions, took vows of celibacy, and engaged in strict religious practices encouraging purity. They took ritual baths all the time—Qumran has more bathrooms per square foot than a McMansion in Phoenix. They were freaky about urine and excrement. They did not relieve themselves on the Sabbath—at all! And they wouldn't relieve themselves inside the city walls, meaning they schlepped up the hill behind a big pile of rocks to do their business.

The Book of Isaiah was the central pillar of his teaching. From Isaiah's prophecies, the Essenes established two goals at Qumran; First, was to prepare for the birth of the Messiah. Holy Women were chosen for purity and set apart for the purpose of being chosen by God to be the Mother of God. (add note here about strict laws of diet, bathing, etc.) It is believed by many that Anna, the mother of Mary, was an Essene. The second, was to prepare for a future Apocalypse that would engulf the entire world destroying it.

After Jesus was born and later crucified, the Essene community disbanded, hiding hundreds of scrolls and fragments of texts copied in their Scriptoriums and sealed in clay jars throughout the desert hills around Qumran including many copies of the five books of Moses, portions of every Hebrew Bible book except Esther and Nehemiah, and several complete copies of the Book of Isaiah. The scrolls also include 14 apocrypha and the sect's own rule books until they were discovered at the very moment the State of Israel was born in 1948.

---

10 See the Damascus Document

Conclusion: When will the last seven weeks of the Great War Begin?

2

The End of Days

Two thousand years ago, the Disciples of Christ had questions. In Matthew 24, Jesus was sitting on the Mount of Olives, when the disciples came to him privately, saying, *"Tell us, when will these things be, and what will be the sign of Your coming, and the end of the Age?"* (Matthew 24:29-36):

In answering them, Christ revealed certain "signs' that would take place before the end of the Age.

But immediately after the tribulation of those days, the sun will be darkened and the moon will not give its light, and the stars will fall from the sky, and powers of the heavens will be shaken. And then the sign of the Son of Man will appear in the sky (Scepter) and then all the tribes of the Earth will mourn, and they will see the Son of Man coming on the clouds of the sky with power and great glory.

Now learn the parable from the fig [author's note: Fig leaves were used the cover the nakedness of Adam and Eve]: when its branch has already become tender, and puts forth its leaves, you know that summer is near; even so you too, when you see all these things, recognize that He is near, right at the door. "Truly I say to you, this generation will pass away, but My words shall not pass away, "But of that day and hour no one knows, not even the angels of heaven, nor the son, but the Father alone."

Then Christ said, *"Verily I say unto you, this generation shall not pass, till all these things be fulfilled.*

This controversial verse is in all three "Olivet Discourse accounts," found in Matthew 24:1-51, Mark 13:1-37, and Luke 21:5-33. For some time, critics of the Christian faith argued that Jesus explicitly said here that all of the events prophesied in the Olivet Discourse, including His return, would happen before the last person living at that time died. It is perhaps why St. John wrote seven times in the Book of Revelation that "the time is at hand." The return of Christ is called *Parousia*, in Greek, meaning "arrival" of Christ's physical, royal personage.

Jesus's royal visit has not happened yet. So, what did Jesus mean "this generation shall not pass until all be fulfilled?" By "generation" Jesus must have meant the end of the *Age of Pisces*, the fish, one of twelve constellations in the Zodiac. If he did not, it would mean Christ was a false prophet. I would prefer the Piscean / Aquarian Age interpretation as we shall see throughout the Armageddon Stones, the prophets before Christ, the Ancient Egyptians, Adam & Eve, Enoch, Noah, Ezekiel, Isaiah, Daniel, St. John, and Mohammed and his visions of the future, all pointed to the SIGN OF MAN- AQUARIUS, A ZODIAC AGE that would arrive 2,160 years after the birth of Jesus Christ, the Ruler of the Piscean Age.

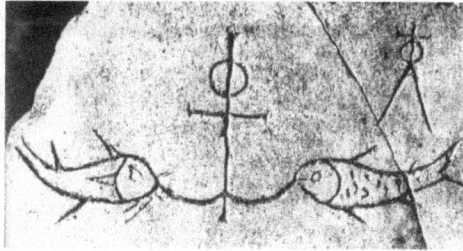

It's 2016. We are on the cusp of entering the Age of Aquarius. But to understand our movement into this "Golden Age" I think it is important to understand what the Bible says about three previous ages personified by prophets and "avatars" to further understand Jesus's statement, "All shall be fulfilled at the end of this generation."

Astronomical preoccupation by the ancients is a pastime not well explained by scholars today. Their passion to determine the length of the year could be explained if oral and written tradition believed in the reappearance of what was often called *"Firefall"* a destroying fire that fell from heaven," and fixing approximate dates for its return. Stonehenge, pyramids, ziggurats, and observatories in Mexico and South America testify to this obsession to record, calculate, and predict celestial events, because it was believed, what happened before, will without failure could be predicted by the precise observations of the stars, sun, moon, stars, and the planet Venus, from their precisely aligned stone temples. It was not a matter of if, or fear, but a knowing, a future catastrophic event will happen again.

Astronomy was on the reason, the fifth-century Roman writer, Nemesis—*a name Loius Alverez dubbed on the dinosaur killing asteroid*—for example, also describes his belief that a certain astronomical alignment, like the biblical prophets, caused a periodically repeated destruction and renewal of the world:

"The planets will be restored to the same zodiacal sign, both in longitude and latitude, as they had in the beginning when the cosmos was first put together; that in stated periods of time a conflagration and destruction of things will be accomplished, and once more there will be a reconstitution of the cosmos as it was in the very beginning. And when the stars move in the same was as before, each thing which occurred in the previous period will without variation be brought to pass again."[11]

Dynastic Egypt, began in between the 34th and 31st century B.C., by Menes / Narmer the first Egyptian pharaoh of Upper and Lower Egypt some 1000 years after the "Age of Taurus"— the Bull began in 4609 B.C. [12]

Two Thousand One Hundred Sixty (2160) years later at the end the Taurian Age, the Earth wobbled into the next constellation, the sign of the ram. The "Age of Aries" dawned, and a great exodus in the evolution of mankind took place. Legends of the Ram persist in almost every culture during the advent of Aries. Moses came up out of Egypt and said, "I am the ram, the ram of God." Exodus.

As our galaxy traveled through the infinite realms of space at approximately one million miles per hour, the Earth slowly wobbled through the Age of Aries, and mankind progressed into the "Age of Pisces," the fishes. Christians were to choose the fish for their symbol in that new age.

The Prophet for the Piscean Age is revealed by the first Gospel of the New Testament, where we read, "Out of Egypt have I called my Son."

---

11 Arthur Lovejoy and George Boas, Primitivism and Related Ideas in Antiquity, Johns Hopkins Press, Boston, 1935, p84.

12 http://earthsky.org/space/what-is-the-zodiac

Jesus, came up out of Egypt and said, "Follow me and I will make you to become fishers of men." To the ancients, each astronomical age held a special meaning, however, the three preceding ages and particularly the Age of Aquarius were special. Why? We find the symbols of the four ages appearing regularly through the Bible. The prophet Ezekiel and St. John in their visions saw 4 beasts, a bull, a lion, an eagle, then ended with THE MAN —Aquarius.

Today, as we wobble into the "Age of Aquarius," the water bearer, the prophets have spoken of an end to wars and the beginning of a golden age of peace and enlightenment. Jesus declared the secret, "I will pour out my spirit upon Men—thus the image of the water bearer pouring out spirit over humanity." Not just Spirit, but a *Flood of water*.

So why all the astrology? Many Christians, Muslims, and Jews may be shocked that all the prophets (there can be no dispute) referred to the "living creatures—angels, cherubim" — the four cardinal points of the zodiac, and in sacred Scripture. (See Prophet Zodiac Table below).

The Naming of the Constellations

As the Earth orbits the sun, the sun appears to move against the background stars (red line). The constellations (green) through which the sun passes define the zodiac. Credit: Tau'olunga (via Wikipedia)

When our ancestors looked up on our ancient skies, the constellations, unlike the astrological signs, were not of equal size and shape. The stars that make up a constellation are not, for the most part, physically related. The constellations are just patterns our ancestors saw as they gazed skyward and tried to make sense of everything.

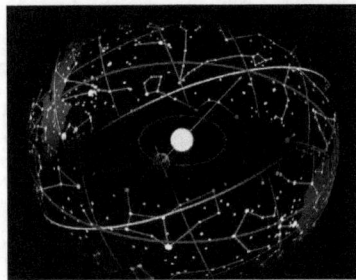

Today's constellations are specific to ancient Greek culture. Most of them were introduced by the Greek astronomer Ptolemy in the 2nd century but he derived them from much older sources, including Sumerian and Babylonian cuneiform texts dating from 3500 B.C.E. Modern constellation boundaries weren't defined until 1930 by the International Astronomical Union. With the current boundaries, there are actually thirteen constellations that lie along the sun's path. The extra one not listed in any horoscope is Ophiucus, the Serpent Bearer, who sits between Sagittarius and Scorpius—a 13th Zodiac Sign.

The original Greek word for zodiac is "zodia," and is translated from Greek in the Latin tongue in the Book of Revelation (KJV) as and "living creatures" in the Revised Kings James Version. We know that the

Babylonian zodiac is also reflected in the Hebrew Bible. E. W. Bullinger interpreted the creatures appearing in the books of Ezekiel and Revelation as the middle signs of the four quarters of the Zodiac, with the Lion as Leo, the Bull is Taurus, the Man representing Aquarius and the Eagle representing Scorpio. The zodiacal signs are an abstraction from 12 physical constellations, and each represents exactly one twelfth of the 360-degree full circle, or the longitude traversed by the Sun in about 30.4 days.

To repeat again, without exception, the prophets Ezekiel, Daniel, and St. John in the Book of Revelation, refer to the same "living creatures" identified by a Lion, (Leo), an Eagle (Scorpio), the Ox or calf (Taurus), and Man (Aquarius).

The question becomes: if astrology was prohibited by Jewish teaching, why were the prophets in the Name of the Lord, using Zodiac signs in their prophecies?

How old is the Zodiac?

The Zodiac is much older than Greece. Many have maintained that Egypt was the first to give shapes and names to the star groups; Dupuis, perhaps inspired by Macrobius of our 5th century, traced the present solar zodiac to that country, and placed its date 13,000 years (11,000 B.C. —an auspicious date that correlates with a previous apocalypse, Firefall / impact, anterior to our era).

Hebrew antiquaries have long recognized Enoch as the inventor of the Dodecatemory divisions of the zodia; and both Berossus, a Chaldean historian of about 260 B.C. and Josephus declare that Abraham was famous for his celestial observations and even taught the Egyptians. Following the opinion of Josephus, Origen said that the constellations were known long before the days of the patriarchs by Noah, Enoch, Seth, and Adam -- indeed were mentioned in the Book of Enoch (before the Great Flood-cite from the Enduring Mystery and Gospel in the Stars) as "already named and divided." In Arabia the zodiac was Al Mintakah al Buruj, the Girdle of the Signs.

In the Mesopotamian *Epic of Creation*, dated 2350 B.C., discovered by George Smith in 1872, the zodiac signs were described as *Mizrata*. A very similar word appears for the Milky Way, -- generally supposed to be the original of the biblical Mazzaroth; or the form Mazzaroth, being the form used in the *Targum's* and later Hebrew writings. This word, although of uncertain derivation, may come from a root meaning "to watch" the constellations, thus marking the watches of the night. Another word, Ezor, meaning "Girdle; while others have referred them to Zahir, or Zohor, from Zuhrah, meaning "a Glittering Star, and so signifying something especially luminous. Still this Bible word has been variously rendered, appearing for the Great Bear, Ursa Major, Sirius, the planets, or even for the constellations in general.

Why 12 Zodiac Signs?

Some authors have linked the twelve tribes of Israel with the twelve signs. Martin and others have argued that the arrangement of the tribes around the Tabernacle (reported in the Book of Numbers) corresponded to the order of the Zodiac, with Judah, Reuben, Ephraim, and Dan representing the middle signs of Leo, Aquarius, Taurus, and Scorpio, respectively.

Each zodiac "living creature Axial precession is the movement of the rotational axis of an astronomical body, whereby the axis slowly traces out a cone. In the case of Earth, this type of precession is also known as the precession of the equinoxes, lunisolar precession, or precession of the equator. Earth goes through one such complete processional cycle in a period of approximately 25,960 years or 1° every 72 years, during which the positions of stars will slowly change in both equatorial coordinates and ecliptic longitude. Over this cycle, Earth's north axial pole moves from where it is now, within 1° of Polaris, in a circle around the ecliptic pole, with an angular radius of about 23.5 degrees.

Hipparchus is the earliest known astronomer to recognize and assess the precession of the equinoxes at about 1° per century (which is not far from the actual value for antiquity, 1.38°)" The prophets saw the living creatures of the zodiac as cherubim or angels connected with "wings" and used them to divide, time, time, time, and half times—used in their prophecies I describe later in the *Armageddon Stones* in a side by side prophetic comparison in chapter—a startling revelation to be sure. This astronomical knowledge was passed down to all the prophets, who knew the zodiac surrounding God's Throne in heaven governed the gears of prophetic time when future catastrophic events would happen.

Ezekiel, Daniel, and St. John Zodiac Living Creatures of the Zodiac:

	Cherubim with wings two above two 2 covering body	Right side	Right Side	Right Side	Right Side	Left side	Left side	Left Side	Left side
Living beasts Creatures		Right side	Right Side	Right Side	Right Side	Left side	Left side	Left Side	Left side
Ezekiel (1: 4-14) at Kebar		Man	Aquarius	Lion	Leo	Ox	Tauru s	Eagl e	Scorpi o
	Wings	4		4		4		4	
Daniel		Lion	Leo	Eagle	Scorpio	Ox	Taurus	Eagle	Scorpio
St. John		Lion	Leo	Eagle	Scorpio	Ox	Taurus	Eagle	Scorpio

Characteristics of Ezekiel's four living creatures: (Ezekiel 4:9-14)

Fire folding in on itself
Brightness/amber (bright orange/brown)
living creatures
likeness of man
Four faces
Four wings each
Feet were straight
Feet like sole of a calf's foot
Feet sparkled like color of burnish brass
Burning coals of fire appearance of lamps: it went up and down among the living creatures, and the fire was bright, and out of the fire went forth lightning.

The vision of the Four Wheels (Ezekiel 4: 15-21)- 4 wheels deffourne` 4 living creatures

1st wheel upon the Earth with four faces
Colour of beryl (green)
One Likeness
The appearance of wheel within wheel
Followed each other wherever they went
Rings above them so high and dreadful
Full of eyes (stars)
And when the living creatures went, the wheels went by them: and when the living creatures were lifted up from the Earth, the wheels were lifted up.

The appearance of the wheels and their work was like unto the colour of a beryl: *(The name beryl is derived (via Latin: beryllus, Old French: beryl, and Middle English: beril) from Greek "beryls "which referred to a "precious blue-green colour-of-sea-water stone. The living creatures were constellations in the "sea" of in sky above, the heavens.*

When they went, they went upon their four sides: and they turned not when they went. As for their rings, they were so high that they were dreadful; *and their rings were full of eyes round about them four.*

Vision of the Divine Glory (Ezekiel: 22)

And the likeness of the firmament upon the heads of the living creature was as the colour of the terrible crystal stretched forth over their heads above. And under the firmament were their wings straight, the one toward the other: everyone had two, which covered on this side, and everyone had two, which covered on that side, their bodies. And when they went, I heard the noise of their wings, like the noise of great waters, as the voice of the Almighty, the voice of speech, as the noise of a host: when they stood, they let down their wings. And there was a voice from the firmament that was over their heads, when they stood, and had let down their wings.

And above the firmament that was over their heads was the likeness of a throne, as the appearance of a sapphire stone (blue sky): and upon the likeness of the throne was the likeness as the appearance of a man above upon it.

And I saw as the colour of amber, (yellow-orange)

Fire from loins downward/downward

Brightness around it

Appearance of bow like the cloud in a day of rain with brightness around it

Brightness like glory of the Lord.

The four living creatures, the four cherubim, the four angels, are the Lion, the Eagle (Scorpio), the ox (Taurus) and Man (Aquarius) are placed univocally and undeniably in Scripture and are reflected later in the same language of the Book of Revelation by the Apostle John. They are the four cardinal points of the zodiac. The Sun moves through each zodiac sign every 2,160 years, totaling 25, 920 years.

2160 years before Christ, Abraham was the light bearer for the Aryan Age, which preceded Jesus the Piscean Age. But just as Jesus was a fulfillment of Abraham's covenant, so Abraham still lives through all of the 12. To say it another way, there is no light bearer of any age that is only for that age. Jesus Christ is the light bearer for the Aquarian Age as well. We move into a new age which is an expansion of consciousness into making application of our understanding in new ways and new directions. In the Piscean Age, it was our responsibility to learn to love one another. In the Aquarian Age it is our responsibility to put that love to work in specific, concrete ways.

So, each Age is expanding one on the other and taking the message a zodiac age further. There will be someone who will be a light bearer to the Aquarian Age, but that light bearer will not be a replacement for Jesus Christ, but will rather be a servant of the Christ who will take the message into another level of understanding.

Zodiac Sign	Animal	Age Time Period
Taurus	Bull	4609 B.C.
Aries	Ram	2540 B.C.
Pisces	Fish	380 B.C.
Aquarius	Man	1759 A.D. +

Depending on the astronomer/astrologer this dividing line between the end of one age, and the beginning of the next may include two or three centuries. Regardless, the difference between 4 ages of the Bull, the Ram, the Fish and the Man is 6,380 years—close to the biblical 6000 to 7000 years of the life space or world age of the Apocalypse...

Have we entered the Age of Aquarius?

In every successive generation, usually at the end of the century, there have been prophets who have proclaimed that their generation was the last, and that generation expected the Battle of Armageddon to occur in their lifetime or just after.

Modernizing the End of Days

Good intentions abound. Many devoted Christian pastors have spent years, sometime decades of meticulous research to decipher the prophetic works of Daniel, Isaiah, the words of Christ and His Disciples, and the Book of Revelation. William Miller, a Baptist preacher, created the "Millerites" a movement who believed, *"That Jesus Christ will come again to this Earth, cleanse, purify, and take possession of the same, with all the saints, sometime between March 21, 1843, and March 21, 1844."* His basis was his interpretation of the prophecy of Daniel 8:14: "Unto two thousand and three hundred days; then shall the sanctuary be cleansed." He first assumed that the "cleansing of the sanctuary" represented purification of the Earth by fire at Christ's Second Coming instead of the sanctuary in Heaven.

Then, using an interpretive principle known as the day-year principle, Miller, along with others, interpreted a prophetic "day" to read not as a 24-hour period, but rather as a calendar year. Miller became convinced that the 2,300-day period started in 457 B.C. with the decree to rebuild Jerusalem by Artaxerxes I of Persia. His interpretation led him to believe and promote the year 1843. Despite the urging of his supporters, Miller never announced an exact date for the expected Second Advent. But he did narrow the time period to sometime in the Jewish year 5604, stating: March 21, 1844. The date passed without incident, but the majority of Millerites maintained their faith.13

Jehovah's Witnesses Judgment Day predictions for 1878, 1881, 1914, 1918, 1925, and 1975. The US radio evangelist Harold Camping built up a ministry worth millions of dollars predicting that 21 May 2011 would be the big one. Camping later admitted he was probably mistaken, and donations to his ministry suffered an apocalyptic fall themselves. And most recently the "annihilation of the world" predicted for October 7, 2015, by Chris McCann's Bible Fellowship passed by without so much as a whimper. Usually, the attempt to nail down the Day of Judgment are based solely on calendar math sifted from Old Testament and New Testament sources. If so, it appears we either don't understand what the prophets have handed down to us, or we don't have all the pieces of the Scripture to accurately predict when the End will come.

Solving the Judgment Day puzzle may be one of the most important endeavors ever made in the history of mankind. To find an answer another approach is needed, one that crosses the borders of religion and faith.

Defining the End of Days

What are we to believe? For the most part, intelligent, educated men regard apocalyptic prophecies with disdain. Such attitudes are predicted by the Apostle Peter, "*Know this first of all, that in the last day's mockers will come with their mocking, following after their own lusts, and saying, "Where is the promise of His coming? For ever since the fathers fell asleep, all continued just as it was from the beginning of creation."*

Every generation has looked for these same signs and continues to predict the end of the world. Of course, one day a prophet will eventually

---

13 Wikipedia. The Millerites.

be right. The prophets call the time *before* Judgment Day: "The End of Days" or "The Latter Days."

Before we can intelligently discuss whether or not we are living in the *End of Days*, we must understand what is meant by the term. It is fairly loosely worded, probably with many definitions, but most often used in Christian circles to indicate the period of time immediately preceding and including a great period of Tribulation, or "Testing." Since the phrase "End of Days" or "End Times" is not found in Christian Scriptures, if one must use this term, I would prefer to say the End Times of this present Age, the Age of Pisces—symbolized by the fish, which began 2000 years ago.

If we were to use a more Biblical term—one found in the Scriptures we would talk about the "last days" or, more precisely, the "last days of this present age." We presently stand at the cusp of a "New Age," Aquarius, represented by a Man pouring water (Spirit) over the Earth. Several terms in Biblical Scripture alert the student of prophecy to Judgment Day, also called "The Day of our Lord," or "The Great and Terrible Day of the Lord," which is the Final Day ending the *End of Days.*

We have some clues from various places in Biblical Scripture, as well as myth and legend from other ancient sources that indicate what will occur. One of these is found in II Acts in the Bible, where Peter quotes from the prophet Joel 16-20:

And it shall be in the last days,' God says, `That I will pour forth of my spirit (symbol of Aquarius) upon all mankind; And your sons and your daughters shall prophesy, and your young men shall see visions, and your old men shall dream dreams; Even upon my bond slaves, both men and women, I will in those days pour forth of my Spirit.

The apostle Paul, in writing to his spiritual son Timothy, gives us a different viewpoint on the difficult times that will be coming in the "last days." He depicts mankind as becoming very self-centered, unholy and treacherous. He says that men will love pleasure more than they love God, even though they might have a form of "religion." We find this in II Timothy 3:1-7:

But realize this, that in the last days, difficult times —will come. For men will be lovers of self, lovers of money, boastful, arrogant, revilers, disobedient to parents, ungrateful, unholy, —"unloving, irreconcilable, malicious gossips, without self-control, brutal, haters of good, - treacherous, reckless, conceited, lovers of pleasure rather, than lovers of God; Holding to a form of godliness, although they have denied its power; avoid such men as these. For among them are those who enter into households and captivate weak women weighed down with sins, led on by various impulses, always learning and never able to come to the knowledge of the truth."

According to the Egyptian seer, Asclepius, our society may have already met the first prerequisites of Armageddon:

Darkness will be preferred to light, and death will be thought more profitable than life; no one will raise his eyes to heaven; the pious will be deemed insane, and the impious wise; the madman will be thought a brave man, and the wicked will be esteemed as good. As to the soul, and the belief that it is immortal by nature, or may hope to attain immortality, as I have taught you, all this they will mock at, and will even persuade themselves that it is false... And so, the gods will depart from mankind, a grievous thing! And only evil angels will remain, who will mingle with men, and drive the poor wretches by main force into all manner of reckless crime, into wars, and robberies, and frauds, and all things hostile to the nature of the soul.

Moral degeneration is a common theme among End Time prophecies. According to Isaac Asimov's Book of Facts (1979), an Assyrian clay tablet dating to approximately 2800 B.C. was unearthed bearing the words *"Our earth is degenerate in these latter days. There are signs that the world is speedily coming to an end. Bribery and corruption are common."* This is one of the earliest examples of the perception of moral decay in society being interpreted as a sign of imminent end.

Modern interpretation of Biblical prophecies has found their way into the hearts and minds of millions. A pre-tribulation view of Armageddon was popularized in large measure by Hal Lindsay's bestseller, *The Late Great Planet Earth*, started weaving the Christian Armageddon scenario, based on his interpretation of the Book of Revelation. Christian writers have sold over a hundred million books describing, in their view, what will happen during the End Times. After the Rapture, the Apocalypse scenario follows chapter for chapter the calamities described in the Book of Revelation, authored by John the Evangelist; the appearance of the anti-Christ, an evil man who is Satan personified. He will mislead even the very elect (spiritual people who should be wise enough to know) into one world government hell..... The pre-Rapture comes first, then the Day of the Lord, then Anti-Christ who runs a one-world government, then Armageddon and then the Resurrection of the Just/Millennial Kingdom of the Lord Jesus Christ.

Modern day prophecy pundits try to show us evidence that the End Times are already here. ISIS, and al-Qaida, believe Mohammed's version of Armageddon has begun or will begin soon. The Terrorists have inserted themselves in the place of the Anti-Christ. They are pushing a theology of Armageddon as a final war between Islam, the Jews, and Christians. They believe their Caliphate will usher in the End of Days, and that the 12th Imam after Mohammad will return, ushering in a new Kingdom.

Islam has an *End of the World* checklist similar to Christians and Jews. This advertised "book" future described by prophecy pundits, is just that...an interpretation of prophecy which may have nothing to do with a future that actually happens. I am not saying the writers of these books and their interpretations of the future are wrong. I do believe; however, things will play out quite differently than they think.

Apocalyptic literature was written in obscure terminology, employing the mystical arts of astrology, mathematics, geometry, numerology, temple measures, and mythical symbolism, to hide the primal secret. In some cases, as with Nostradamus, this was done to protect life and family from religious persecution, but more often was to conceal the inner meaning of the prophetic tradition. Why? The knowledge was bluntly considered to be dangerous. What if you knew the precise day and hour of your death? How would it affect the living of your life?

A Turkish Scholar Bediuzzaman during the 1940's, stated to Said Nursi, about knowing the secret a planetary "appointed time."

If the appointed hour of death was specified, the first half of life would be passed in absolute heedlessness, and the second half in absolute terror, as every day a further step was taken towards the gallows. This would destroy the wise and beneficial balance of hope and fear. Similarly, if the end of the world, its death and appointed hour, had been specified, the Early and Middle Ages would have been virtually unaffected by the idea of the Hereafter, and the later ages would have been passed in terror. No pleasure or value would have remained in the worldly life, nor, as an act of will, would the worship of God, between hope and fear, have held any importance or purpose.

Also, if the death of the world had been specified, some of the truths of belief would have been clearly obvious and everyone would have affirmed them willy-nilly. The mystery of man's accountability and the wisdom and purpose of belief, which are tied to a man's choice and will, would have been negated. It is for numerous benefits such as these that matters related to the Unseen remain secret...It is because most hidden cosmic events are tied to such incidences of wisdom that to give news of the Unseen or to foretell events has been prohibited. In order not to be disrespectful and disobedient in the face of the principle None knows the Unseen save God, those who with dominical leaven even, give news of the Unseen other than concerning man's accountability and the truths of the belief, have done so allusively and indirectly.

In fact, the good tidings of the Prophet Muhammad (PBUH) I the Torah, Gospels, and Psalms, are veiled and obscure, in consequence of which some of the adherents of those scriptures put various meanings on those passages and did not believe them. However, since the wisdom in man's accountability necessitates that the questions included among the tenets of belief are communicated explicitly and repeatedly, the Qur'an Miraculous of Exposition and its Glorious Interpreter (Peace and blessings be upon him) tell of the matters of the Hereafter in detail, and of future worldly events only in summary fashion."[14]

The cyclic nature of the Apocalypse has only been revealed to the initiates of the secret mystery school traditions of the Ancient World. The most famous of these were the Greek schools of Pythagoras, the Egyptian Schools of Temple Building, including the Masonic Order and Freemasons, and perhaps the most ancient, the Chaldean Schools of Astronomy, where the patriarch Abraham, learned of the secret traditions from the descendants of Noah, the survivors of the World Deluge.

That it is an "appointed" time is echoed in Mark 13:33:

Take heed, keep on the alert; for you do not know when the appointed time is. "It is like a man, away on a journey, who upon leaving his house and putting his slaves in charge, assigning to each one his task, also commanded the doorkeeper to stay on the alert.

Therefore, be on the alert for you do not know when the master of the house is coming, whether in the evening, at midnight, at cockcrowing, or in the morning, lest he come suddenly and find you asleep.

The Dead Saints and the prophets exhort. Stay alert. Pay attention. It is clear when the "event" occurs, it will be sudden, unseen, and according to the prophecies, it will happen during a time of peace. I think that this is pointed out very beautifully in 1 Thessalonians 5:1-3:

Now as to the times and the epochs, brethren you have no need of anything to be written to you. For you yourselves know full well that the day of the Lord will come like a thief in the night. While they are saying, 'Peace and safety!' then destruction will come upon them suddenly like birth pangs upon a woman with child; and they shall not escape.

Prophets and Dead Saints believe a future Apocalypse will happen because a similar Apocalypses destroyed Earth *many times before* and *will happen again* when the Apocalyptic clock strikes Midnight. Ancient Egyptians held a similar understanding of periodic cosmic catastrophe.

---

14 Nicola Costa Adam to Apophis (The Rays – The Fifth Ray pp.99-100, Turkish Scholar Bediuzzaman Said Nursi (1878-1960). pp. 27-28.

Over 2,500 years ago, Solon, the lawmaker and sage of Athens, in his youth had visited the High Priest Senehis in the temple of Vulcan in the city of Sais in Egypt.

The following narrative from Plato's Timaeus, describes the conversation between Solon and the venerable priest about recurring catastrophes:

"O Solon, Solon, you Greeks are but children, and there is never an old man who is a Greek."

Solon, hearing this said: "What do you mean?"

I mean to say that in mind you are all young, for there is no opinion handed down among you by ancient tradition, nor any science that is hoary with age. And this is the cause thereof: There have been and there will be again, many destructions of mankind arising out of varied causes, the most violent these caused by fire and water, and lesser ones in a thousand other ways. You have preserved a story that Phaethon, the child of Helios, the sun, having yoked the steeds of his father's chariot, because he was not able to drive them in the path of his father's usual path, burned up all that was on top of the earth and was himself destroyed by a thunderbolt.

Now this has the form of a myth, but really it signified a deviation from their courses of the bodies moving around the earth, and in the heavens, and indeed a great conflagration of things on earth by fierce fire which recurs at long intervals. From this calamity, the fact that we live on the low-lying land by the Nile, who is our savior, we are delivered. When, on the other hand, the gods purge the earth with a deluge of water, among you herdsmen and shepherds on the mountains are the survivors, whereas those of you who live in cities are swept away by the waters into the sea. In this country the waters come not from the heavens but up from below and these are predictable and help the crops, for which reason it is said that the libraries preserved here are the oldest.

The fact is, that whenever the winter frost or summer sun does not prevent, the human race is always increasing, and when nature does interfere it diminishes. Whatever has happened to your country or to ours or to any other region of which we are informed, if any actions which are noble or great, or in any other way remarkable have taken place, it has been written down of old in histories, and is preserved in our temples. Whereas, you and the other nations having just provided yourselves with letters and other things which states required, when after the usual interval of years, like a plague, the flood from heaven comes sweeping down afresh upon your people, and

*leaves none of you but the unlettered and uncultured, so that you
become young as ever, with no knowledge of all that happened in old
times in this land or in your own.*

*As for those genealogies which you related just now, Solon,
concerning the people of your country, are little better than children's
tales, for in the first place you remember but one deluge only, where
there have been four with many lesser ones.*

It is clear from this account passed on by Solon, that the
Egyptians believed Firefall —Armageddon fire— periodically
destroyed the Earth in regular, predicable intervals. The ancients
believed Comets were the source of danger, and that they returned,
regular predictable intervals. In many ways it typifies our
understanding today of the periodicity of comets, howbeit, an
unusual one. Most comets have orbits with periodicities less than
125 years, yet some do not return after several thousand years,
sometimes millions of years.

Yet, the Egyptians were not referring to cycles of millions of
years, but shorter periods, perhaps only a few thousand years in
length.

While Plato may not have totally comprehended the true nature
of the information passed down to him from the Egyptian sages, he
grasped the essential facts: the descent of a world-consuming fire
which periodically destroyed men, plants, and animals here on earth
at regular intervals which could be predicted, much in the same
manner we predict the return of a periodic comet.

The Prophet Amos reveals the same knowledge; "Woe unto you that
desire the day of the LORD! to what end is it for you? the day of the
LORD is darkness, and not light." (Amos 5:18)

In the *Book of Enoch*, which is widely considered to be a collection of
works by different authors, some corroborated by the Dead Sea Scrolls
written around 200 B.C., but some parts are still believed to be missing.
The first chapter starts by informing the reader that a great judgment
will be sent upon the Earth in form of a deluge (Noah's flood), and only
a handful of select members of the human race will survive. Enoch warns
*of another judgment yet to come after the deluge, a prophecy referred to
by Jude in the New Testament:*

*'I saw seven stars like great burning mountains, and to me, when I
inquired regarding them, the angel said: 'This place is the end of heaven
and earth; this has become a prison for the stars and the host of heaven.
And the stars which roll over the fire are they which have transgressed
the commandment of the Lord in the beginning of their rising, because
they did not come forth in their appointed times. And He was wroth*

with them and bound them till the time when their guilt shall be consummated (even) ten thousand years.'

(In a dream) Enoch watches the future apocalypse happen:

'And again I saw with mine eyes as I slept, and I saw the heaven above, and behold a star fell from heaven, ...And behold I saw many stars descend and cast themselves down from heaven to that first star...and behold all the children of the Earth began to tremble and quake before them and to flee from them...'

Signs of the End of Times?

#1 Sign: Wars and rumors of wars. The threat of Nuclear War/WW III remains an uneasy possibility because Russian incursions into Ukraine, Crimea, and Syria. ISIS and other terrorist groups are seeking nuclear, chemical, and biological weapons. China is confronting the US on its border and in the South China Sea. On January 6, 2016, *The Telegraph* 15 reported, "North Korea claimed to have achieved the "next level" of "nuclear might" yesterday after conducting the underground explosion of a "hydrogen bomb." This was the country's fourth nuclear test since 2006, but the first allegedly to involve a hydrogen weapon – also known as thermonuclear device. Bombs of this kind, which rely on nuclear fusion, release far more destructive power than other weapons based on a fission reaction.

If North Korea's claim was true, the country would have joined the ranks of the world's thermonuclear powers".

Then, on February 7, 2016, CNN Seoul 16 reported "North Korea launched a satellite into space Sunday. The launch triggered a wave of international condemnation and prompted a strong reaction from an emergency meeting of the U.N. Security Council. Though North Kora said the launch was for scientific and "peaceful purposes," it is being widely viewed by other nations as a front to test a ballistic missile, especially coming on the heels of North Korea's purported hydrogen bomb test last month. Pyongyang carried out both acts in defiance of international sanctions." 17

~ISIS announced formation of the Caliphate covering a third of Syria and Iraq. The announcement was on the first day of the month of

---

15 Julian Ryall in Tokyo and David Blair in an article written in the telegraph.

16 By Ralph Ellis, K.J. Kwon, Tiffany, AP, and Tim Hume, CNN.

17 U.N. Security Council condemns North Korean rocket launch, Updated 5:15 PM ET, Sun February 7, 2016

Ramadan 2014. The Shura Council of the Islamic State of Iraq and al-Sham (ISIS) decided to bring back the caliphate, declared its rule once again, and proclaimed Abu Bakr al-Baghdadi as the first "caliph of Islam" in this new era. This is a dramatic, I believe, prophetic development. ~Chronicle 380

February 12, 2016. *The New Cold War. Russia invades and takes over Georgia in 2008. Russia invades Crimea in 2014, after hosting Olympics. NATO alliance is treaty bound to defend. If Russia invaded the 3 Baltic states, Estonia, Latvia, Lithuania. They would be defeated in 3 days. US use to have 300,000 troops in the 1980's. Now only 30,000. Russia has ramped up their military maneuvers since the Cold War; aggressive close aircraft flyby's (within 15 feet)*

February 13, 2016. *Obama ramping up asking for Quadruple the amount of funding for expanding troops in Europe. "Rand points out" points out a need for an increased commitment of troops. If the United States and NATO ramp up and match Russian troop size and ability to mobilize, currently 3 to 1, this MA.D. thinking has initiated a new Cold War...yesterday.*

#2 Sign: Earthquakes, Volcanic Eruptions, and Great Tsunami's

We have experienced many horrific global catastrophes since Year 2000: (add)

2007 Chinese Earthquake – 50,000 dead
2010 South Pacific Tsunami – 250,000 dead
2012 9.0 Japan Earthquake & Tsunami- 30,000 dead.

Are these catastrophes signs of the End of Days? As terrible as they are, these catastrophes pale in comparison to ancient and modern prophecies about the millions or billions who will perish during the End of Days.

#3 Sign: Celestial

2014 - 2015 Jewish Holiday Eclipse Schedule

John Hagee has brought the Four Blood Moon theory and its relevance to the Nation of Israel into millions of US Christian households. He believes, "According to the Biblical prophecy, world history is about to change dramatically. As for what that "hugely significant event" might be, no one really knows. In all likelihood, nothing will happen — or, more likely, given the annoying human capacity of finding patterns where there are none, *something* will happen somewhere, and we'll attribute it to the lunar eclipse, even though there's absolutely no causal link.

If you do believe in *signs*, though, this celestial alignment (or *syzygy* to give its scientific name) combined with a tetrad is a sign that something significant is about to happen. Pastor John Hagee, who has written a book on the topic, *Four Blood Moons*, believes that these rare tetrad moons are astronomical signs concerning events that may fulfill Biblical prophecies concerning Israel. Every time [this alignment has happened] in the last 500 years, it has coincided with tragedy for the Jewish people followed by triumph.

In 1493, a tetrad coincided with the expulsion of the Jews by the Spanish Inquisition; in 1949, the establishment of Israel occurred during a tetrad; and the last tetrad, in 1967, coincided with the Six-Day War.

"And once again, for Israel, the timing of this Tetrad is remarkable," said Hagee. The first lunar eclipse of this tetrad occurred on April, 2014, and coincided with the Jewish festival of Passover the same evening. "Second will be on October 8, at the time of the Feast of the Tabernacles." On April 4, 2015, during Passover, we will have another blood moon. Then finally, on September 28, during next year's Feast of the Tabernacles, the fourth blood and final moon will dawn."

The last Blood Moon occurred on September 28, 2015. While Israel had a devastating War with Hamas during the summer of 2015 that cost several thousand lives, it was not a global war. I would not identify these blood moons or this war as a sign revealing that we are nearing the End of Days.

Total Solar Eclipse: August 21, 2017

As predicted by astronomer's decades in advance, the total eclipse of the sun begins with the shadow arriving with perfect accuracy and touches down in the north Pacific Ocean at 16:48:33 UT*, at local sunrise. (At that spot, the sun will actually rise while eclipsed. This is a sight few people - even veteran eclipse chasers - have seen, and from what we hear, it is quite uncanny.)

Woman Crowned with 12 Stars

Precisely thirty-two days later, on September 23, 2017, another wonder will appear in the Heavens. It is spoken of in Rev. 12:1-2:
Now a great sign appeared in heaven: a woman clothed with the sun, with the moon under her feet, and on her head a garland of twelve

stars. Then being with child, she cried out in labor and in pain to give birth."

On September 23, 2017, the constellation Virgo will be clothed in the sun, the moon will be under her feet, and on her head will be the twelve stars of Leo the Lion. She then gives birth to Jupiter. This heavenly configuration only happens on September 23, 2017. (Someone needs to check with an astrologer to verify how often this astronomical configuration occurs). Further, the Woman crowned with stars, occurs the day AFTER Yom *Teruah—Rashashah*, Feast of Trumpets (Why important? Define)

Are these the heavenly signs spoken of by the Apostles, Christ, and John in the Book of Revelation? Biblical and Islamic Scripture indicate we should look for signs in the skies, including Sun turning black as sackcloth (solar eclipse) and the moon turning blood red (lunar eclipse), but these occur nearly every year, so the prophecies by Christ only point generally to the skies.

Astronomical signs revealing signs of Armageddon are numerous throughout mythology and sacred scripture;

The sun will be turned to darkness and the moon to blood before the coming of the great and dreadful day of the LORD. Joel 2:31. Of course, a dark sun (solar eclipse) and red blood moon (lunar eclipse) occur every year, fairly often, so dating Joel's prophecy is nearly impossible from these few clues.

Genesis of the Angry God

In Book I of the Dead Saints Chronicles, I make the statement. (No Angry God here!) The Creator is never angry. However, the Old Testament describes Judgment from God raining down from Heaven, beginning from the Garden of Eden for Adam and Eve's disobedience, Noah's Flood because all of men had become evil, Sodom and Gomorrah for their immorality and sins, Moses invoking God's Wrath, upon Pharaoh, smiting him with the Plagues of Egypt, Elijah calling down fire from Heaven to destroy the Priests of Baal. The keys to understanding evolve from Fire descending from the Heavens above, destroying humans, animals, and plant life locally, regionally, and sometimes globally, as in the case of the Great Flood, or in Judaism Noah's Flood.

It's important to convey that it is not a Loving God sending a rain of fire to destroy evil men. It is not an act of Judgment. Any space rock or comet fragment that pierces the atmosphere raining fire on the Earth below, has been out there for billions of years. Just as volcanoes explode, or devastating earthquakes crumble our cities, or tsunami wash over the land killing all in its path, it is not God's wrath employed against humanity.

All of these destructive events are managed by our Creator. The Sun, the Moon, the orbits of the planets, and ...comets and asteroids— four-billion-year-old debris that sometimes strikes our planet. It is only humanity's interpretation of these destructive events that give them negative meaning, even the Final Judgment. Death and destruction may be "unpleasant," but it is NOT an angry God Judging us.

Judgment Day

Falling stars epitomizes our fears of Judgment Day...A Day when the heavens roll up like a scroll; the stars fall like rain towards the earth; loud explosions are heard in the sky; the ground shakes; buildings collapse; there is confusion and fire everywhere. As Abraham Lincoln witnessed during the Leonid Meteor storm of 1833, the sky filled with millions of falling stars. Immediately, in the mind of Lincoln, it invoked Biblical verses from the Book of Revelation. I'm sure Lincoln thought for a moment, "Is this the DAY? Is this the Day of the coming of our Lord? The Day of His promised return? Am I ready?

The meteor storm of November 12–13, 1833, caused an innkeeper to awaken Abraham Lincoln and exclaim that the "Day of Judgment" had come. This illustration of the 1833 Leonid's by R. M. Eldridge, depicting the shower as a fulfillment of Biblical prophecy, appeared in an early 20th-century book, Bible Readings for the Home. Right: A manuscript page in the handwriting of Walt Whitman preserves Lincoln's eyewitness description of a great meteor shower — almost certainly the 1833 Leonid's.

In the talk after business was settled, one of the big Dons asked Mr. Lincoln if his confidence in the permanency of the Union was not beginning to be shaken — whereupon the homely President told a little story. "When I was a young man in Illinois," said he, "I boarded for a time with a Deacon of the Presbyterian church. One night I was roused from my sleep by a rap at the door, & I heard the Deacon's voice exclaiming 'Arise, Abraham, the Day of Judgment has come!' I sprang from my bed & rushed to the window and saw the stars falling in great showers! But looking back at them in the heavens I saw all the grand old constellations with which I was so well acquainted, fixed and true in their places. Gentlemen, the world did not come to an end then, nor will the Union now."

The great Leonid meteor storm on the morning of November 13, 1833, dazzled North Americans with a celestial spectacle they would never forget. From Canada to Mexico, the streaks of light filling the sky and the shouting in the streets roused people from their beds to witness what many assumed was the end of the world. Recently we discovered a document indicating that one of the eyewitnesses was a future U.S. president, Abraham Lincoln. Startled from his slumber early one morning, a young Lincoln beheld the sky filled with falling stars and fireballs. Decades later the battlefield flames of the Civil War rekindled in him a memory of those fiery heavens. During one of the darkest times of the war, President Lincoln related an anecdote about the meteor shower and the fixed stars as a metaphor for the state of the Union.[18]

Every year the return of a bright comet brings out our religious fervor, because 2.2 billion Christians know the prophecies made by Jesus Christ, his apostles, and especially those written in by John the Evangelist in the Book of Revelation. 1.6 billion Muslims know the prophecies of Mohammad Qur'an 82 verse, The Day. Muslims believed in the *Qiyamah* (Last Judgment) during which time Jesus will come to earth, end all wars, and kill *ad-Dajjal* - the Muslim anti-Christ. Then every person who ever lived will be bodily resurrected, before being judged by God. The faithful go to heaven, and the rest to hell. Apparently there's also room for some "People of the Book," i.e. Jews and Christians.

Their prophets are selling apocalyptic fervor like candy. Judgment Day is near— even night upon us. A DAY that begins with fire falling like rain from the sky accompanied by clouds and loud explosions. It's in our bones. It's in our DNA. And this fear goes far back in time, many thousands of millennia and we don't know why.

In 1991, my original title to the Armageddon Stones was the Final Judgment (Scan in old David Book Picture). Ancient mythologies describe the future destruction of our world as a final judgment at the end of each Age because of man's disobedience, declining morality, and evil actions. Each judgment is accompanied by stones that fall from heaven destroying the world by fire and water. Mythology and religion have integrated the Judgment and Planetary Death with our own personal death. Barbara Walker writes in *The Crone*, "The hidden connection between doomsday and the dissolution of the individual, in ordinary death, is plainly shown in the Manichaean hymn called *The Ship of God*, where the body and cosmos are interchangeably mingled:

All the comets quivered, and the stars whirled about, and each of the planets turned awry in its course. The earth shook, my foundation beneath, and the height of the heavens sank down above. All the rivers,

---

18 Review and Herald Publishing Association Library of Congress/Charles E. Feinberg Walt Whitman Collection: Sky & Telescope November 1999 35 nascences" in a volume titled Specimen Days & Collect (1882).

*the veins in my body, dried up at their source. All my limbs have connection no longer... The reckoning of my days and months is ended. Harm befell the course of the zodiac's wheel. The seal of my feet and the joints of my toes—each link of life of my soul was loosed. Each joint of my hands and fingers—each was loosed, and its seal taken off. All the grisly parts—their life grew feeble. And cold became each one of my limbs.*19

It appears Isaiah ben Amoz also knew about the prophecies contained in the Manichean hymn, also written during the 6th century B.C. in Iceland, who composed his writings in 515 B.C. during the Babylonian captivity. It is one of the most popular works among Jews in the Second Temple period (c. 515 B.C.E - 70 CE), and in Christian circles it was held in such high regard as to be called "the Fifth Gospel." It is clear the Prophet Isaiah understood 2,800 years ago the Apocalypse we have yet to experience today: He writes in Isaiah 28:9-22:

Whom shall he teach knowledge? and whom shall he make to understand doctrine? them that are weaned from the milk, and drawn from the breasts. For precept must be upon precept, precept upon precept; line upon line, line upon line; here a little, and there a little: For with stammering lips and another tongue will he speak to this people. To whom he said, This is the rest wherewith ye may cause the weary to rest; and this is the refreshing: yet they would not hear. But the word of the Lord was unto them precept upon precept, precept upon precept; line upon line, line upon line; here a little, and there a little; that they might go, and fall backward, and be broken, and snared, and taken. Wherefore hear the word of the Lord, ye scornful men, that rule this people which is in Jerusalem.

Because ye have said, We have made a covenant with death, and with hell are we at agreement; when the overflowing scourge shall pass through, it shall not come unto us: for we have made lies our refuge, and under falsehood have we hid ourselves: Therefore thus saith the Lord God, Behold, I lay in Zion for a foundation a stone, a tried stone, a precious corner stone, a sure foundation: he that believeth shall not make haste. Judgment also will I lay to the line, and righteousness to the plummet: and the hail shall sweep away the refuge of lies, and the waters shall overflow the hiding place. And your covenant with death shall be disannulled, and your agreement with hell shall not stand; when the overflowing scourge shall pass through, then ye shall be trodden down by it.

From the time that it goeth forth it shall take you: for morning by morning shall it pass over, by day and by night: and it shall be a vexation only to understand the report. For the bed is shorter than that a man

---

19 Barnstone, Willis, E.D. 1984, The Other Bible, San Francisco: Harper and Row. P. 321.

can stretch himself on it: and the covering narrower than that he can wrap himself in it. For the Lord shall rise up as in mount Perazim, he shall be wroth as in the valley of Gibeon, that he may do his work, his strange work; and bring to pass his act, his strange act.

Now therefore be ye not mockers, lest your bands be made strong: for I have heard from the Lord God of hosts a consumption, even determined upon the whole Earth.

—————— **3** ——————

Something BIG This Way Comes

She that hath borne seven languisheth she hath given up the ghost; her sun is gone down while it was yet day the residue of them I will deliver to the sword before their enemies.
~Jer 15:9

The Dead Saints prophecy something BIG is coming soon.

Very soon.

The event appears to be destructive, causing planetary destruction within a span of 1 minute—or less. It could happen anytime. It could be tomorrow; but according to the prophets, it will likely occur in the next 20 years or less.

What will cause such great destruction? Will millions perish in a destructive major event? Is this event set in stone?

According to Carl, who was pronounced clinically dead of smoke inhalation says a significant event is coming where we all right now have an important role to play;

I collapsed, passed out. The next thing I knew I was in an operating room where people were working on my body. It's like I was there, but I wasn't there. I died, or at least they said I'd died – I was clinically dead "I was shown "some things" and others were "hidden" temporarily. I'm not some kind of prophet, nor was I expected to give anyone warnings. I was asked to "help," nothing more, if I could. It was as if I've become a "tool." THEY told me if I failed, another or others would be sent. I can be replaced, but I wasn't meant to feel replaceable. THEY didn't make little of my volunteering, but THEY also wanted me to be at peace if I was unable to accomplish what I was sent to do. I think THEY were being kind, but now that I'm back, and thinking like a human again, I sometimes wonder if THEY lacked belief in my ability to do what they asked of me. When I think of HOME, I know this cannot be true. THEY would never ask something you're incapable of, nor would THEY attempt to bring you harm, or pain of any kind. Something important and vital is coming. THEY felt that those who wanted to help should be allowed to do so, if it was their choice."[20]

Sybel agrees. Her heart stopped and she saw her spirit come out of her body. During her NDE, she visits with Jesus and sees people busy preparing for something really BIG:

Then suddenly my heart stopped. I could feel my spirit come out of my body and I saw my body lying there as if asleep. I was carried as if my angels, though I did not see them far up into the heavens and at a great speed. I was taken to a place surrounded by music. It was like no music I have ever heard. More beautiful than all the best music my ears have heard and it came from everywhere. I stood there and suddenly I saw Jesus. It was quite an awesome shock to be standing in His presence. I was in such awe, I could not speak. My eyes were fixed upon Him. His hair was white like wool and hung down to His shoulders. His skin was like brass without one wrinkle. His eyes were like flames of fire and when He spoke, it was with great authority. Yet when He spoke it was kind and gentle and loving. When He spoke it sounded like thunder rolling across the North Carolina skies but much louder than any I had ever heard. You would think I would have been afraid but I was not. Quite the contrary. I felt happy! I felt so happy, a feeling like none other I had ever felt...."

...I saw Jesus. I know He is real. I don't have to doubt. I know! I also know I was the most happy person alive at that moment. I have never seen hell nor do I want to, but I have seen Jesus and I have been to heaven. You can think I am crazy but I am as sane as a person can be. You might think it was a dream but your body temperature doesn't drop in a dream. All I can tell you is Heaven is real. **There are people there and they are very busy getting ready for something really big"**[21]

20 Carl D NDE, #3163, 10.16.12, NDERF.org
21 Sybel S's NDE, #214, 03.02.03, NDERF.org

What is the Big Event?

Will the BIG EVENT be WWIII (nuclear, biological, chemical) solar flare, earthquake, tsunami, polar shift, volcanic eruption, asteroid, or comet impact? Or all of the above? What do the prophets in the Bible say? Do other religions prophesize Armageddon and the Apocalypse? How do ancient prophecies match up with modern prophecies of the Dead Saints?

Dead Saint Vision of Future Mass Destruction

James was born in 1939 and had his NDE in 1947, which was recorded on the NDERF.org website in 2010. His wife, seven years younger, was born in 1946. Currently they are aged 76 and 69 respectively as of 2016. The background is important because if James' vision of a "mass extinction" event in the future received during his NDE is accurate as claimed, it will most likely occur in the next 20 years if his wife lives to 89 years old. If his wife lives to an average of 84 years, she will die in the year 2028:

> *I was eight years old. In the late summer of 1947, I contracted a fatal illness that resulted in my death approximately 11 hours later. When I came out of my body I watched and heard the Doctor and Nurses as they declared me dead…*
>
> *At that time an Angel approached me and I asked him what was happening. He told me that I had died and that as soon as the other Angel arrived we would go to Paradise. We spent several minutes there and watched as the Doctor talked with my parents.*
>
> *When the other Angel arrived we departed through the wall traveling what most describe as a tunnel until we reached the gates of Paradise. Arriving there I first saw someone that looked a little like me and when I asked who this person was I was told it was my brother. At that time, I never knew I had a brother that had died. Soon we were joined by two others that were introduced as another brother and sister, a set of twins that had also died. I met and saw many of my blood line down through the ages.*
>
> *When I was told why we were waiting, I demanded to see Jesus. Jesus asked me to return to my body because my family needed me. I argued a while, but Jesus promised me an additional 10 years to my life and that he would watch over me all the years of my life if I would return.*
>
> *Still objecting, He instructed that I be shown what would happen if I did not return. When I saw the vision of my father being taken to jail and my mother in a casket I was devastated. Then Jesus allowed me to*

be shown the future of the world. Wars, strife, no rain, death, earthquakes, etc.

Upon returning to my body, which was a unique experience itself, I was instructed by the Angels to sleep and when I awoke I would be free of the illness, except that I may not remember some things about my early life. At the moment that did not sink in until I could not even recognize my parents and siblings when they came to visit me in the hospital.

The hospital staff began called my "Miracle Boy". The local newspapers picked up the story and when I was discharged from the hospital there were several photographers and other people to greet me and wish me well. They also gave me some flowers.

I soon learned that telling this story was not the thing to do because, in those days and in that community, people did terrible things to you and thought you were crazy or a devil. So, after several encounters at school my parents moved us from the mountain to where no one knew us and I was instructed my parents never to speak of the event again. I told no-one of the event and was shocked 20 years later when I married that my sister told my wife about the experience which caused a little problem. I told my wife that my sister is a little strange and dismissed what she told my wife.

(As I said) I was shown a vision of my future wife, who at the time would have been only four months old, the events leading up to the destruction of our world, a vision of what would happen to my family if I did not return. So far, everything has been totally accurate even including the current political condition of the world and of our President. (Obama).

For over sixty years I have had to lead a double life, hiding what I know. I have had five heart attacks and survived two terminal cancers. My additional ten years will be up soon. When the Doctors told my wife and I that I would live less than six months my wife then told my daughters about what my sister had told her and my daughter pumped me to tell her the whole story. Her I cannot resist, so about ten years ago I told my daughter and, well, let's say the jury is stills out on her opinion. When I tell my wife about events that will occur she is astounded that I have never been wrong. I can't tell her too much because she will be frightened and it will have a very bad effect on her.

What would happen if I told [everyone], and everyone believed the events to come? It has been hard enough living as long as I have knowing the terrible things that my wife and children will endure."[22]

Sobering, isn't it? James doesn't say what causes the future mass extinction. *What event will deal the deadly blow?* Nuclear War? Disease?

22 James C NDE, #2137, 06.20.10, NDERF.org

Pestilence? Natural calamities such as earthquakes, volcanic eruptions, Tsunami? Exploding comet fragments? Bolide/asteroid impacts?

How bad could it get? According to Scripture, very bad:

For then shall be great tribulation, such as was not since the beginning of the world to this time, no, nor ever shall be. And except those days should be shortened, there should no flesh be saved: but for the elect's sake those days shall be shortened." Matt 24:21.

Part II

Paradise Lost

—— **4** ——

10,500 B.C: Atlantis, Mexico, Egypt & South America

Ancient Egypt

Dead Saint Memory of Ancient Egypt

~*A wetter ancient Egypt has support from Zach, "Somewhere between 10pm and 3am the following events took place. I was asleep and at the same time very aware that I was physically travelling very fast through time. I was shown my life starting at the Stone Age, I was aware that I was hunting and interacting with other people, I was aware that I was a young man. I could see the landscape and saw my life being revealed in a minute to minute manner. This period ended in a flash of White and Orange. I then moved to the next life, I was in the Mountains and an Ocean where I was riding horses. I then moved to Europe and again aware of old buildings and Cities. I wasn't here for long and no there was no specific city I was aware of. I then came into the pyramid age and was approximately 20 to 30 years of age and was employed to assist in the building of the pyramids.*

At the time of the pyramids the area was not a desert, it seemed very lush, I saw in detail how the pyramids were built. Again I saw my life in a moment to moment manner and was aware of eating dinner with other workers. There was no sense that we were slaves or in the situation under duress. I left this at approximately 40 years of age."[23]

October 5-19, 1992

Paul Solomon teamed up with John Anthony West, a rogue Egyptologist who had garnished world-wide attention regarding a much "older" Sphinx. Mr. West was truly onto something earthshattering. USA TODAY 1991 headline, "If the Sphinx predates dynastic Egypt, this would have major archeologic implications" says Egyptologist John Anthony West. "Quite simply, we would have to rewrite the history of when advanced civilization began." The trip was billed as the trip of a lifetime. I remember thinking, "This is great!" My mind reeled with possibilities.

John Anthony West

The Sphinx.

Many believe its form was carved by the Gods to be a guardian of the pyramids. The Sphinx symbol wasn't limited to Egypt, but was also found in ancient Phoenician, Syrian, and Greek societies. In Greek legend, the Sphinx devoured all travelers who could not answer the riddle it posed: "What is the creature that walks on four legs in the morning, two legs at noon and three in the evening?"

The hero Oedipus gave the answer, "Man," causing the Sphinx's death.

The lion was a powerful symbol in ancient Egypt as it represented strength and courage. The great cat was also considered the supreme guardian and tamed lions sometimes accompanied kings into battle. Not just as a mascot, but as the physical presence of a god meant to protect troops. The Sphinx was the combination of two symbols, LION GOD—ZODIAC LION and KING PHAROAH / ZODIAC MAN-AQUARIUS into one icon. But whose HUMAN FACE sculpted into the other icon? History believes it bears the face of the ruling pharaoh at the time of its construction: Chephren.

---

23 Zack J's Probable NDE, #2438, 10.11.10, NDERF.org

But was it Chephren?

Some Egyptologists and Archaeologists believe the Sphinx is much older than we think, and dates to a time, many thousands of years earlier. If this were the case, the Head of the Sphinx could not be Chephren. Conventional science has held that the Sphinx was carved out of an outcropping during the reign of King Khafre around 2500 B.C., but other Egyptologist and archaeologists think differently.

One such rogue Egyptologist is John Anthony West. In 1979, he wrote a book entitled *Serpent in the Sky*. In the book West suggested that the Sphinx was far older than the pyramids and its severe erosion was the result of rain, not blowing sand. Therefore, concluded West, the Sphinx must have been built thousands of years earlier when the land was much wetter.

Nobody gave West's theory much attention until West brought in a trained geologist from Boston University named Robert Schoch. Schoch examined the Sphinx and thinks some of the fissures in the rock were indeed created by running water or rain. His conclusion is that the front and side of the Sphinx dated from 5000 to 7000 B.C. and was remodeled during Khafre's era to give the likeness of the pharaoh. Other Egyptologists argue that the original estimate is still right and that the fissures found by Schoch were the result of wet sand being blown up from the Nile River, not rain. As of 2016, both Schock and West believe the Sphinx is yet older than their original estimates and have pushed back its construction another 3000 years to 10,500 B.C. or older. Other megalithic sites around the world are being dated to this ancient time, a period when a great catastrophe occurred destroying many world civilizations.

"The architects of the Sphinx could well be one of those legendary lost civilizations that appears in the mythology of people all over the globe." he says. "But I hate to use the "A" word because it drives everybody so crazy."

That's "A" for Atlantis.

In fact, the Chephren temple complex in front of the Sphinx, attributed to Pharaoh Chephren of the 4th dynasty, supposedly built about the same time as the Sphinx in 2,200 B.C. appears to have been *built by two distinct cultures,* although archaeologists will defiantly argue they are of one culture. Giant limestone blocks, some weighing 450 tons appear tremendously ancient, with granite ashlars cut to fit over ancient worn limestone. The same applies to the Pyramids, especially the Chephren Pyramid, where huge limestone blocks can be seen to underlie the granite and limestone coverings. It is part of a megalithic culture which can be found world- wide. The Chephren temple complex, like Tel Megiddo in Israel, appeared to have been built over another more ancient, and even more impressive sacred site. Something that looked as if it had been built by giants.... a belief upheld by Arabic legends and curiously enough in the Bible in the book of Genesis, "In those days there were giants..."

The Sphinx has a secret- The Hall of Records

Naturally, the question arises, "Who left behind such an incredible technology, and why should it be discovered after being lost for thousands of years?" According to the sleeping Solomon, the Hall of Records was preserved by a dying race, priests/scientists from "Atlantis".

"These instruments, words, teachings, ideas, understandings were preserved by a group of people that you might call a dying race, or a finishing of their time. And these instruments and records [contained in the Hall of Records] related as much to their own mistakes as to their discoveries. And they were hidden in such a way, or preserved in such a way, to be awakened at a time when the needs of earth were similar to the time of their creation-- as if these masters, teachers, were saying, "In the evolution of mankind and the development of his technology and such, he will again arrive at the point we have arrived at, and will make some of those mistakes that we have made, and could very well destroy his earth, his land, his race, as we have destroyed so much. But let that we've discovered through our own mistakes and our growth, as well, be made available to that next race that should reach such a point. So you see, it is not as if an ego minded, superior race is handing down a gift of understanding or knowledge to a lesser people. But rather recognizing that they had learned and realizing that there would be a time of need for this understanding left it to become available and preserved it in such a way that it was not likely to be discovered until the race reached that state of consciousness and level of technology, as would allow for its discovery."

There have been rumors of passageways and secret chambers surrounding the Sphinx and during recent restoration work several tunnels have been re-discovered. One, near the rear of the statue extends down into it for about nine yards. Another, behind the head, is a short dead-end shaft. The third, located midway between the tail and the paws, was apparently opened during restoration work in the 1920's, then resealed. It is unknown whether these tunnels were constructed by the original Egyptian designers or were cut into the statue at a later date. Many scientists speculate they are the result of ancient treasure hunting efforts.

Several attempts have been made to use non-invasive exploration techniques to ascertain if there are other hidden chambers or tunnels about the Sphinx. These include electromagnetic sounding, seismic refraction, seismic reflection, refraction tomography, electrical resistivity and acoustical survey tests. Studies made by Florida State University, Waseda University (Japan), and Boston University, have found "anomalies" around the Sphinx. These could be interpreted as chambers or passageways, but they could also be such natural features as faults or changes in the density of the rock. Egyptian archaeologists, charged with preserving the statue, are concerned about the danger of digging or drilling into the natural rock near the Sphinx to find out if cavities really exist.

Are these "anomalies" secret chambers? And is it worth risking damage to such a work as the Sphinx in order to find out? In their research using seismographs to calculate the depth of erosion in the body of the Sphinx, the team of archaeologists have uncovered the existence of three cavities under the limestone floor on the side of the Sphinx and between the paws. The existence of such cavities has been rumored for centuries and are thought to contain the records of the society that built the Sphinx. Arabic legends further describe that beneath the Sphinx is a record of a race of beings now extinct encapsulated in a monument called the Hall of Records.

Arabian authors of the early Christian era, passed on several interesting bits of legend that refer to the Great Pyramid, its history, and its ancient contents. Mohammed Ebn Abd Al Hokm stated, *"The Pyramids were constructed by Sheddad Ben Ad before the deluge."*[24] Abou Mohammed Al Hassan Ben Ahmad Ben Yakub Al Hamadani said, *"The Pyramids were antediluvian, and they resisted the force of the great flood."*[25]

In the Long Road Back, pp 137-138, the author writes, *"In about 390 A.D., the Roman Ammianus Marcellinus noted: "Inscriptions which the ancients asserted were in the walls of certain underground galleries in the pyramids were intended to prevent the ancient wisdom from being lost in the Flood. Abou Balkh, in 870, preserved the same tradition, writes: "Wise men, previous to the Flood, foreseeing an impending judgment from heaven which would destroy every created thing, built upon a plateau in Egypt pyramids of stone in order to have some refuge against the calamity."*

The most detailed pyramid tradition recounted, however, was by Ibn Abd Alhokim, and it can still be read today in the Akbar Ezzeman Manuscript, kept at Oxford. Pyramidologist John Greaves, in 1646, offered this translation of the tradition in his work.

---

24 Operations Carried On at the Pyramids of Gizeh in 1837, Volume 2

By Howard Vyse

25 Publisher:مطبعة دار الكتب والوثائق القومية بالقاهرة, [Cairo], 2004 / Maṭbaʿat Dār al-Kutub wa-al-Wathāʾiq al-Qawmīyah bi-al-Qāhirah, [Cairo], 2004

From the Enduring Mystery:

"He which built the Pyramid was Saurid Ibn Salhouk, who lived three hundred years before the flood. The occasion of this was because he saw in his sleep, that the whole earth was turned over, with the inhabitants of it, the men lying upon their faces, and the stars falling down and striking one another, with terrible noise, and being troubled with this, he concealed it. Then after he saw the first stars falling to the earth, in the similitude of white fowl, and they snatched men up, and they carried them between two great mountains, and these mountains closed upon them, and the shining stars were made dark. And he awakened with great fear, and assembled the chief Priests of all the provinces, a hundred and thirty priests, the chief of them called Aclimun. He related the whole matter to them, and they took the altitude of the stars, and made their prognostication, and they foretold a deluge. The king said, "Will it come to our country?" They answered, "Yes, and it will destroy it."

While these stories are surely mixed with preconceptions and wishful thinking, they are probably founded on historical fact. Just as the city of Troy was once considered myth, the legends surrounding the Great Pyramid continue to direct our attention to an "antediluvian" race who left the great pyramid and a yet undiscovered "Hall of Records" as a library of knowledge for the education of future generations.

As early as 700 years ago, the Pyramids were covered in solid white limestone casing stones, however, in A.D. 1303, *"a massive earthquake loosened many of the outer casing stones, which were then carted away by Bahri Sultan An-Nasir Nasir-ad-Din al-Hasan in 1356 to build mosques and fortresses in nearby Cairo. More casing stones were removed from the great pyramids by Muhammad Ali Pasha in the early 19th century to build the upper portion of his Alabaster Mosque in Cairo not far from Giza. Later explorers reported massive piles of rubble at the base of the pyramids left over from the continuing collapse of the casing stones, which were subsequently cleared away during continuing excavations of the site."*[26]

The key to solving the true riddle of the Sphinx must be found. Secular Humanists from many disciplines scoff at solving the puzzle. Why look for answers when the question is irrelevant in the first place? Destroyed Ancient civilizations are myths. They never existed. Prophecies of the End of the World are the manifestation of our fears of death.

---

26 Wikipedia.

Abydos - The Strange Story of Omm Sety

1989 Trip – Japanese Contingent
- 20 people on the trip
- We did not go to Abydos
- Emiko translator.

The purpose of the Great Pyramid and Sphinx. Built 10,500 B.C. Edgar Cayce. The King's Chamber is commonly believed to be Cheops' burial chamber. The granite coffer in the King's Chamber, according to Mendelssohn and others, most likely was never intended for the mummy of Cheop's. Some have proposed that initiation candidates were sealed into the coffer by placing a stone lid over it, creating an airtight space that slowly suffocated them until they died. The intent of the ceremony was to produce a near-death experience. The soul during the dying process would travel up the air passages of the pyramid to the star Sirius and bring back afterlife revelations of the Light. Candidates who brought back visions of the afterlife, like modern Dead Saints, were transformed by sacred visions, and often gained healing and prophetic abilities. They became wise counselors to ancient kings. Source of the writings of the Egyptian Book of the Dead.

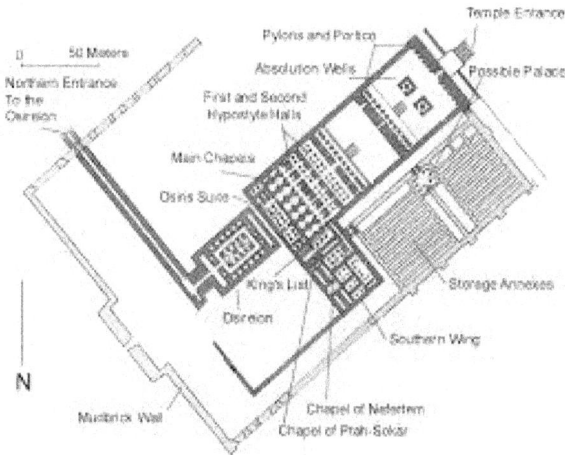

Budge examined the ancient Egyptian text of the Book of the Dead. He observed that ancient Egyptians appeared to be Monotheistic. They worshipped the Light.

The Egyptian ceremony is similar to ancient and modern shamanistic rituals, where the initiation candidate dies from an ingested poisonous herb taken during a prepared ceremony. If they survive the ritual, they often come back with healing and prophetic abilities, and are promoted to leaders of the tribe.

Arrival in Egypt – October 1989

Thirty reserved a spot to make the trip, which provided enough profit to bring George Everett (George Everett Photographer/Director) Virginia Beach, VA, a professional camera man to record the trip on Hi-8, (note: Have DVD's posted on You Tube channel) and two students from Hearthfire, Paul Edward and me.

Excited doesn't quite describe how we felt.

Staying at the Mena House, Paul Edward and I set out to search the nearby Egyptian desert around the pyramids early the next morning. Of course, we knew about the prophecies of John Peniel, spoken of in Paul's Readings, "*There is not a great deal of reason or purpose to your attempting now to move stones, or find hidden places, or open doors for they shall be discovered, opened by this one. And this is not to say there will be no further discoveries until that time, for those who dig and search and seek in this area will find many interesting things.*"

Tremendous shafts dotted the landscape, some over 40 feet straight down, and tombs for the royal families of the pharaoh over a three-thousand-year period. So many that there are still uncovered remains today. Camels, coke, and Baksheesh.

Searching Stella and ancient graves, I remembered what John Anthony West had said about the Chephren Complex and the discontinuity between the Granite Ashtars and the limestone blocks. They certainly were not from the same time period. That is not debatable. Any intelligent mind could see that the 4th dynasty stones were of a much more recent time period than the megalithic limestone blocks, obviously worn by water and extremely long periods of weathering. But if you ask any orthodox Egyptologist, "Why the discrepancy in the appearance of age, if they were built by the same pharaoh," he will not say, "it is an anomaly." But instead shrugs off the suggestions as "an illusion." It does not fit the current model of Egyptian history.

It is such a nuisance to rewrite history!

After several hours of poking, using our divining rods, and realizing that 60 feet of sand probably separated us from the ancient glory of Egypt, we gave up and returned quite thirsty to the Mena House.

We were on our way to Abydos. I knew nothing about Omm Sety. I was about to find out.

Paul Solomon & Omm Sety (1977)

Paul Solomon told me about a remarkable encounter he had with Omm Sety[27] in 1977, then caretaker of the Temple of Seti I, in Abydos, Egypt. Paul had read Omm Sety's exceptional life story, detailed in Jonathan Cott's, *The Search for Omm Sety* (Doubleday, 1987), and had made arrangements for his Egypt tour group to meet her.

Omm Sety, born Dorothy Louise in England in xxx, had a near-death experience when she was 3 years old. She accidentally fell down a flight of stairs and was announced dead by the physician summoned. Laid to rest in her own bed while the family proceeded to grieve, Omm Sety, were shocked when she suddenly "came alive" after apparently being dead for hours. Dorothy's parents were only too delighted to welcome her back into the land of the living, believing the old doctor had made a mistake. She perfectly recovered and proceeded to play with her toys. It was from this point onwards that she began to have recurring dreams of being in an ancient building with huge columns, interpreted by her as a temple.

When she was 4, her parents took her with them on a visit to the British Museum, and it was here in the Egyptian galleries that the little girl suddenly became aware that she was 'home'. Dorothy, at first, didn't cause too much concern, but as she grew older, her love and her memories of the ancient Egypt grew stronger. Friends and family could only wonder at her stories. According to Cott, Dorothy recalled 3500-year-old memories of *"her lifetime with Pharaoh Sety, during which time she served as a priestess in the Temple of Isis, their subsequent love affair and her resulting suicide to save him face since she was consecrated to the goddess."*

The temple at Abydos, erected by Seti I in the 13th century B.C., was a place of constant devotion for Dorothy. During her teenage years Dorothy spent every available moment studying Egyptology. This course of study included being taught to interpret hieroglyphics by the legendary Sir Ernest Wallis Budge, Keeper of Egyptian Antiquities at the British Museum.

At age 14, Dorothy saw a photograph in a magazine showing 'The Temple of Sety the First at Abydos', in Upper Egypt, and immediately connected it with the large columned building of her recurring dream. It was mainly through her dreams, recorded in detail in her diaries that she came to believe that she was the reincarnation of 14-year-old virgin priestess called Bentreshyt, who had lived at the Temple of Abydos

27 Dorothy Louise Eady, also known as Omm Sety, was Keeper of the Abydos Temple of Seti I and draughtswoman for the Department of Egyptian Antiquities. Born: January 16, 1904, London, United Kingdom. Died: April 21, 1981.

during the reign of Seti I. She told her father that the Temple was her home, the place where she had once lived, but was confused as to why the buildings were in ruins and there was no garden.

But it would not be until she was 29 years old, as the wife of an Egyptian student that Dorothy would travel to Egypt, where she became the first woman ever to work for the Egyptian Antiquities Department. Dorothy had a son from her marriage, whom she named Sety, much to the annoyance of her husband. In keeping with the Egyptian custom of not referring to women by their first name, Dorothy was then known as Omm Sety, 'mother of Seti'. Many years afterwards, in 1956 after her marriage had ended, she finally made her way 'home' to Abydos, where she remained, living in a small peasant house until her death in 1981.

Cott writes, "Although her life was an inexplicable adventure in which the barriers of time were blurred, one event, which occurred in 1958 while she was working in the Hall of the Sacred Boats is especially significant. It is detailed in Cott's book and written in her own words as she describes a fall into a hidden chamber/hall, which she later denied being able to find again, passing it off as a dream or hallucination due to an illness."

Paul's Story

"I met Omm Sety several times in the late 1970's during my tours of Egypt with small groups in Abydos. She was a quaint English woman. Quite charming, powerful, woman. In 1977 she met with our little rag-tag group for about 30 minutes, telling her stories about Ramses II as if they had occurred yesterday. She described her famous story of her encounter with a hidden room under the Seti Temple, and of the treasures she saw there. When she finished talking, she was a little tired, and her assistants, the people who were looking after her, took her away to rest. I asked if I could see her again and she said, "Of course."

After the rest of the Tour Group left Cairo for their destinations, Paul returned a week later with a friend to have a private meeting with Omm Seti.

Paul remembered, "When I arrived in her village, I asked someone to send for her. When she saw me, she looked at me impishly out of the corner of her eye and said, "You want to see the room, don't you?"

I said, "Of course I do!"

Although Omm Seti was in her late seventies but could still get around fairly well. I followed her to the back of the temple, the portion that straddles the Osirion, an ancient temple of megalithic proportions, to a corridor in the "L" section of the temple. At the end of that corridor, she leaned forward with her hands on the wall. Suddenly, I felt strange talking about it, it seemed as if we were going through walls and time. These things I cannot explain.

I heard rumbling sounds as if gigantic stones were moving over one another, and the sound of rocks falling. When she pushed against the wall, the floor gave away under us as if a large, hinged stone opened downward, and we fell forward on our hands and knees. She may have swung a wall open, but the transition from just standing in the hallway, to suddenly getting up on a sandy floor, having fallen, is just missing from my consciousness. I'm sorry I can't describe it sensibly.

I reached for Omm Sety afraid that she had been hurt in the fall. Amazingly, she was laughing. She thought it was a delightful adventure. She told me that we must move quickly before she was missed, and her companions started to look for her. The first thing I noticed was the inexplicable light source in the room. It revealed a room not tangled with cobwebs as she described it in Cott's book, but clean and in order -- a little wonder, since she had 32 years during her care of Sety's temple to clean it up!

Omm Sety seemed to know about everything in the room which seemed to me to contain hundreds of objects. I was asked not to touch anything because everything was so old. There was one object... a five-foot pillar made of black stone, diorite or black granite, with four disks which appeared to be made of gold centered on the pillar toward the top. I had seen many pictures of this object in hieroglyphics throughout Egypt, even on the walls of the Sety temple. I specifically remembered the back wall of the Sety Temple describing in graphic detail the mystery play of the death and resurrection of Osiris, the only one like it in Egypt. Egyptologists have always thought this depiction of Osiris was just a symbol, but this object looked like a machine. (Osiris, Pharaoh of the underworld, was a mythical King of Pre-historic Egypt, and represented by the Djed column, often thought to represent the "backbone" of man or an analogy of death and resurrection)

Another object, almost life-sized, I immediately recognized as an Apis Bull which was made of solid gold and appeared to be the light source in the room. "Omm Sety confirmed the fact that the Bull was the source of light in the room and further explained that the golden Apis statue was a type of "battery" and hence, the "energy" source for the light. The light was not bright but afforded enough light to read the hieroglyphics in the room. The interesting thing is the Apis Bull seemed to light the room without casting shadows. It was not like you had a bulb in the middle of the room...While in the chamber, which was only a short time, Omm Sety spoke of many interesting things, among which was the possibility that the same "batty charge system" that provided energy for the golden Apis could have been the energy source of the Ark of the Covenant[28]

"Both were kept in a sealed room with no other light source that allowed seeing in the room." As Sety explained, she felt these sacred instruments stimulated the photons present even in a darkened room, but it did not emit photons or send out any kind of rays from itself. It is more like drawing energy to itself rather than sending it out."

Paul continued his story. Of significant interest in the room and commanded my attention was a green disk displayed on a rectangular, black lacquered table, a smooth stone about the color of lapis lapilli. It appeared to be worn smoothly either by handling or by water and had hieroglyphics carved in a circle around its circumference, almost like a horoscope. Sety told me that it was used to maintain a constant temperature and needed no energy source to activate it. She took my hand and put it near the stone so I could feel the heat coming out. "It was responsible for the near perfect condition of the room and its contents."

The walls drew my attention almost immediately because they were entirely covered with hieroglyphics in full color. Every inch was covered with paintings and glyphs.

This is the only room I've seen in which you could see work as it must have looked originally. The ceiling was painted dark blue, and filled with white, five-pointed stars.

We quickly moved down a hallway piled high on one side with golden artifacts and treasures and walked through a square archway with no door. We passed through a maze of small rooms until we walked through a doorway into the courtyard in front of the Sety temple. There I saw a pool with clear, running water, like a spring. It wasn't fancy...but it seemed like a clear pool.

I caught up with Sety, who was just a few steps ahead, trying to hear everything she said. I happened to look back to see the clear pool again and got the shock of my life. Instead of seeing clear running water, the water was stagnant and filled with algae.

I didn't say anything to Omm Sety as we left the temple complex and began to walk slowly to her little home. Just before we parted, she turned to me with a look that could melt iron, "You're wondering why I haven't contacted the department of antiquities aren't you?

28 The Ark of the covenant is sometimes described as perfect cube, 20 cubits by 20 cubits (40' x 40') (I-Kings, Ch:6), and contained within it the 10 commandments written on two tablets of stone. Two gigantic angels ten cubits, (20' high), overlaid with gold, rested on the olive wood Ark also overlaid with gold, with an outspread of wings of 20 cubits (40') The Ark of the Covenant was an adaption by Moses and Aaron of the Egyptian Ark, even to the kneeling figures on its lid (goddess maat). Bas-reliefs on the several temples, including the temple at Edfu, show Egyptian priest carrying their ark, the sunboat of the gods, -- which closely resembled the Ark of the Hebrews -- upon their shoulders by means of wood staves like those described in Exodus.

"Well, it had crossed my mind."

"What do you think they would do with all of this? Those narrow-minded idiots in Cairo could study these instruments for years and never figure them out. Besides, they would probably just stick these sacred devices in a museum where they don't belong. These treasures are Sety's. They belong here and should be used appropriately by an initiate." Omm Sety took a long breath, "One or two others know about the treasury's existence, but you're the first to see the treasury besides myself. I needed to pass this information into the hands of someone I trust, someone who understands the importance of this discovery. Paul, I've not long to live. I've done my job caring for this treasure."

In early 1981, Paul returned to Egypt see Omm Sety again with a friend. She had fallen the previous year and broken her hip, and now required a walker to get around. I never found out what they talked about. She died a few months later.

Seti I Temple & the Osirion

The long four-hour bus trip to Abydos. Camels and the Egyptian poor. More Baksheesh.

The bus pulled into the parking area in front of Sety's temple. I turned to George, our camera man sitting behind us, who had been oblivious to our conversation, "We're going to get this on tape."

"Paul, do you remember the reading you gave in the sarcophagus of the King's chamber back in 1981 concerning the entrance into the Hall of Records?" I pulled out the reading excerpt, "And were the temples dismantled you would discover ideas and glories you hadn't dreamed of. And they are available -- not so hidden that you could not find them -- through great effort! For researchers who attempt to dig to enter into the Hall of Records of the Great Pyramid will find themselves somewhat disappointed. *But those who discover the manner of passing through stone, will pass into the chamber and find their instruments which would amaze those who consider themselves superior in technology in this time.*"

"What you told me about the Seti treasury must be a key to finding the lost Hall of Records near the Great Pyramid! *The ability to pass through stone...*It must have been something Omm Sety did when she pushed against the wall. Did she say any words, use any sounds to trigger the stone's movements?"

"Not that I remember," Paul raised his eyebrows. "She just pushed forward. It's all very confusing. It sounds like we fell through a doorway when we entered because I heard the sound of stones grating against one another, but when we left the treasury, we walked through a door opening into daylight, and when I looked back there was a solid stone wall. I know it doesn't make sense, but that's what happened."

"Paul, I can't believe you've seen a priceless treasure room, a find greater than Tutankhamen, and you haven't been back since?" I said exasperated.

"We'll I haven't been exactly up to it, have I", he looked at me with a grin. I'd nearly forgotten that he had been laid out five years since his Pancreatitis operation in 1984.

Hearing Omm Seti's story, Paul Edward separated from the main group, and went to the Western area to the Osirion. To our astonishment, we discovered all the water had been pumped out. It had been flooded for decades. A rickety wooden ladder went down a hundred feet into the bottom of the Osirion. We pointed at the ladder and asked one of the workers if we could go down for a few minutes and take a look around. He understood our body language and the Baksheesh. Works every time. Our backpack had the Hi-8 camera and flashlights. Hearts beating fast, we scuttled down to the muddy floor of the Osirion, open to the sky. One entrance disappeared underneath the Temple. The Osirion in Abydos had been flooded since the completion of the Aswan Dam in 1970, because of the rising water table.

Paul Edward and I were one the first tourists to see the floor of the Osirion in nearly two decades. What a rush. We headed through the dark opening of the Eastern Red Granite entrance, the mud up to our ankles. Flashlights on, we plunged into the pitch-dark room that appeared to go under the Seti I Temple. Of course, there were no secrets to find. Plain granite walls. No bas reliefs. We were too excited to be disappointed. The ten minutes squishing around with flashlights found no treasure, but it was exciting, none-the-less.

Back to Abydos

I was prepared. I studied the Seti I temple blueprints preparing for the day when I could visit the temple and look for the room. It seemed unreal. These things only happen in the movies. Paul, dressed in an Egyptian cotton striped Galabia, and me in jeans, searched the southern Wing of the Seti I temple adjacent to the Hall of the Kings lists, which is a jaunt up a flight of stairs that lead to the western view overlooking the Osirion. Alas, it was now flooded again. I wouldn't be tromping in mud to look there.

Going back down the stairs, we walked on the massive stones on the walls and massive stones below of the Southern Wing, all decorated with colorful bas reliefs of Pharaoh Seti I in ceremony with Osiris, Isis, and Horus. We came to the one room that was locked with massive steel bars that looked like a jail cell. It was the Hall of Boats Sanctuary. I asked one of the Seti guides, "Why is that room locked and barred?"

In his broken English he said, "Office. Office." I looked at the blueprint I had copied from Omm Sety's book. It was the same location as Omm Sety's library. Locked out. Paul stood there quite coy in his Egyptian Gallabiyah staring at me, smoking his meerschaum pipe. I had no idea I would encounter locked doors. Looking through the steel bars, I could see bookshelves on the opposite wall. I couldn't see into the rest of the room. Heck, the shelves probably covered the part of the wall we needed to push against anyway. Downtrodden, I accepted that I would not see Seti's secret Treasury Chamber today.

We were not done. After dark, taking me with him, Paul sought out a local Arab contact to arrange to obtain the rest of Omm Sety's personal belongings from the remnant of her estate; mostly letters and photographs contained in two boxes. How did he know of their existence and who to ask?

Once we returned to the United States, I had a chance to sift through the boxes. It is a treasure trove of old photographs and letters with historical significance, but alas nothing related to the Treasury Room.

Meeting with Hanny El Zeiny

The next day, when we returned to Cairo, I arranged a meeting between Hanny El Zeiny, former friend and confidant of Omm Seti to meet with John Anthony West, Paul Solomon and me, to inquire about the Treasury Room and the locked office. I always believed Mr. Zeiny must have known about this. He was Omm Sety's personal friend for many years. He's just not talking about it. He's not letting on. But El Zeiny genuinely denied possessing such knowledge from Omm Sety. He described the same story about Omm Sety's fall into the treasury room while walking on top of the Western part of the Temple, described in Jonathan Cott's book. *"I went up the stairs and started walking along the top of the wall, when I suddenly became very dizzy and fell, twisting my ankle and hurting my left shoulder. I remember hearing a loud grating sound, like that of a grindstone at work, and I rolled down a fairly steep slope; the grating sound was renewed, and I found myself in darkness."*

That was that. I poised the question. Was Seti I unaware of the Osirion until he had already begun building the temple? That is why the L leg on the back side of the temple. I pointed out that this could not be so, because the Osirion was perfectly aligned as if, Sety had found the Osirion first, and aligned his 19th dynasty Temple to be integrated into the entire master temple plan. Perhaps Seti I found something old, very old there when he built his Temple.

Solving the Omm Sety Treasury Room Mystery

Henny said, "Omm Sety told me I have a job to do.

Maybe my being with her when I was, was precisely the time she may felt the need to share the chamber with someone else, for the contents may be important to mankind. I find it hard to believe that she did not have opportunities to do the same with others, though I probably will never know.

But the one thing I do know. Omm Sety trusted me, and she didn't want to die and not share its contents with someone, and it would be conceited of me to think that she did not show it to others. But the one thing I do know is the individual who is able to accept the challenge and enter into that room has a tremendous responsibility. The treasury room under Sety's temple is only a small portion of that contained in the Great Hall of Records at Giza. And there is one who has been chosen already to open those records."

"The herald, John Peniel." I said.

He nodded.

"But when will that happen?"

"It's up to him." Perhaps, he's waiting for the key this temple offers to open the Records in Giza. Maybe this is a key we can pass on."

"First, we've got to find him again, you know what that's like."

"We won't find him. He'll find us. That's the way he works."

As much as I wanted to know the truth, we were soon back on a plane to the US, ending my search. I hoped to return with Paul Solomon to Egypt to make arrangement to look in that locked room, but Paul died in March 1994.

Before Paul Solomon died, he typed a letter to Marcel Vogel, a respected nuclear physicist, attempting to describe where he fell through the wall, but six months later, Marcel died suddenly. Marcel Vogel. (get letter copy) and find chapter written.

Maybe Paul revealed its secret to someone unknown to me, but for now, the secret died with him.

I wouldn't return to Egypt until February 2009.

Egypt 2009 with Rob Pennington & JAW

The Treasure Room notes (quoted from the Search for Omm Sety)

During roof work on the Sety I Temple in 1958, and even though she had the keys to all the doors, 54-year-old Omm Sety found it easier to get in and out of the building by going up the stairs to the roof, walking along the top of the southern wall of the unroofed Western Corridor and down the scaffolding at the west of the temple. Her experience is detailed in *Abydos: Holy City of Ancient Egypt,* by Omm Sety and Hanny El Zeiny, which I quote in its entirety as I believe every word to be important: (note: Both Omm Sety and Hanny El Zeiny are deceased. Referencing the quote should be sufficient without requesting permission. The L.L. Company no longer exists. The only books in circulation are dated to 1983.)

"At the time, I was working in the Hall of the Sacred Boats cataloguing and fitting together about 3000 pieces of inscribed stone that had once formed the doorways, columns and window grills of the magazines and Audience Hall of the Temple. At that time, work was also in progress in the roofing of the temple, and although I had a *key to all the doors*[29], *it was easier to get in and out of the building by going on the roof, walking on top of the southern wall of the unroofed Western Corridor and down the scaffolding at the west of the temple.*

We then had Asiatic flu in the neighborhood, and I caught it. One morning I was feeling rather bad, and as a couple of aspirins and a short rest had no effect, I decided to call it a day and go home. I went up the stairs and started walking along the top of the wall, when I suddenly became very dizzy and fell, twisting my right ankle and hurting my left shoulder. I remember hearing a loud grating sound, like that of a grindstone at work, and *I rolled down a fairly steep slope*; the grating sound was renewed, and I found myself in darkness."

After a while the dizziness passed off enough to allow me to stand up and grope for the wall. I touched some smooth limestone blocks and stood there wondering what to do next. Presently, I sensed very faint threads of light filtering down from above, as though through cracks in the roof [of the Treasure chamber], and as my eyes become accustomed to the gloom, I found that I was standing in a narrow passage less than three meters (ten feet) wide. A narrow path, perhaps about fifty centimeters (20 inches), ran along the base of the wall; but the remainder of the width of the passage appeared completely filled with boxes, offering tables,

29 The Main Entrance to the temple, both doors of the Corridor of Kings, both doors of the Western Corridor, and that of the Hall of Sacred Boats are fitted with iron gates that are always locked.

cases, bales of linen; and everything had the gleam of gold. Feeling my way along the wall, but the remainder of the width of the passage appeared to be completely filled with boxes, offering tables, vases, bales of linen; and everywhere there was the gleam of gold. Feeling my way along the wall, I limped along.

The passage seemed endless, and, to my left, crowded with objects. I stumbled and fell, and, on trying to rise, saw what I took to be the God Horus himself bending over me, his hands raised as through 'in astonishment.' From his waist down, he was standing upright, but from the waist upward he was bending over, his fierce falcon's face peering down at me, and his Double Crown [representing Upper and Lower Egypt] sticking out at right angles without any means of support. There I squatted meekly, thoroughly embarrassed and trying to think how one should address a god under such circumstances. Then I suddenly realized that 'Horus' was only a painted wooden statue, life size, and originally standing upright with the arms bent at the elbows, and the hands raised. Insects had eaten away part of the front of the body, causing the upper part to lean over. Scrambling to my feet, I noticed similar life-sized statues of Osiris and Isis leaning against the far wall, apparently uninjured.

Near where I stood was a gold vase about 25 centimeters (10 inches) high. It had an oval body, a long neck, and a trumpet-shaped mouth, and stood in a wooden ring stand. By the faint light I could see a cartouche engraved on its body, but it was too dark to read it. But by the length of the frame, I knew it was not the cartouche of Sety, but one of the later kings, perhaps from the Twenty-sixth Dynasty. I picked it up. It was very heavy, and at first, I thought I would take it with me as evidence of what I had discovered by accident, but I finally decided against it and put it down in its place.

I began to feel very ill again but continued to limp along, half unconscious. Suddenly, I found myself out in the open air, almost blinded by the sunlight. I was standing beside the well in the Second Court of the temple, and approaching me was a young man, a stranger. He stared at me in frank astonishment and asked if I knew where the architect in charge of the restoration was. I told him and, still staring in surprise, he thanked me and left. I went to enter the temple by the main door, only to find that the keys [to unlock the gates] were not with me. I went around to the back of the temple I order to reenter it by the Hall of the Sacred Boats by the way I had left it, when I met two of the gaffers (watchmen.) They all cried out, "Where have you been? You are all dirt and cobwebs!" And so I was; no wonder the young man had stared at me so hard!

I replied that I had fallen down and hurt my ankle and had forgotten my keys. But I did not tell them anything more. I managed to crawl back to the Hall of the Sacred Boats; and there were the keys just as I had left them, on the table.[30]"

Did it Really Happen?

'"Two days later the Chief Inspector of Antiquities Department for this area came here, and I told him about this experience. He was very astonished and interested, but neither of us could decide if it was just a hallucination caused by the fever, and to this day I do not know for certain. All that I am sure of is that I really fell, my ankle was swollen and painful for a week and I had a big bruise on my shoulder. Also, I was covered in cobwebs, and having left the keys behind, there is no way in which I could have reached the front part of the temple except by going around from the outside, where the gaffers (watchmen) would have seen me, and where moreover, I would not have collected any dust or cobwebs all over my clothes.

'If this really did happen, then there is only one possible explanation! I must have hit a stone with my shoulder as I fell, which turned on a pivot and opened into a sloping passage. This would account for the grating sound. But how did I get out again? ALL that I can suggest is that a deserted hyena's lair in the side of the wall may have communicated with the "Treasure Passage." Later, the lair caved in, but its place remains, still clearly visible. The chief inspector got interested and told me to try to find the supposed pivot stone. I looked for it in every possible place in the temple, including the "Blind Rooms" [two rooms— one on top of the other – without windows or doorways that are located immediately behind the Inner Chapels of the Osiris Complex], pushing and butting against the walls of these rooms, but with no results, except some more bruises!

'There is some significant point, however. The paving stones of some of the aisles in the hypostyle halls are large, single slabs that resemble those in the upper "Blind Room." These could well be roofing subterranean passages. But in all other places, the paving stones are smaller and irregularly shaped.

One thing I am certain of -the Temple of Sety still holds some secrets... as a matter of fact, lots of them. One day a patient archaeologist may come to Abydos to investigate all its unknown and fascinating possibilities. And maybe he will stir the enthusiasm and admiration of the entire world with something bigger and more

30 Omm Sety and Hanny El Zeiny, Abydos: Holy City of Ancient Egypt, © L.L. Company, 1981, pp 176-178

important than Lord Carnarvon's discovery of Tutankhamun's mortuary treasures in 1922.'[31]

History validation Age of Pyramids & Sphinx?

"I asked his Majesty, about the Osirion (ancient temple behind Temple of Seti I) and he said, "No, I didn't build it, it is very much older and existed long before my time. (1279 B.C.). She told King Sety that according to researchers, the Great Sphinx was built by Khafra (Chephren], and he said, "Oh, no.... it's older, it's older... "I was told that it was a monument to Horus (the Child of Osiris and Isis) himself."

I repeated, "It's a monument to Horus, but they say it was built by Khafra. "And again, he said, "oh, no...."[32]

MEXICO-10, 500 B.C.

One indication that a far earlier civilization existed in the Valley of Mexico comes from the tantalizing discoveries made in the Mexican highlands by a Scotsman, William Niven. As a mining engineer working for a Mexican corporation, Niven described coming upon the remains of *two separate prehistoric civilizations* at depths of from six to thirty feet below the present level of the Valley of Mexico. These discoveries, made between 1910 and 1930, have been casually brushed aside by archaeologists, primarily because they do not fit the orthodox historical model.

When he died in 1937, his obituary in the *American Historical Review* listed him merely as a professor who had been engaged for several years in mineralogical research, an honorary life member of the American Museum of Natural History and various other such scientific societies who had become involved in Mexican archeology, nothing more. Nothing to indicate he might have made the most controversial archeological discovery of the western world.

Actually, William Niven had been exploring in Mexico since 1889. While digging among the ancient, ruined cities in the unknown and uninhabited portion of the state of Guerrero, southwest of Mexico City in the Acapulco area, he began to receive periodic visits from local Indians who came to him with terra-cotta figurines and other objects for sale. Though the Indians pretended to have found these objects at the pyramids of the Sun and Moon at San Juan Teothuacan, Niven realized that the source of the artifacts must be nearer; with a bribe of five pesos ($2. 50) at that time) he managed to discover the actual spot Between

31 Omm Sety and Hanny El Zeini, Abydos: Holy City of Ancient Egypt, © by L.L Company, 1981, pp. 176-178

32 Jonathan Cott, The Search for Omm Sety, A Story of Eternal Love, Studio 33 Books, Random House Group, Ltd., pg. 93

Texcoco and Haluepanta, hamlets just north of Mexico he came across hundreds, if not thousands, of pits dug into the sand, clay, and tepetate used for material by the builders of Mexico City for more than three hundred years. Exploring these pits, which Niven says cover an area of about ten by twenty in the northwest corner of the Valley of Mexico, he came across vast layers of what appeared to be very ancient ruins, whole prehistoric cities Lying as deep as thirty feet below the plain, which appeared to have been overwhelmed by a series of cataclysmic tidal waves, perhaps at several thousand-year intervals, which, as Niven described them, had left telltale strata of boulders, sand, and pebbles. By their depth beneath the surface, Niven estimated the oldest remains might go as far back as 50,000 years.

Four to six feet below the first pavement, Niven says he encountered a second concrete floor," but in the intervening space failed to find a single piece of pottery or other trace to indicate that humans had once lived there. Beneath the second pavement, he describes coming upon what he considered "the great find of my many years' work in Mexican archeology."

Niven discerned, beneath a well-defined layer of ashes from two to three feet thick, analyzed as being of volcanic origin, traces of innumerable buildings, large, but regular in size, the remains of a vast city which appeared uniformly at the same level throughout more than a hundred clay pits. In one of the houses, most of which were crushed and ruined, filled with ashes and debris he says he found an arched wooden door which had turned to stone. The walls of this house were bound together with white cement, harder than the stone itself. In one uncrushed room, about thirty feet square, full of volcanic ash, with a flat roof of concrete and stone, Niven says he came across many artifacts and human bones, which "crumbled to the touch like slaked lime."

According to his detailed report, a complete goldsmith's outfit was still on the floor with some two hundred models of figures and idols molded in clay turned to stone, each model thickly coated with iron oxide, bright and yellow, presumably there to prevent the molten metals adhering to the patterns while in the casting pot. Niven says the ornaments were unlike any found in Palenque or Mitla or anywhere between. The work was fine, beautifully polished, demonstrating an advanced degree of civilization. On the walls Niven found paintings in red, blue, yellow, green, and black, which he says compared favorably with the best he had seen from Greek, Etruscan, or Egyptian works of a similar sort. The ground color of the wall was a pale blue; six inches down from the fourteen-foot ceiling a frieze painted in dark red and black ran around the room, glazed with some native wax which had perfectly preserved the color and pattern, which depicted the life of some person, apparently a shepherd, from birth to death. Beneath the floor Niven found a tomb three feet deep, lined with cement, in which were seventy-five pieces of bone, all that was left of a skeleton. A large fragment of the skull contained the blade of a hammered copper axe, which appeared to have been the cause of death, for it had not been removed.

Niven also found in the tomb 125 small terra-cotta idols, manikins, images, and dishes, some with features strongly Phoenician or Semitic, one sitting cross-legged with a hollow movable head set on its neck by a cleverly devised truncated Tenon fitted into a mortise at the base of the skull. Less than three miles away Niven found an ancient riverbed in the sands and gravel of which he says were thousands of terra-cotta and clay figures with faces representing "all the races of southern Asia."

Then, in 1921, in the course of excavations at Santiago Ahuizoctla, a hamlet contiguous to Amantla, about five miles northwest of Mexico City, Niven came across a discovery so startling he says it opened up for him a whole new field of archeological research. At a depth of twelve feet Niven described coming across the first of a series of stone tablets with very unusual pictographs. Systematically exploring other clay pits and temperate quarries within an area of twenty square miles, he claimed he was able to unearth during the course of the next two years 975 more tablets.

In the end he says he found more than 2600. Though there was nothing in these tablets by which he could determine their exact or even approximate age, Niven deduced from the depth at which they were buried and the accumulation of debris on top of them that they were over 12,000 years old and more likely closer to 50,000. The tablets, which Niven carefully numbered in the order in which he found them, had no particular shape. They appeared to be water-worn stones with smooth surfaces on which the figures had been carved, often to conform with the shape of the stone, much like the so-called Cabrera stones found almost contemporaneously in Peru, which depict strange human beings with four-fingered hands in combat with dinosaurs, though such prehistoric creatures were thought to have been extinct many millennia before the appearance of man on the planet.

Niven's showcase number 6 containing a portion of his collection of carved stones from the Valley of Mexico. When William Niven died in Austin, Texas, in 1937, the New York Times described him as distinguished mineralogist and archeologist who had discovered buried prehistoric cities beneath the Valle of Mexico. He was also noted as the discoverer of four new minerals including cytrialite, thorogon, and nivenite. According to the Times, Niven donated to the Mexican government the best of the relics he found in Mexico, keeping for himself some, which he sold to finance further archeological expeditions. With what was left over there were enough pieces to establish in Mexico City a private museum of 30,000 exhibit. Note: In lava fields dated over ten thousand years, Niven discovered thousands of figurines and tablets in a foreign language unrelated to the Mayan or Olmec cultures.

At last, some serious attention is being paid by Niven by Robert Wicks of the University of Washington in Seattle, who is preparing a book on Niven and his tablets. In correspondence with leading academicians in the field of Mexican antiquities, Wick's has managed to obtain their admission that negligence of Niven's work may have been in error.[33]

Over 2,500 years ago, the lawmaker and sage of Athens, Greece, Solon, in his youth had visited the Temple of Vulcan, the oldest temple of the city of Sais (Port Said today) Egypt. Solon had sought out this temple because he knew the priests apparently had access to an incredibly ancient collection of papyrus manuscripts which made up the Alexandrian library, and that somewhere in its hidden recesses lay the ancient books carrying the history of the ages.

South America

A haunting tale and one not easily dismissed describes a tribe of Indians named the Urgha Mongulalas living in the Amazon jungle of South American have carefully recorded their past in the Chronicle of Akakor, translated by Carl Brugger, who relates a fantastic story about "shining ones" who descended from the skies around 13,500 B.C. in Brazilian/Peruvian Jungle, who encountered a tribe who lived in caves like animals without reasoning intelligence.

The tale describes these "shining ones" descended in golden ships and taught them the laws of nature and the heavens; how to read and write, how to till and cultivate their land, how to build their homes, towns and cities. They remained with the tribe for three thousand years and then they returned to their constellation in the sky which they called Schwerta.

But before they left, they told the tribe that a great catastrophe was about to befall the world because of the evil of the majority of its inhabitants. They advised the people to take shelter, when the disaster struck, in the thirteen underground cities which they had built for them. Promising that they would return one day, the gods flew into the heavens on their golden wings.

It is recorded in The Chronicle of Akakor that two major catastrophes nearly wiped out the whole of mankind. The year given for

---

33Peter Thompkins, Mysteries of the Mexican Pyramids, Harper and Row, New York, 1976 pp.356-361.

the first one is 10,468 B.C. and a terrible flood is spoken of. A change of climate occurred caused by a shifting of the earth's axis:

That is the news of the downfall of man. What happened on the earth? What made them tremble with fear? Who made the stars dance? Who made the waters gush out of the cliff? It was terribly cold and an icy wind swept over the earth. It was terribly hot and men burnt up in its breath. Men and animals fled in panic-stricken fear. They ran hither and thither in desperation. They tried to climb the trees, but the trees hurled them far away. They tried to creep into caves and the caves collapsed on top of them. What was below was pushed up above, and what was above sank into the depths.

But there were many survivors among the Mongulala's for they had followed the directions of the gods and had hidden in the caves and underground cities. It is said that there was a second dreadful disaster about 4130 B.C. The Mongulala's were told that the world disaster recurs every 6,000 years. Again the tribe survived by descending into the underground cities. At last the time came for him to leave them but he told his people that there would be a third world cataclysm before the gods returned which is expected to occur toward the end of this century.

"And finally, after many years the gods returned as they had promised. They came in their golden ships and the tribe welcomed them gladly. The gods did not remain long with the people but stayed only three months. However, two brothers remained behind, one called Lhasa and the other Samon. The latter went to the East and the former remained with the tribe. The people looked upon Lhasa as the divine master and the Chronicle says this of him: Lhasa was one of the gods. For thirteen days he rose into the sky, for thirteen days he walked to meet the rising sun, for thirteen days he assumed the shape of a bird and was truly a bird (tale of Quetzecotl) for thirteen days he changed into an eagle."

He remained with the Mongulala's for several hundred years and led them into prosperity united with right living. At last the time came for him to leave them but he told his people that there would be a third world cataclysm before the gods returned which is expected to occur toward the end of this century.

He commanded them, "The people must prepare at that time to go once more into the caves and underground cities which had been built so long ago for the; thus, they would survive and be amongst those who would be the remnant left after the earthquakes and tidal waves had devastated the world. Before Lhasa departed,he gave the tribe certain machinery and artifacts which the latter was to guard carefully; one of these was a box sealed tight. He said that when the box began to sing that would be the signal for the return to earth of the gods from Schwerta* (*German word for "sword" the constellation Orion*).

The present chief of the tribe announced in 1978 that the box started to sing sounding as though it were full of bees. [authors note: The Sword of Orion also reflects the Egyptian creation of the Giza plateau reflecting the three stars of Orion's sword in the design of the three great pyramids. The Pit may represent the entrance to the Underground Cities built in South America.]

Then Lhasa left them, flying in his golden ship that sped as fast as any arrow to his distant home among the stars. There is an interesting fact regarding the legends of South and North American Indians.

Although the many tribes have differing names for their gods there are one or two similarities in the stories concerning these deities that are too far from being mere coincidence to be ignored.

This story begins on March 3, 1972, when Karl Brugger, a German journalist, met an Indian chieftain named, Tatunca Nara. After Brugger taped an interview with him, he accompanied Tatunca Nara into the Amazon jungle in an attempt to return to the secret city of Akakor. (Bragger's story can be read in his book, *The Chronicle of Akakor*.) Tatunca Nara was a tall man with long dark hair and brown eyes that gleamed with suspicion, and for good reason - he was a Mestizo, a half-breed, the stigma that would one day come to haunt him. Tatunca Nara told Brugger the tale of his tribe, the Ugha Mongulala, a chosen people by the Gods 15,000 years ago, long before academia says the first humans set foot on the North American continent. He said the language they spoke was Guechua, a written language of 1,400 symbols, each yielding a different meaning depending on their sequence. According to Tatunca Nara, the tale of Akakor was recorded in a tribal book entitled Chronicle of Akakor (not to be confused with Brugger's book THE Chronicle of Akakor), and begins at the year zero, which corresponds to the year 10,481 B.C. on the Gregorian calendar. He went on to say that during year zero "...the Great Masters, left the Ugha Mongulala...Before the year zero, men lived like animals, without laws, without clothing..." and that the Great Masters brought "the light" (here we have shades of Prometheus' story of bringing light, or knowledge, from the heavens to humankind.) Before year zero, the continent was "...still flat and soft, like a lamb's back...the Great (Amazon) River still flowed on either side." He continued to explain that sometime before year zero (Tatunca Nara guessed about 3,000 years before) "...glimmering golden ships appeared in the sky. Enormous blasts of fire illuminated the plain. The earth shook and thunder echoed over the hills." The strangers in the golden ships looked like humans with fine features - white skin, bluish-black hair and thick beards.

"They (the Ugha Mongulala tribe) had no tools as they did which, as if by magic, SUSPENDED THE HEAVIEST STONES, FLUNG LIGHTING AND MELTED ROCKS." These strangers civilized the tribe and built three great cities of stone called: Akanis, Akakor, and Akahim. The city of Akakor was built up the Purus River in a valley of mountains between Brazil and Peru. Likely locations are the Madre de Dios Province of Peru and the Acre Province of Brazil. "The whole city is surrounded by a high stone wall with thirteen gates...The gates are so narrow they give access only to one person at a time, and the plain in the east is guarded by stone watchtowers where chosen warriors are always on the lookout for enemies. "Akakor is laid out in rectangles. Two intersecting main streets divide the city into four parts corresponding to the four universal points of our Gods...

The Temple of the Sun and a stone gate sit on a wide square in the center. The temple faces due east, toward the rising sun, and is decorated with symbolic images of our Former Masters..." Tatunca described the city in great detailed - temples of artfully hewn stones, golden mirrors, life-size stone figures flanking the entrance to the temple, the temple's interior walls covered with relief, and that there is a large stone chest sunk into the front wall of the temple where the first written laws of their Former Masters is located. He said there were another 26 stone cities around Akakor, the largest being Humbaya and Paititi in Boliva, Emin (near the lower reaches of the Great [Amazon] River) and Cadira in the mountains of Venezuela. "...all these (cities) were completely destroyed in the first great catastrophe 13 years before the departure of the Gods."

"...The ancient Fathers also erected three (3) sacred temple complexes: Salazere, on the upper reaches of the Great River, Tiahuanaco, on the Great Lake, and Manoa, on the high plain in the south." A giant pyramid was erected in the center of these sacred temple complexes and a broad stairway led up to the platform where ceremonies were conducted. Interestingly enough, Tiahuanaco is the only place named that is known today and does indeed have a pyramid located in its center.

Tatunca Nara said he had seen only Salazere with his own eyes, which lies at a distance 8 days' journey from Manaus at a tributary of the Great River. Its palaces and temples have now become completely overgrown by the Liana jungle. Only the top of the great pyramid rises above the canopy of the forest. According to Tatunca, there are also underground cities inside the mountains we call the Andes. Tunnels link "lower Akakor" with the underground cities. These tunnels are large enough to accommodate for five men walking upright and they are so extensive that it takes many days to reach one of the other cities.

Tatunca said these underground cities were artificially illuminated by vertical shafts that ascended up to the surface where an enormous silver mirror dispersed light over the whole city.

These tunnels and subterranean cities were built by the Former Masters. From Tatunca Nara's memory - quoting the written Chronicle of Akakor: "And the Gods ruled from Akakor -- They ruled over men and the earth. They had ships faster than birds' flight, ships that reached their goal without sails or oars and by night as well as by day. They had magic stones to look into the distance so that they could see cities, rivers, hills, and lakes. Whatever happened on earth or in the sky was reflected in the stones. But the underground dwellings were the most wonderful of all. And the gods gave them to their Chosen Servants as their last gift. For the Former Masters are of the same blood and have the same father." In year zero, the Former Masters left, but before they left there was some kind of "War between the Gods." This war was horrible and devastating. Afterwards, the Former Masters left and a global catastrophe ensued. The Ugha Mongulala and the surrounding tribes lapsed into 6,000 years of barbarism. In the year 13 (10,468 B.C.) the course of the rivers was altered, and the elevation of the mountains and the strength of the sun changed. (Could this be an account of a polar shift - with massive amounts of volcanic ash in the sky?) During this time, "...Continents were flooded.

The waters of the Great Lake (near Tiahuanaco?) flowed back into the oceans. The Great (Amazon) River was interrupted by a new mountain range and now flowed swiftly toward the east (and into the Atlantic Ocean.) Enormous forest grew on its banks. A humid heat spread over the easterly regions...In the west, where giant mountains had surged up, people froze in the bitter cold of the higher altitudes..." All this would be known as the "First Great Catastrophe." After this first Great Catastrophe, the empire was set in ruins. Many of the passages that linked the borders of the empire were blocked; the mysterious light that illuminated the subterranean dwellings was extinguished; the twenty-six cities were destroyed by a tremendous flood; and "the sacred temple precincts of Salazere, Tiahunanaco, and Manoa lay in ruins, destroyed by the terrible fury of Gods." This would NOT be the last. In 3166 B.C. a second catastrophe occurred. This catastrophe ended the "Years of Blood" - the 6000 years of barbarism that had ravaged the land since the First Catastrophe. Just after this 2nd catastrophe, the Gods returned to Akakor, but only a few.

Their stay was short. Only two brothers stayed: Lhasa and Samon. Lhasa stayed with the Ugha Mongulala and Samon flew off to the east. Lhasa, now king of the Ugha Mongulala, fortified the kingdom and supposedly had Macchu Picchu built as an outpost of the empire. "Lhasa was the decisive innovator of the Ugha Mongulala Empire. During the 300 years of his rule, he laid down the basis for a powerful empire. Then he returned to the Gods. He convened the elders of the people and the highest priests and passed his laws on to them. He ordered the people to live according to the Gods' bequest forever and to obey his commands." After establishing this powerful empire, "he ascended the Mountain of the Moon, which looms over Machu Picchu, and in his flying disk forever withdrew from the humans.

So what happened to Samon? Tatunca Nara explained that Samon's empire was a mirror image of Akakor, located by a mighty (Amazon?) river. Lhasa often visited his brother with his flying disk to form a strong link between the two nations. In 3056 B.C., he commanded the construction of a great city at the mouth of the Amazon River named, Ofir. Ofir became a powerful seaport. Here, Samon's empire docked with their valuable cargoes of gold, silver, ancient scrolls, rare woods, fine fabrics, and unknown green stones. Soon, Ofir became one of the wealthiest cities of the empire and also a target for attacks from the tribes in the East. In repeated attacks, these eastern tribes stormed the city, raided the ships at anchor and disrupted the communications with the interior.

One thousand years after Lhasa's departure, the empire disintegrated. The savage tribes of the East had succeeded in conquering Ofir and burned it to the ground. Subsequently, "the Ugha Mongulala yielded the (Atlantic) coastal provinces in the east and withdrew into the interior of the country. And the connection to Samon's empire was severed." Believed to be located on the borders of Brazil and Venezuela, Akahim was/is "a gigantic stone city shaped like an outstretched finger" that lies behind a great waterfall. It has "lain in ruins for 400 years, though it was in close alliance with Akakor for thousands of years." When the "White Barbarians" began to advance into their territory, the former residents of Akahim sought refuge underground. Apparently, Akahim and Akakor were/are linked together via a subterranean passage... Tatunca Nara goes on the say that circa 2470 B.C. the Inca founder of legend; Viracocha (an Ugha Mongulala who apparently had been banished from the tribe for breaking some law) founded the Inca dynasty and built Cuzco. (The Inca Empire would later become a sister nation of the Ugha Mongulala.)

Believed to be located on the borders of Brazil and Venezuela, Akahim was/is "a gigantic stone city shaped like an outstretched finger" that lies behind a great waterfall. It has" lain in ruins for 400 years, though it was in close alliance with Akakor for thousands of years." When the "White Barbarians" began to advance into their territory, the former residents of Akahim sought refuge underground. Apparently,

Akahim and Akakor were/are linked together via a subterranean passage... Tatunca Nara goes on the say that circa 2470 B.C. the Inca founder of legend, Viracocha (an Ugha Mongulala who apparently had been banished from the tribe for breaking some law) founded the Inca dynasty and built Cuzco. (The Inca Empire would later become a sister nation of the Ugha Mongulala. Then, in 570 A.D., white, bearded strangers sailing in long ships with a fierce dragon at the bow came up the Amazon River. They called themselves "Goths" (Goths?) and allied themselves with the people of Akakor. Their sailing ships consisted of iron armor, black sails, and colorful dragon heads and could carry up to 60 men. But the Germanic tribe of Ostrogoths (a warrior race that conquered Italy within 60 years) were defeated by the East Roman General Narses at the battle of Vesuvius in 522 A.D.) and the last survivor of this tribe had disappeared without a trace. Linguists claim to have found traces of their language in Southern France and Spain, but there is no definite proof where they migrated.

Well, according to Chronicle of Akakor, the Ostrogoths made an alliance with bold sailors of the north (North?) and ended up in South America. The subsequent union of the Goths and the Ugha Mongulala again strengthened Akakor. The Goths built new defenses and showed the Ugha Mongulala how to make iron and armor. Eventually, "The White Barbarians" invaded South America and conquered the Incas. Five years later, the Ugha Mongulala withdrew into the inner recesses of Akakor, and according to Tatunca Nara, they departed from Macchu Picchu and ordered their frontier cities abandoned and destroyed. This helps explain Macchu Picchu.

As more and more Spanish and Portuguese landed at the mouth of the Amazon and continued to conquer more and more of South America, Akakor's sister city, Akahim (note the similarity of the name with city of Arakim in Siberia, Russia?), was attacked by hostile tribes and abandoned. While the men wanted to retreat, the women insisted on fighting the White Barbarians. And so the legend of the Amazons was born. According to Tatunca Nara, in 1920, the Spanish captured fifteen Inca nobles and held them prisoner in Lima.

Tatuncas father, Sinkaia, sent 80 worriers through the Lhasa's underground tunnels in an attempt to free them. For three moons they stealthily made their way to the capital of the White Barbarians. When dawn broke, they stormed out from hiding and attacked the White Barbarians. An ensuing battle raged. 120 White Barbarians were killed. But the White Barbarians were overwhelming. None of Sinkaia's warriors returned to Akakor. All had given their lives "as faithful servants of the Gods for the Chosen People." Later, in 1932 A.D., the Ugha Mongulala attacked a white settlement on the upper reaches of the

Santa Maria River. They killed all the men and took four women captive. Three of the women drowned in their attempt to escape, but one survived. Her name was Reinha, a German missionary. Reinha found her way to Akakor and grew fond of the city and its people. She eventually married Prince Sinkaia and gave birth to Tatunca Nara.

NOTE: The main points in the story from the records of Slavic Aryan Vedas:

1. The Ugha Mongulala means Mongolian Ear (but does not relate to modern Mongolian races in ancient Slavic Aryan language), a chosen people by the Gods in that area 15,000 years ago.

2. The language Gods spoke was Guechua in Slavic Aryan the neurolingual meaning of word Guechua is RECHA (means - LANGUAGE).

3. Guechua (RECHA), a written language of 1,400 symbols, each yielding a different meaning depending on their sequence SLAVIC ARYAN RUNIC ALPHABET consists of ~1400 RUNES, each yielding a different meaning depending on their sequence.

4. Ugha Mongulala tribal book entitled Chronicle of Akakor begins at the year zero, which corresponds to the year 10,481 B.C. on the Gregorian calendar (after the Great Masters started new calendar system and left the Ugha Mongulala tribe). IN SLAVIC ARYAN VEDAS THIS IS A TIME OF GREAT COOLING - 10908 B.C.

5. At ~ 14000 B.C. glimmering golden ships appeared in the sky. Enormous blasts of fire illuminated the plain. The earth shook and thunder echoed over the hills." The strangers in the golden ships looked like humans with fine features - white skin, bluish-black hair and thick beards in Slavic Aryan Vedas there is a description of such people and also a description of the year 10908 B.C. as TIME OF GREAT COOLING.

6. The Ugha Mongulala tribe had no tools as Gods did, which, as if by magic, SUSPENDED THE HEAVIEST STONES, FLUNG LIGHTING AND MELTED ROCKS." These strangers civilized the tribe and built three great cities of stone called: Akanis, Akakor, and Akahim (AS RUSSIAN ARAKIM).

7. In year zero (10,481 B.C.), the Former Masters left, but before they left there was some kind of "WAR BETWEEN THE GODS." This war

was horrible and devastating. Afterwards, the Former Masters left and a global catastrophe ensued.

8. In 3166 B.C. a second catastrophe occurred. This catastrophe ended the "Years of Blood" - the 6000 years of barbarism that had ravaged the land since the First Catastrophe. Just after this 2nd catastrophe, the Gods returned to Akakor, but only a few - Lhasa and Samon.

9. Lhasa supposedly had Macchu Picchu built as an outpost of his empire.

Four years after Tatunca Nara was born, Reinha (his mother) returned to Germany as an ambassador to Hitler's Third Reich. A year later she returned with three German leaders and negotiated an agreement with the Ugha Mongulala. The Ugha Mongulala and the Germans would be allies in a plan that would rule Brazil. The Nazis would invade Brazil in 1945, occupy the large coastal cities and the Ugha Mongulala would attack the white settlements in the interior. After the expected victory, the Germans would rule the eastern provinces along the coast and the Ugha Mongulala would reclaim the region of the Great (Amazon) River.

According to Tatunca Nara, the first Nazi soldiers reached Akakor by U-boat in 1941; the last soldiers arriving in 1945. For years the Germans lived with the Ugha Mongulala, arming them and training them for a war that never came. But in 1963 fighting erupted between the Germans and the Ugha Mongulala and Peru. The Germans and the Ugha Mongulala killed a number of white settlers in the Madre Dios region, but when the Peruvian government counter-attacked, the Ugha Mongulala retreated back to Akakor. In 1968, a plane crashed near Akakor.

Sinkaia ordered his son, Tatunca, to go to the crash site and kill the survivors, who were being held captive by another tribe. But instead of slaying the survivors, Tatunca was able to release them from their captives and led them to Manaus. As it turned out, the 12 survivors were officers of the Brazilian government. Tatunca Nara eventually became the new tribal leader of the Ugha Mongulala and in 1972 he went to Manaus to negotiate with the White Barbarians in an effort to secure peace with them (he felt it was useless to fight them any longer.)

It was during this trip that he met Karl Brugger, the German journalist who documented the story you are reading here. Karl Brugger checked out what elements of Tatunca Nara's story he could and found them to be true. For example, Natunca claimed he saved the lives of 12

Brazilian officers (whose plane had crashed in the jungle) by obtaining their release from the Haisha Indians who had held them captive. He then led them to Manaus, where he originally met Karl Brugger. According to Brugger, Nara's story has been documented in the archives of Rio de Janeiro, Brasilia, Manaus, and Rio Branco.

Independent newspaper documentation of the tale is available beginning in 1968, which mentions "a white Indian chieftain who saved the lives of 12 Brazilian officers by obtaining their release from the Haisha Indians and leading them to Manaus.

Witnesses said he spoke broken German, a number of Indian languages from the upper Amazon, and a little Portuguese."

It was during Tatunca and Karl Brugger's second meeting that Brugger accepted Tatunca's offer to accompany him up into the dangerous and forbidden rainforest to see the secret city of Akakor for himself firsthand. On September 25, 1972, with a Winchester rifle, two revolvers, machetes, food, hammocks, jungle attire, medicine and other provisions and equipment, Tatunca Nara, Karl Brugger and a Brazilian photographer departed Manaus by river and motored up the Rios Purus to the secret city. Once they reached the Rio Yaku, their plan was to continue by canoe as far as they could and then proceed on foot through the foothills of the Andes to Akakor.

Tatunca estimated that it would take six (6) weeks. Once they reached the Rio Yaku, their plan was to continue by canoe as far as they could and then proceed on foot through the foothills of the Andes to Akakor. Tatunca estimated that it would take six (6) weeks. On October 5, ten days after they left Manaus, Brugger reported they had abandoned their boat at Cachoeira Inglesa (English waterfall) and began their final journey to Akakor by canoe. As they neared their destination, Brugger and the photographer became uneasy. Tatunca Nara began painting red stripes on his face and yellow stripes on his chest and legs, as if he was preparing to return to his people. The snow-capped mountains of the Andes mountains towered before them.

At this point they must have traveled far up the Rio Yaku and into Peru. On October 13, they lost their canoe over dangerous rapids. Most of their food and medical supplies were lost, and their camera equipment was destroyed. This was the excuse Brugger and the photographer needed to abandon the expedition and return to Manaus. They were only 10 days away from Akakor. But Tatunca Nara did not follow. With a bow, a small quiver of arrows, and a hunting knife, Natunca Nara disappeared into the forbidden jungle. Tatunca Nara was never seen or heard from again, or so the story was written. Some say he is mentally unstable and will do tours for interested parties.

Some have accused of Karl Brugger and Von Daniken of fraud, solely based on the Edgar Cayce-ite date he writes about of 10,481 B.C. BUT there are too many underlying details that convey truth to the story. What can we deduce from Tatunca Nara's story, and Karl Brugger's account?

Could Brugger have made up the whole story? Could he have combined existing legends with 20th century history? It's possible but not likely. Although some parts of the story may seem far-fetched, for the most part they are true, as far as Brugger knew it. There are some parallels between the Chronicle of Akakor and the Peruvian legends of Gran Paititi. Was Paititi where the last Incas fled? Was Akakor the Inca Paititi?

Tatunca Nara spoke broken German. If he lived in the jungle all his life, where did he learn German? From Reinha, his mother? Since the Chronicle of Akakor was published, several explorers have lost their life searching for Akakor. Gregory Deyermenjian, an American explorer who had spent a lot of time in Peru in search of Gran Paititi, reportedly told David Childress (the author of Lost Cities and Ancient Mysteries of South America) about a young American from a wealthy family who came to Cuzco (Peru) in 1977 obsessed with finding Akakor.

He said the young American hired a hotel owner to escort him to the headwaters of Rio Yaco where he planned to meet an Indian guide who would lead him to Akakor. But the headwaters of the Rio Yaco is located in a very remote area nearly impossible to reach from Peru. The hotel owner escorted him as far as Cosnipate. The young aristocrat was never seen or heard from again. It seems quite plausible that the Ugha Mongulala are real people, and with that, we must accept the reality that they have real traditions and real cities. If Tatunca Nara really had a German mother, then it is quite plausible that Germans could have had an influence on their traditions and their mythology.

Karl Brugger investigated whether the Nazis actually occupied Akakor and whether or not they had a plan to invade Brazil. According to Brugger, the Third Reich believed it was essential to keep Brazil neutral in order for the German U-boats to have uncontested control of the South Atlantic, and an invasion of Brazil was the natural expansion plan of the Third Reich.

But the United States ruined their plan by persuading the Brazilian government to align with the Allies. Unhappy with Brazil's decision to side with the Allies, Germany retaliated by sinking as many as 38 Brazilian ships by U-boats from 1942 to the end of the war. But was it possible for Germany to land soldiers in Amazonia and make their way to Akakor?

Apparently, there are eye-witness accounts of German U-boats landing on the coast of Rio de Janeiro, and in 1938 a Nazi U-boat established contact with the German colony at Manaus and made a geological survey including a documentary film of Amazonia, which has been preserved in the East Berlin archives.

In May 1945, on the day before Germany surrendered, two U-boats, U-530 and U-977, departed from northern Germany and headed to South America. About three months later, they surrendered to Argentina, at different times. When the captains of the U-boats were turned over to the Americans, under thorough interrogation, they said they had had no important passengers onboard and that the reason they surrendered to Argentina was because they did not want to surrender to the British.

 But where had they been prior to their surrender? And why did it take them nearly three months to surrender?

Natunca Nara supposedly knew nothing of the war in Europe. His knowledge of the war was limited to what the Germans told him, and yet, according to Natunca Nara, Germans were still sending soldiers to Akakor in May 1945.

As if that wasn't suspicious enough, Karl Brugger was murdered outside his apartment in Manaus by an unknown assailant (assassin?). Why? Did his murder have anything to do with his book, and/or his knowledge of Akakor? The answer will probably never be known. And so the question begs to be repeated: Does an ancient stone city or cities exist on the Brazilian-Peruvian border? On December 30, 1975, the Landsat II satellite photographed an area of southeastern Peru at 13° S latitude, 71° 30" W longitude. The photographs revealed 12 pyramids covered with trees. Unfortunately, all attempts to reach these pyramids have led to the death and disappearance of many explorers. Could these pyramids be a part of the ancient Akakor complex?

It is known that a group of people calling themselves "representatives of the German Consulate," stormed his hotel room and confiscated all documents, as well as the entire material Brugger about the three lost cities. The same happened in Germany and what little was left, or rather survived, was in the hands of friends of the journalist and happily this way could not be confiscated or censored!

WHO therefore be interested in eliminating Karl Brugger and along with him to destroy all your precious records on important archaeological repositories Brazilian great vestiges of a lost civilization evolved? Why exactly prevent the publication of the original drafts of his book, which were held by a German editor? For what exactly the Nazis had much interest in the Amazon jungles...?

Warning & Disaster

Talmud says, "Seven days before the deluge the Holy One changed the primeval order and the sun rose in the West and set in the East (Tractate Sanhedrin), and another rabbinical source declares that in the time of Moses, *"the course of the heavenly bodies became confounded."*[34] [author's note: This would have been inaccurate, because a pole shift this large, a Flood and Tsunami would have occurred within 7 hours, not 7 days.

---

34 Leket Misrash 2A, Perek 8, quoted from Rabbi Elieser, a contributor to the Talmud, quoted by Louis Ginz

5

Apocalypse: 10,850 B.C

An ominous sunset marked the close of the Pleistocene, as tenacious humanity hung perilously close to extinction near the 7000-foot summit of the Dome of the Gods, one of the highest peaks of the Great Smokey Mountains.

Two Clovis[35] Indian survivors', Quetzlte and his wife, Nura[36], exhausted to the bone, leaned heavily against the cool granite facing, when another brilliant flash burst overhead, followed instantaneously by a horrific explosion.

When would the sky stop burning? Another brilliant fireball raced across the heavens, followed by another and another in quick succession, each ending in a deafening explosion that shook the ground. Quetzlte wiped the burning sweat from his eyes and surveyed the overhanging ledge above him. They were almost to the summit. Gathering his courage, Quetzlte struggled to find a secure handhold in the stone crevice over his head. Finding one, he pushed Nura ahead of him over the ledge, and quickly strained to follow her, pulling himself up and over the overhang, safely landing next to her. They laid prostrate for a moment staring straight up into the burning red sky. Last night his grandfather warned his people that the Great Day of the apocalypse had come, and this time, the old shaman told

---

35 Clovis Indians, Ancient North American race

36 CNN reported on May 18th, 2014 of the discovery of an early American Indian skeleton in a Yucatan Mexico cave dated 13,000 years ago. According to the archeologists, they believe the teenage girl may have gotten lost in a cave looking for water, and fell to her death in a deep, dark pit. Curiously, the team of archeologists named her "Naia," as she's known today, a name similar to my own fictional of "Nura." I wrote about in 1995. (I am sure there is no correlation, but it's an interesting coincidence.). Why did Naia retreat into the cave 13,000 years ago? Was she running and hiding from our impacting comet?

them, the world would burn and then drown in a great deluge. He pleaded with his people to quickly seek the highest mountain tops, so they might survive and carry on the ancient traditions. Few listened to the old man, and fewer still took action. Life had continued in much the same manner for thousands of years, why therefore, should it change now? Quetzlte had implicitly trusted his grandfather, and so, without hesitation, he and his wife fled their hearth and home, to seek out the refuge of the mountains -- but nothing could have prepared him for this.

Countless streaks of light flashed through the encroaching darkness, each ending in deafening explosion that seemed to pound the earth like a cosmic hammer. The quickening tempo of the firestorm served Quetzlte a warning -- something terrible was about to happen. He cowered under a nearby stone ledge and searched frantically for a cave hidden against the mountainside. He knew it was nearby, but where? Nura saw it first. Pointing to a nearby shadow, she pulled him off his knees and together they charged through the tempest towards the cave entrance. In the mad dash for safety, they hardly noticed the comet approaching from the Northwestern horizon. It quickly grew in size and brilliance, until the

night sky became as bright as day. Nura screamed and stumbled, but Quetzlte caught her mid-stride, and kept running on until they passed the threshold of the cave. Scrambling quickly into its depths, the brave antediluvians braced one another, closed their eyes, and waited for impact.

A deafening roar shook the mountain, as the fireball rushed over the Smoky mountains and continued briefly eastward before exploding into millions of fragments, raining fire and death on the inhabitants below. Twelve-thousand-nine-hundred years ago a massive fall of meteors from a disintegrating comet flashed across the American Midwest, smashed off part of the Blue Ridge Mountains in Virginia and North Carolina and slammed into the coastal regions of North America at speeds in excess of 150,000 miles per hour. Travelling nearly tangentially to the Earth's surface, barely two minutes could have elapsed between the moment when the comet first flared up on the horizon and its catastrophic explosion over the Atlantic. During its descent, the gases hurled backward from the flaming giant must have attained a luminosity to 20-100 times that of the sun, blinding permanently any curious eyes sighting it. After entering the lowest and densest part of the atmosphere, the heat and stress on its surface would have reached a critical mass, causing the cometary body to burst, resulting in several explosions, shattering it into innumerable lethal fragments, gouging a trench of deadly devastation across the southeastern part of North America.

I wrote the above fictional reenactment with *Nura* and *Quetzlte* in 1995 to dramatize the 10,850 B.C. impact recounted in ancient legends and myths. Back then I did not have the impressive geological evidence that clearly shows a mass extinction level event that occurred 13,000 years ago in North and South. Millions of woolly mammoths, giant ground sloths and sabre-tooth tigers were wiped out.

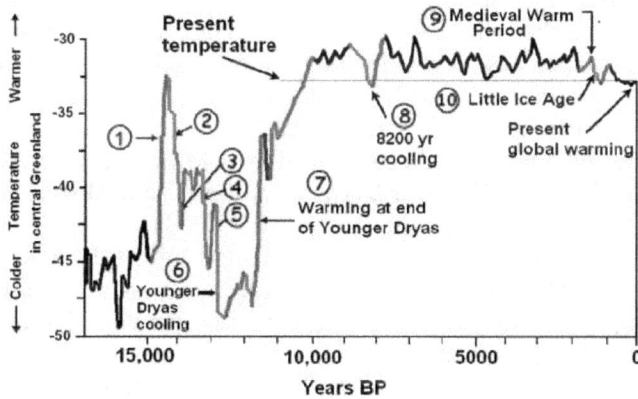

How? By the impact of a devastating comet/meteor 12,900 years ago.

A controversial 2007 study by California Professor James Kennett, suggests that an ancient cosmic triggered a vicious cold snap called the 'Younger Dryas impact theory' that lasted nearly 1,500 years. Another research team from the University of California, Santa Barbara, went a step further. They proclaimed to have further evidence to back up Professor Kennett's 2007 'Younger Dryas impact theory'. Their separate study last year found further evidence that a cosmic impact 12,900 years ago could have led to the demise of the 'Clovis' people of North America.

There was more evidence. A layer of platinum from an ice core taken in Greenland dated back to the time of a known abrupt climate transition, known as the 'Big Freeze' that led to the demise of both humans and the Megafauna. Another study by researcher Michail Petaeve and his Harvard colleagues at Harvard University, writing in the journal Proceedings of the National Academy of Sciences, found evidence of a 100-fold increase in the platinum concentration in the ice-cores that were 12,890 years-old. 37

Their conclusion? A comet impact tipped the world into its colder phase, making dozens of species extinct.

The Younger Dryas, or the 'Big Freeze', saw a rapid return to glacial conditions in the higher latitudes of the Northern Hemisphere between 12,900–11,500 years ago. Researchers claim that debris thrown into the atmosphere from an impact may have tipped the Earth into global cooling, wiping out mega-fauna along with native cultures such as the Clovis people. The theory is partly based on the presence of nano-diamonds along Bull Creek in the Oklahoma Panhandle, which is one type of material that could result from an extra-terrestrial collision.

North America's woolly mammoth, giant ground sloths and sabre-tooth tigers are thought to have been wiped out by a devastating comet 12,900 years ago. The research team re-examined the distribution of nano-diamonds in Bull Creek's geological record to see if they could reproduce the original study's evidence.

'We were able to replicate some of their results and we did find nano-diamonds right at the Younger Dryas Boundary,' said Alexander Simms, an associate professor at the University of California. Researchers analyzed 49 sediment samples representing different time periods and environmental changes. They identified high levels of nano-diamonds immediately below and just above Younger Dryas deposits and in late-Holocene near-surface deposits.

The research team re-examined the distribution of nano-diamonds at Bull Creek's geological record to see if they could reproduce the original study's evidence at Bull Creek that showed the paleosol—the ancient buried soil; the dark black layer in the side of the cliff that formed during the Younger Dryas and ended in the late Holocene which began at the end of the Pleistocene 11,700 years ago and continues to the present. The researchers believe that the presence of nano-diamonds was not caused by environmental changes, soil formation, cultural activities, or the amount of time in which the landscape was stable.

They believe whatever process produced the elevated concentrations of nano-diamonds at the onset of the Younger Dryas sediments may have also been active in recent millennia in Bull Creek. A 'recent' meteorite impact did occur near Bull Creek but scientists aren't sure exactly when. However, they claim this provides compelling evidence that the nano-diamonds were the result of a comet collision.

---

37   http://www.dailymail.co.uk/sciencetech/article-2550915/Did-comet-kill-woolly-mammoth-Study-backs-theory-impact-13-000-years-ago-triggered-extreme-cold-snap.html#ixzz3RGUBjBxR

Megafauna Disappearance- Were they killed off by Man or Comet?

Many researchers have come to the conclusion that these pre-historic animals, including the ferocious carnivores, were killed off by man. I am willing to admit that man with his superior cunning probably killed all of these animals, even the mammoth, cave bear, and lion, on rare occasions, but to suppose that man could have killed millions or even in great numbers, is not at all reasonable or likely.

History has taught us that the American Indian was never able to kill the buffalo in any numbers with the bow and arrow, or by driving a herd over a cliff or into a river on some rare occasion. In fact, the Indians did not kill more than they needed, as the white man did, and there is no reason to suppose that prehistoric man could or did.

The violent circumstances under which the Clovis people and the mammoths and many other large Mammalia met their doom 13,000 years ago point to a sudden, massive inundation and drowning by water. The new science of taphonomy, a study of the dying process from its onset until the animal is embedded in a geological stratum, has recently proven that the mammoths died suddenly by asphyxiation, or drowning.

They perished not only in Siberia, but in every corner of the globe. In nearly every instance, death was sudden and catastrophic. Quadruped species from both arctic and tropical regions have been found buried together in caves and excavations in the United Kingdom and Europe, and literally millions of animals, mammoths, extinct bison, horse, woolly rhinoceros, camel, and saber-tooth tigers, along with great quantities of bones and tusks, are mangled and mingled among splintered masses of torn and uprooted trees. They have been mined out of miles of excavations along stream valleys near Fairbanks, Alaska, *extending in a huge arc perhaps 1,500 to 2000 miles long and sometimes 1020 feet high.* Mammal remains are for the most part, dismembered and articulated, even though in the permafrost of the north, some fragments yet retain in their frozen state, portions of ligaments, skin, hair, and flesh. At least four layers of volcanic ash may be traced in these deposits although they are extremely warped and distorted.

More species died out in Africa and North America at the end of the last ice age than had vanished during all previous Pleistocene ice ages combined. And whereas the life-forms that had been lost during the earlier ice ages had been replaced by related species, the ecological niches that were emptied by the last wave of extinctions remained empty. In North America, for example, the mammoths, horses, camels, ground sloths, peccaries and giant beavers that once roamed the Great Plains disappeared virtually overnight without replacement." After an exhaustive study of past and present fauna, Russel Wallace, the eminent zoologist of the late 19th century, declared: "We live in a zoologically impoverished world, from which all the hugest, and fiercest, and strangest forms have recently disappeared.

It is a marvelous fact, and one that has hardly been sufficiently dwelt upon, this sudden dying out of so many large mammalia, not only in one place but over half the land surfaces of the globe." [38]

Scientists have known for many years of the mammoths in Siberia, but generally supposed their icy demise was a freak aberration of nature, and thus only a very rare occurrence. In 1946, Dr. Hibben published a book called *The Lost Americans* which went into great detail concerning his unique observations of the great quantities of animal remains found frozen in central Alaska. He (Kelly) "tells of the tremendous quantities of bones found all over North America and especially in Alaska. Dr. Hibbon estimates that some *forty million animals* met their death in this continent and probably an equal number in Asia. He discusses the several causes that scientists have advanced for the sudden extinction of these Pleistocene animals but concludes that none fill the requirements and that "Who or what killed them is still a mystery." He tells us that the whole of central Alaska is underlain with the bones of such animals as mammoth, mastodon, bison, wolves, bears, lions, and others.

"Throughout the Yukon and its tributaries, the gnawing currents of the river and eaten into many a frozen bank of muck to reveal bones and tusks of these animals protruding at all levels. Whole gravel bars in the muddy river were formed of the jumbled fragments of animal remains. The Alaskan muck is like a fine, dark-gray sand. It is very moist, is eternally frozen, and apparently has been so ever since the glacial age and the times of early man. Even in summer the ground thaws only about three feet down from the surface. Eskimo dogs in the warm Alaskan summers habitually dig shallow holes in the ground so that they may lie on the frozen muck beneath to keep cool. Within this mass, frozen solid, lie the twisted parts of animals and trees intermingled with lenses of ice and layers of peat and mosses. It looks as though in the middle of some cataclysmic catastrophe of ten thousand years ago the whole Alaskan world of living animals and plants was suddenly frozen in mid-motion in a grim charade." (Hibben)

38Ibid, Ice Ages, pg 55

Depopulation of Humans and Animals

The Younger Dryas comet theory has been criticized for a number of reasons, including suggestions that the human population didn't decline at the time.

The evidence begs otherwise. According to Historian Daniel Warner:

"Scholars who study human population patterns have come up with some interesting figures for the period between 6000 B.C. and today. They estimate that 2,000 years ago there were about 250,000,000 people on Earth -- not very many considering that intelligent human beings have been around for nearly four million years! In 4800 B.C. the global population, they say, was 20,000,000. In 5000 B.C., we are told, there were only 10,000,000 people living on all the continents of the world. A thousand years before this -- in 6000 B.C.-- the experts claim that there were only half that many people on Earth. Based on these figures, the total world population about 10,500 B.C. would only number about 250,000, less than the population of a small city."

Traditions say that the only survivors were those who were able to climb to the tops of high mountains. People living in valleys must not have had time to make such a climb, which ordinarily took several hours or even days, since we know that the flood happened suddenly. "One story tells how many inhabitants of the interior of South America had to flee to a very high mountain in order to escape the water and to find food. While they were attempting to climb the mountains some of them fell in the mouth of a great fish and were swallowed. Others reached the top in safety. In their stories they have preserved the names of some of the heroes of ancient times in the days of catastrophe and flood. One name reminds us of the Patriarch Noah from the account in Genesis of the Deluge or Flood. This name is *Noeaha.*"[39] *(not local, North and South America too)*

The Masco Tribe says in the time of the great destruction, which destroyed a resplendent and ancient empire that was ruled over by white kings, it took twelve hours to go around a certain tree. Some of the people wanted to climb to the top of this tree to escape the catastrophe. But some of them were unsuccessful and fell off the tree *into boiling water which covered the Earth.*"[40]

"All of the tribes of the Madre de Dios area speak of a time when tremendous cataclysms shook the Earth and how in the interior of South America nearly everyone perished." [41]

---

39 Roy Norville, The Road in the Sky, pg. 123-124
40 Ibid, The Road in the Sky, pg 124.
41 Ibid, The Road in the Sky, pg. 124

The Huachipari say that the 'Rocks of Writing' are very ancient and existed before the Spanish, and before the Incas, and even before the Pre-Incas. They, and other tribes, say that those who reached the highest points of the mountains were saved in the days of the catastrophe, and such represent the direct descendants of the 'remnant that remained.'

The Bible tells us that Noah landed on the mountains of Ararat; Xisuthrus, in Chaldea, brought his boat to rest on Mount Korkura; the Manu of the Hindus went to the Himalayan plateaus; Bochica in South America took refuge on the Altiplano of the Andes; and Coxcox, the Mexican Noah, did the same in the Sierra Madre. The North American Indians spoke of the Rocky Mountains and African tribes stated that their own survivors gathered on the Ethiopian plateau.

All these mountain ranges have peaks higher than twelve thousand feet. Mammoths caught by an abrupt upheaval of the ground have been found frozen at altitudes of twelve thousand feet in central Asia.

Five Plateaus Survived the Deluge Worldwide

If we imagine the earth flooded and ravaged by waves breaking at altitudes of up to twelve thousand feet, what can we see behind the reach of the cataclysm? We see five high plateaus, in Mexico, Peru, Ethiopia, Iran and the Himalayas.

These plateaus were practically the only parts of the globe that were spared by the Deluge, though there were probably other places that might have served as a refuge for people, such as Greenland and Egypt, where the topography of the land spared a massive inundation by water, although they did not escape it entirely.

Very Few Survivors:

The Universal Deluge nearly wiped out the whole human race, since traditions mention the following numbers of survivors:
Eight according to the Hebrew tradition
Two after the deluge of Deucalion in Greek mythology
Two thousand nine hundred and one in Persian mythology.
Two to six in Assyro-Babylonian deluge
Ten to a hundred in the Chaldean deluge of Xisuthrus.
One in the deluge of Manu of the Catapartha Brahmana.
Four or five in the Greek version in which the deluge of
Ogyges is older than that of   Deucalion.
Eight in the deluge of Manu from the Mahabharata.
Eight in the deluge of Satyavrata.
Two in the deluge of Kymris (Belgian Celts).

Two in the Eddas of the Scandinavians.

Two in the dulge of the Lithuanians.

Two according to the traditions of the Canaris in Ecuador.

Fifty to a hundred in the deluge of Bochica (Columbia)

Fifty to a hundred in the deluge of the Chichimecas in the first age, known as the 'Sun of the Waters." This deluge is specifically said to have been worldwide.

Two in the Mexican deluge of Coxcox.

Four in the traditions of Brazil.

Noah's Ark: A metaphor of RESCUE and a story of mass extermination of humans, plants, and animals. The Noah narrative does not truly explain the survival of the animals. According to the biblical narrative, between two to seven of each species were saved, which does not provide enough of a gene pool to repopulate the planet from two to seven of each species to save them from extinction. Furthermore, to preserve a multitude of sensitive species from excessive cold, heat, dampness and dryness, and to give each species proper temperature and food, would have taken resources such as electricity which were likely unavailable to the people of that time period.

Noah's ark is predominantly an allegorical representation of the Deluge. Noah's age, the dimensions of the Ark, the period of the deluge, the number of animals, the number of human survivors cannot be taken literally, although the story is likely based on historical fact.

The animal kingdom appears to have met the same destiny as their more intelligent human cousins during this period. Between 15,000 B.C. and 6,000 B.C., seventy genres of large mammals became extinct including all North American members of seven families, and one complete order, the Proboscidea. Staggering losses of 40,000,000 animals, the vast majority of extinctions occurred between 11,000 B.C. - 9,400 B.C. To put this in perspective, during the previous 300,000 years only about twenty genres had disappeared.

Why did so many animals and humans die off?

If the animal and flora kingdom were so abundant, wouldn't the human species, being more intelligent and adaptable, have flourished and multiplied after four million years of evolution?

The Megafauna Mystery

What could have caused the sudden, worldwide extinction of the giant mammals that recently inhabited the earth? Baron Georges Cuvier (1769-1832), the father of comparative anatomy and paleontology and one of the great geologists of all time, suggested that the mass extinction of the gigantic Ice Age beasts was caused by a sudden, catastrophic inundation by ocean water and freezing temperatures. He reasoned from the evidence he saw around him that the earth had undergone many sudden and violent revolutions. He saw the evidence of world-wide extinctions and survivals and the overlapping of some species from one age into the next; and although he did not invent the word "evolution" he certainly described its effects.

Darwin is usually credited with having originated the theory of evolution, but in reality, he only popularized the word and enlarged upon the ideas of Cuvier and others. The strange historical fact arising out of this situation is that the theory of evolution could be born out of two such divergent views of world cosmology as the catastrophism of Cuvier and the uniformitarianism of Darwin."

In speaking of the sudden nature of the inundation, Cuvier made this statement:

"Repeated irruptions and retreats of the sea have neither all been slow nor gradual; on the contrary, most of the catastrophes which have occasioned them have been sudden; and this is especially easy to be proved with regard to the last of these catastrophes, that which, by a twofold motion, has inundated, and afterward laid dry, our present continents, or at least a part of the land which forms them at the present day. In the northern regions it has left the carcasses of large quadrupeds which became enveloped in the ice and have thus been preserved even to our own times, with their skin, their hair, and their flesh. If they had not been frozen as soon as killed, they would have been decomposed by putrefaction. And, on the other hand, this eternal frost could not previously have occupied the places in which they have been seized by it, for they could not have lived in such a temperature. It was, therefore, at one and the same moment that these animals were destroyed and the country which they inhabited became covered with ice.

This event has been sudden, instantaneous, without any gradation, and what is so clearly demonstrated with respect to this last catastrophe, is not less so with reference to those which have preceded it."[42]

According to Flint, a mammoth was unearthed miles within the Arctic Circle and more than 2000 miles north of the present range of the elephants, but unlike any living elephant, the animal bore a "heavy coat of yellowish-brown wool interspersed with long black hair," and had "four instead of five toes on each foot." Its head was high crowned and very short, giving the animal a profile quite distinct from that of any living elephant. The specimen was, indeed, that of the extinct wooly mammoth, a species that ranged widely over the snow fields of Europe

---

42 Cuvier, Essay on the Theory of the Earth, pg. 14-15.

and North America during the last part of the Pleistocene ice age. According to Flint, "Unmistakable pictures of such creatures were left by prehistoric cave dwellers on the walls of some of the caverns in France, but the species has been extinct so long that we find no allusions to it either in historical records or in the legends[43] of any living people."[44]

The violent circumstances under which the mammoths and many other large mammalian met their doom 12,800 years ago point to a sudden, massive inundation and drowning by water. The new science of taphonomy, a study of the dying process from its onset until the animal is embedded in a geological stratum, has recently proven that the mammoths died suddenly by asphyxiation, or drowning.

They perished not only in Siberia, but in every corner of the globe. In nearly every instance, death was sudden and catastrophic. *Quadruped* species from both arctic and tropical regions have been found buried together in caves and excavations in the United Kingdom and Europe, and literally millions of animals, mammoths, extinct bison, horse, woolly rhinoceros, camel, and saber-tooth tigers, along with great quantities of bones and tusks, are mangled and mingled among splintered masses of torn and uprooted trees. They have been mined out of miles of excavations along stream valleys near Fairbanks, Alaska, *extending in a huge arc perhaps 1,500 to 2000 miles long and sometimes 1020 feet high*. Mammal remains are for the most part, dismembered and articulated, even though in the permafrost of the north, some fragments yet retain in their frozen state, portions of ligaments, skin, hair, and flesh. At least four layers of volcanic ash may be traced in these deposits although they are extremely warped and distorted.

Cited from Richard Noone, Ice Ages - More species died out in Africa and North America at the end of the last ice age than had vanished during all previous Pleistocene ice ages combined. And whereas the life-forms that had been lost during the earlier ice ages had been replaced by related species, the ecological niches that were emptied by the last wave of extinctions remained empty. In North America, for example, the mammoths, horses, camels, ground sloths, peccaries and giant beavers that once roamed the Great Plains disappeared virtually overnight without replacement." After an exhaustive study of past and present fauna, Russel Wallace, the eminent zoologist of the late 19th century, declared: "We live in a zoologically impoverished world, from which all the hugest, and fiercest, and strangest forms have recently disappeared. It is a marvelous fact, and one that has hardly been sufficiently dwelt

43 Contrary to Flint, the Eskimo's legends of the mammoth are quite common. The Greenland Eskimo's legend of the mammoth Is - "The animal with the iron tail (tusks) that runs backwards".

44 Flint, Glacial Geology and the Pleistoce Epoch, John Wiley and Sons, Inc.

upon, this sudden dying out of so many large mammalia, not only in one place but over half the land surfaces of the globe." [45] Hibben further states:

"Throughout the Yukon and its tributaries, the gnawing currents of the river and eaten into many a frozen bank of muck to reveal bones and tusks of these animals protruding at all levels. Whole gravel bars in the muddy river were formed of the jumbled fragments of animal remains. The Alaskan muck is like a fine, dark-gray sand. It is very moist, is eternally frozen, and apparently has been so ever since the glacial age and the times of early man. Even in summer the ground thaws only about three feet down from the surface. Eskimo dogs in the warm Alaskan summers habitually dig shallow holes in the ground so that they may lie on the frozen muck beneath to keep cool. Within this mass, frozen solid, lie the twisted parts of animals and trees intermingled with lenses of ice and layers of peat and mosses. It looks as though in the middle of some cataclysmic catastrophe of ten thousand years ago the whole Alaskan world of living animals and plants was suddenly frozen in mid-motion in a grim charade."[46]

Dr. Hibben gives us a picture of catastrophic destruction that cannot by any stretch of the imagination be attributed to ordinary floods.

(Kelly) *"The presence of bones, trees, peat, and other debris all mixed together down to a depth of nearly 100 feet, points to a cataclysmic flood of tremendous proportions that must have moved across the land, grinding the bodies of the animals with stones and trees and spreading the whole out over the Yukon Valley. Nor ordinary floods as we know them have accomplished this great depth of debris at one time, nor is it at all likely that so many kinds and numbers of animals could have been captured by ordinary flood waters, even in a temperate climate. The Yukon Valley has very little fall and even if the river had expanded to fifty times its present greatest known size, the rising of the water would still have been slow enough for most animals to escape to higher ground or swim to shore. Elephants, or almost any four-legged animal for that matter, can swim for very long distances without tiring, and it is very unlikely that they could have been drowned or torn to pieces in any ordinary river."* [47]

Mammoths Drowned in Ice Age Floods?

"In 1876 Alfred Wallace, after collecting enough bones and fossils to permit him to catalogue the long list of vanished fauna, proposed that a

45 Ibid, Ice Ages, pg 55

46 F. C. Hibbn, The Lost Americans, Thomas V. Crowell Company, publishers, pg. 91.

47 Kelley, Hibben

sudden rise in temperature caused the ice packs to melt, initiating massive flooding which swept away the poor unsuspecting animals. Wallace reasoned that the elimination of so many species must have been the result of some exceptional event that had occurred almost simultaneously in many parts of the world.

Citing evidence that the northern portions of both Europe and North America had been covered with ice when these large animals were disappearing, Wallace maintained that the ice had probably "acted in various ways to have produced alterations of level of the ocean as well as vast local floods, which would have combined with the excessive cold to destroy animal life."[48]

Recently, this hypothesis has become the most prevailing view among paleontologists and geologists. In a pair of olive-grey, silty clay cores dredged up from beneath the Soto Canyon, which cuts through the floor of the Gulf of Mexico off the western coast of Florida, there is evidence of a universal flood. The cores contain within their sediments the skeletons of tiny marine creatures called foraminifera. Radiocarbon dating determined that these creatures died around 9,600 B.C. The skeletons also told Dr. Cesare Emiliani, a professor of geology at the University of Miami, that very suddenly the waters of the Gulf of Mexico had received a massive infusion of fresh water. So great, in fact, was this infusion that it raised the level of all the world's oceans by 131 feet, a rise so great it flooded all of the earth's coastal regions. (over thousands of years).

The infusion of freshwater into the Gulf of Mexico coincided with the Valders Advance, a southward surge of the North American glaciers. According to Dr. Willard Libby, who received the Nobel Prize for developing his method of radiocarbon dating of biological samples, there is evidence that ice was advancing around the world. Radiocarbon dated samples from Europe, South America and New Zealand all show movement at the same time as the Valders Advance. The melting waters from all these glaciers caused a tremendous rise in the world's oceans. Dr. Emiliani in the September 26, 1975, issue of Science, stated categorically, "In spite of its great antiquity in cultural terms, this could be an explanation for the deluge stories common to many Eurasian, Australasian and American traditions."

Were the Megafauna killed off by Man?

Many researchers have come to the conclusion that these pre-historic animals, including the ferocious carnivores, were killed off by

---

48 Ibid, Ice Ages, pg. 57

man. I am willing to admit that man with his superior cunning could have killed all of these animals, even the mammoth, cave bear, and lion, on rare occasions, but to suppose that man could have killed millions or even in great numbers, is not at all reasonable or likely.

History has taught us that the American Indian was never able to kill the buffalo in any numbers with the bow and arrow, or by driving a herd over a cliff or into a river on some rare occasion. In fact, the Indians did not kill more than they needed, as the white man did, and there is no reason to suppose that prehistoric man could or did.

Catastrophic evidence associated with the extinction of mammoths and many other large species have caused great consternation among geologists and paleontologists who marvel at their sudden disappearance.

It is apparent that life has often been disturbed on this earth by terrific events. Nonetheless, living beings have been the victims of these catastrophes. Some which inhabited the dry lands have been swallowed up by inundations. Others which peopled the waters have been laid dry, the bottom of the sea having been suddenly raised. Many races have been extinguished forever and have left no other memorial of their existence than some fragments that the naturalist can scarcely recognize. Cuvier knew these events occurred, but did not know why they occurred, or what caused them. We have a pressing problem in geology, but orthodoxy has no answers for them.

What is the cause the deaths of so many humans and animals?

A collision flood from neighboring oceans. They were torn limb from limb and mixed with the muck and debris of the flood. At the same time the area was moved 2000 miles closer to the newly formed polar cap which contains the present North Pole. Within a few hours or days, they were frozen solid and have remained in that condition to this day.

It has appeared strange and unnatural to us that such great, rich, and varied land masses as the two American continents remained to be discovered by the land-contact sailors of Europe. Now we see that these continents could have been stripped clean of whatever human culture that might have existed before the collision, and that the recovery was very slow compared to the recovery in the Mediterranean basin or in some of the Asiatic areas. At the time of discovery, the men of the Americas were at the primitive level of the stone-tipped arrows of the Indian and imbued in the excessive and barbaric mysticism of the nations of central America while the Europeans were stirring with the a burgeoning world-wide commerce. The history and archaeology of pre-Indian times will bear close study for it is *very likely that a high level of culture existed that was wiped out at the same time with the mammoth.*

6

Impact, Pole Shift, Oceanic Flood

*Then the sun will suddenly shine by night and the moon by day. Blood will trickle forth from wood and stone speak its voice. People will be confounded, and stars change course.*49

One of the most controversial theories surrounding the Apocalypse is the pole shift; a sudden and cataclysmic movement of the planet in which it flips end-over-end in space, or in the view of some pole shift theorists -- the crust of the planet slips around the molten core. In either case, the result is worldwide destruction.

Scientists are generally rigid in their belief that the world has never suffered an abrupt geological shift of the poles. Any movement of the geographical poles, they say, occurs at a nails pace, coinciding with continental drift of tectonic plates in the polar regions crawling along about two inches per year. Have the poles always been tilted at 23.5 degrees? Hidden among the symbols of mythology and theology are hints of this world-shaking event. They tell of global floods, lost civilizations, and reversed celestial orientation of the stars, sun and moon. Ancient antiquity saw the blending of history and allegory.

Pole Shift-Egyptian Records

The Egyptian priests of Memphis, for instance, told Herodotus that there had been 341 pharaohs and 341 high priests in 341 generations during the past 11,000 years. As if to back up the genealogy, 9 pthe high priests told Herodotus (c.485-c.425 B.C.) that the sun had "four times risen out of his usual quarter," and that he [sun] had "twice risen where

---

49 (RH. Charles, The Apocrypha and Pseudepigrapha of the Old Testament, vol. 2 (Oxford: Clarendon Press, 1913, pp 569

he now sets and twice set where he now rises.[50] The Egyptians believed the poles had reversed 180 degrees; south became north; summer became winter. The reversal of the poles is seen on the ceiling of the tomb of Senmut, architect to the Egyptian Queen Hatshepsut. Painted on an astronomical panel, hieroglyphs show the Orion-Sirius star cluster in the night sky proceeding in the reverse direction from what they appear to move today.[51] H. A. brown states in *Cataclysms of the Earth* that in a polar shift greater than 80 degrees, people at the 'pivot points' would see the sun appear to stand still and then move backward and set where it had risen. The stars would also appear to deviate from their usual course.

Several Egyptian documents mirror a similar reversal theme. "Papyrus Anastasi IV and Papyrus Ipuwer refer to the *Earth being turned upside down*, the seasons reversed, time disordered, and the sun

failing to rise."[52] The Harris Magical Papyrus reiterates the pole shift motif, "South becomes north and the Earth turns over."[53] *And in Breasted's Ancient Records of Egypt* III, Sec. 18, the reference "Horakhte, he riseth in the West," appears. Horakhte, was the name given to Horus, the sun god, when he rose in the morning, eastern sky."

"Sura LV of the Qur'an speaks of the "Lord of two Easts and of two Wests."

A recent study published in the scientific journal Nature of ancient ice extracted from the Greenland ice sheet, reveals, in fact, that between ice ages the Earth's climate has changed dramatically in the span of less than a decade with global temperatures plunging as much as twenty-five degrees, sometimes for hundreds of years.[54]

Following each world age, "in a general convulsion of nature, the sea is carried out of its bed, mountains spring out of the ground, rivers

---

50 Herodotus, The Histories (Harmondsworth, U.K.: Penguin, 1985), Book Two, p. 186.

Note: While some Egyptologists suggest this passage describes 1.5 Zodiacal rotations (one rotation is 25,920 years) equaling 39,000 years, it must be remembered that these changes in solar station occurred within 341 generations. How much time can we allow for a generation? Given that the Egyptian year was 360 days (with 5 unlucky days added on at the end of the year), with 30 day months, it is plausible that they considered a "generation" to be 30 years. 341 generations of 30 years is 10,230 years, which from the time of Herodotus in 450 B.C., establishes the beginning of Egyptian civilization at 10,680 B.C., fairly close to the previous magnetic dipole reversal of 10,400 B.C.

51 A. Pogo, The Astronomical Ceiling Decoration in the Tomb of Senmut (XVIIth Dynasty) (New York: Isis Press, 1930), p.306

52 Adolf Erman, The Literature of the Ancient Egyptians (London: Methuen, 1927) p.309.

53 H.O. Lange, Magical Papyrus Harris (Copenhagen: Det. Kgl. Danske Viden Skabernes Selskab, 1927), p.58.

54 Richard G. Fairbanks, 'Flip-Flop End to Last Ice Age," Nature, v362, n6420 (April 8, 1993), p. 495

change their course, human beings and everything are ruined, and the ancient traces effaced."[55]

Conversing with Egyptian priests on his visit to Egypt, Herodotus recounts their claim that four times since Egypt became a kingdom, "the sun rose contrary to his wont; twice rose where he now sets, and twice he set where he now rises."[56]

Drawing exclusively from Egyptian written sources, the first century Latin scholar Pomponius Mela affirms this ancient tradition, stating: 'The Egyptians pride themselves on being the most ancient people in the world. In their authentic annals ... one may read that since they have been in existence, the course of the stars has changed direction four times, and that the sun has set twice in the part of the sky where it rises today."[57]

In fact, the Magical Papyrus Harris specifically refers to a cosmic cataclysm of fire and flood when "the south becomes north, and the Earth turns over."[58]

Other ancient Egyptian records, including certain pyramid texts, corroborate the belief that the sun "ceased to live in the occident, and shines, a new one, in the orient."[59]

Likewise, Plato writes in "The Statesman": "I mean the change in the rising and the setting of the sun and the other heavenly bodies, how in those times they used to set in the quarter where they now rise, and used to rise where they now set...At certain periods the universe has its present circular motion, and at other periods it revolves in the reverse direction .... Of all the changes which take place in the heavens this reversal is the greatest and most complete." Later in this same work the great philosopher writes of the cataclysmic consequences of such a reversal: "There is at that time great destruction of animals in general and only a small part of the human race survives."[60]

Bellamy, in *Moons, Myths and Man,* states that "The Chinese say that it is only since a new order of things has come about that the stars move from east to west." ". . . The signs of the Chinese zodiac have the strange peculiarity of proceeding in a retrograde direction, that is, against the course of the sun."[61]

55 G. Schlegel, Uranographie chinoise (1875), P. 740. Quoteks, Immanuel Velikovsky, Worlds in Collision (New York: Pocket B.C.10. 1977), p. 48.

56 Herodotus, Bk. ii, 142 (Trans. A. D. Godley, 1921). Quote by, Velikovsky, op.cit., p. 118.

57 Pomponius Mela, De situ orbis, i.9-8. Quoted in Velikovsky, Op-cit., P. 119.

58 H. 0. Lange, "Der Magische Papyrus Harris," in K. Danske Videnskabernes Selskab (1927), p. 58. Quoted in Velikovsky, op.cit., p.120.

59 L. Speelers, Les Textes des Pyramides (1923), 1. Quoted in Velikovsky, op.cit., p. 120.

60 Plato,TheStatesman, or Politicus(Trans.H.N.Fowler,1925),pp. 49, 53. Quoted in Velikovsky, op. cit., p. 122.

61 Bellamy, Moons, Myths and Man, p. 69. Quoted in Velikovsky, op.cit., pp. 124-125.

Other accounts, too numerous to mention, could be cited in defense of cataclysmic theory and the shifting Earth thesis.

Firefall in the land of Canaan-Story of Joshua & Moon & Sun Standing Still   1623 B.C. or 10,890 B.C.?

"As they [the Canaanite kings] fled from before Israel, and were in going down to Beth-ron... the Lord cast down great stones from heaven upon them unto Azekiah, and they died; there were more which died with hail stones [stones of barad] than they whom the children of Israel slew with the sword."[62]

The author of the Book of Joshua was surely ignorant of any connection between the two phenomena.: the sun standing still and the fall of aerolites, or meteorites. Was it an improbable story invented as simply an allegory describing God's retribution against the enemies of Israel? The meteorites must have fell in torrents, for they struck down more warriors than swords of the adversaries. Hundreds of thousands of stones must have fallen. Such a cataract of stones would mean that the earth had passed through a meteor swarm.

Joshua emphasizes that the radius of the heavenly wrath was widespread, and as in other descriptions of meteorite falls, earthquakes accompany the tempest:

> Sun and moon stood still in heaven,
> and Thou didst in Thy wrath against our oppressors...
>
> All the princes of the earth stood up,
> the kings of the nations had gathered themselves together...
>
> Thou didst destroy them in Thy fury,
> and Thou didst ruin them in Thy rage.
>
> Nations raged from fear of Thee,
> kingdoms tottered because of Thy wrath...
>
> Thou didst pour our Thy fury upon them...
> Thou didst terrify them in Thy wrath...
>
> The earth quaked and trembled from the noise of Thy thunders.
>
> Thou didst pursue them in Thy storm,

---

62 Joshua 10:11

Thou didst consume them in the whirlwind...

Their carcasses were like rubbish.[63]

31 kingdoms were destroyed; not a plant, animal, or soul is left alive throughout Goshen. The descending hordes of large stones, an earthquake, a whirlwind, a disturbance in the movement of the earth, these phenomena belong together, and match closely reports given in the late 19th century concerning meteorite falls, but on a much larger scale.

We are reminded of the Egyptian high priest who taught Solon, the Greek philosopher, "But the truth of it lies in the occurrence of a shifting of the bodies in the heavens which surround the earth, and the destruction of things on earth by fierce fire;

Then the earth shook and trembled; the foundations of the hill moved and were shaken, because he was wroth. There went up a smoke out of his nostrils, and fire out of his mouth devoured; coals were kindled by it. He bowed the heavens also and came down; and darkness was under his feet. He made darkness his secret place; his pavilion round about him were dark waters and thick clouds of the skies. At the brightness that was before him, his thick clouds passed, hail stones and coals of fire."[64]

We must assume that at some time during the middle of the second millennium B.C., perhaps 1623 B.C., the earth was interrupted in its regular rotation by a comet followed by a meteor storm. The implications of a shifting of the geographical poles are staggering. What could have caused such a disturbance in the momentum of the crust? A strong magnetic field, or a disturbance of the same, so that the sun appeared to lose for hours its diurnal movement.

What Happens during a Pole Shift?

If you can accept the theory that the earth's surface was changed in relation to the rays of the sun by asteroid or cometary impact, then he will understand that all of the oceans of the earth and the seas and the lakes, due to inertia, stood still momentarily and then appeared to move as one body over the earth in a tremendous boiling motion, the water on the impact side of the earth rolling north, over and over, like a giant breaker rolling ashore on a beach. On the other side of earth, the water moved south and in the mid-Pacific, where the opposite moving bodies of water joined, there must have been tremendous turmoil. One needs to understand that all of the water moved at once, because all of the surface

63 Ginzberg, Legends, IV, 11-12.

64 Psalms 18: 7

of the earth moved under it at once. With all of these millions of cubic miles of water moving as one body, the water behind pushed the water ahead over all of the lower land surfaces of the earth, hundreds of miles inland and into valleys as much as 5,000 feet above sea level, seemingly, beyond the power of any oceanic flood to penetrate -- yet the physical evidence is there to see. The physical evidence left behind by this greatest of all cataclysms is real. It can be seen, touched, and photographed.

Nearly 80,000 deluge legends scattered over the face of the earth, describe the descent of fire, the erratic movement of the stars, sun and moon, the sudden the rush of the oceanic-flood, the refuge on mountaintops, and the struggle to survive the subsequent darkness, hunger, and cold that followed.

We will discuss these legends in the next chapter.

Are flood legends, recalling a local tsunami or pole shift caused by meteor impact in the ocean or just a local flood?

A school of mythologists believe that the Great Flood was indeed an actual occurrence, but of local extent and of diminutive dimensions only, some inundation which became exaggerated to a gigantic, heroic scale, as the original disaster receded into the past.

Some mythologists suppose a few flood myths have spread from a single center and were handed on by the missionary activity of some prehistoric colonial empire, much as the dissemination of the Jewish-Christian mythology. But these tales must have been truly overwhelming cataclysms, not inundations caused merely by prolonged rain (although this is a regular feature of one type of deluge myth) which set the low-lying parts of some country under water, or by earthquakes (another striking trait of many deluge tales) which caused huge waves to sweep over some island or surge far inland on some seacoast. The events must have been tremendous, universal, and rather sudden convulsions, to impress themselves so deeply upon man's mind and memory.

After the Flood: Change in Seasons of heat and cold. Change in the atmosphere. Rainbow. Mists rising from the ground.

IMPACT ICE AGE – Younger Dryas

Radiocarbon and cosmogenic dating of glacial moraines in regions all over the world and abrupt changes in oxygen isotope ratios in ice cores indicate that the Younger Dryas cooling was *globally synchronous.* Evidence of Younger Dryas advance of continental ice sheets is reported from the Scandinavian ice sheet, the Laurentide ice sheet in eastern North America, the Cordilleran ice sheet in western North America, and the Siberian ice sheet in Russia. Alpine and ice cap glaciers also responded to the abrupt Younger Dryas cooling in both the Northern and Southern hemispheres, e.g., many places in the Rocky Mts. of the U.S. and Canada, the Cascade Mts. of Washington, the European Alps, the Southern Alps of New Zealand, and the Andes Mts. in Patagonia of South America.

MOUNTAINS OF REFUGE

Ice Age to The Great Warmup: 9,500 B.C. & impacts every 1500 years over 6000 years.

Impact #1. 9500 B.C. – The Great Warmup & the Carolina Bays

11,500 years ago, a disintegrating comet fragment flashed across the American Midwest and exploded over coastal regions of North America at speeds in excess of 150,000 miles per hour. Travelling nearly tangentially to the Earth's surface, barely two minutes could have elapsed between the moment when the comet first flared up on the horizon and its catastrophic explosion. During its descent, the gases hurled backward from the flaming giant attained a luminosity 20-100 times that of the sun, blinding permanently any curious onlookers. After entering the lowest and densest part of the atmosphere, the heat and stress on the comet's surface reached critical mass, causing the cosmic fragment to burst, shattering it into innumerable lethal fragments, gouging a trench of deadly devastation across the southeastern part of North America.

Any humans living in the central and eastern United States at that time would have seen a "blinding flash followed by seismic shocks approximately four minutes later. It would have been almost impossible to avoid being hit by the barrage of enormous ice boulders seven minutes after the flash. Any survivors would have found their habitat destroyed. A local extinction event could have easily occurred under these circumstances, even without considering fires and other effects associated with the meteorite impact." [65]

The great scars left by this catastrophe were brought to light in 1931, when the governments of North and South Carolina decided to carry out a new photometric survey of the coastline and commissioned a company specializing in aerial photography to undertake this task. The aerial survey, which included the stretch between Florida and Cape Hatteras, was startlingly exposed by the eagle eye of the camera. When these films

65 Interpreting Carolina Bays as Glacier Ice impacts, Antonio Zamora, Abstract

were put together and examined, they created an unexpected sensation. All of the pictures showed thousands of peculiar elliptical depressions, or oval craters, locally called "bays," thickly scattered over the Carolina coast and more sparsely over the entire Atlantic coastal plain from southern New Jersey to northeastern Florida. These huge marshy depressions, sometimes overlapping, could be clearly identified as enlarged mud-filled craters appeared to be caused by the impact of gigantic boulders.

This was the only conclusion that could be drawn from the unbiased evidence of the aerial photographs. The survey camera had thrown a completely unexpected light upon a section of the Earth's history. Its keen and incorruptible eye had discovered the so-called Carolina Meteorite.

In 1933, geologists Melton and Schriver of the University of Oklahoma aroused world-wide attention when they announced the bays were depressions left by a "meteoric shower of a colliding comet."[66] Current estimates based on actual counts in limited regions conservatively indicate that half a million bays[67] may have been gouged out by catastrophic meteoric explosions. These detonations, in the air and close to the ground, created shock waves which formed shallow, sand-rimmed depressions. Each exploding fragment produced a trough of unknown depth, circular or oval, according to the angle at which the fragment struck the ground. A remarkable feature about these bays is that all of the longitudinal axes are parallel extending from northwest to southeast. The fragments must therefore have been traveling parallel to each other. Around the bays are thrust walls, mounds of earth thrown up from impact or air explosion, and elevated on the southeastern end, indicating the direction of cosmic impact coming from the northwest. Measurements made on more prominent craters show that the large bays average 2200 feet in length, and in several cases exceed six miles in length.

W. F. Prouty, a geologist who originated the "air shock" theory, published several papers on the Carolina Bays in the *Geological Society of America*. According to Prouty; "No bays have been found outside of the sand-covered coastal plain, and none should be expected according to the 'air-shock' wave theory. Large, rimmed depressions would not have been formed as readily in the clay soil or rocky terrain of the Piedmont as in the sandy Coastal Plain, and such rimmed depressions in clay soil would be more readily destroyed by both erosion and sedimentation. On the other hand, if meteorites did survive the passage to the ground, they

66 F. A. Melton and W. Schriver, "The Carolina Bays--Are They Meteorite Scars?", Journal of Geology, XLI (1933).

67 W.F. Prouty, The Bulletin of the Geological Society of America, Vol. 63, March, 1952, page 167.

would be better preserved in the clay than in the sandy soil."[68] As predicted, areas west of the Carolina Bays, including the southern Appalachian region, the Great Smokey Mountains, Mississippi, Tennessee, Kentucky, and Arkansas, have yielded more meteorites than any other equal area in the United States.[69]

Meteoric fragments may yet be found under the Carolina Bays on the Coastal Plain. According to Prouty, "practically every bay, large or small, surveyed by magnetometer has a well-defined *spot magnetic high or highs* associated with it,"[70] indicating the possibility of a deeply buried mass of iron-meteoric debris somewhere below each of the bays. Locating a mass of iron beneath the bays would indicate fragments of the Carolina meteorite actually impacted the surface *en-masse and* would point to the disintegration of an iron-nickel asteroid rather than an air explosion from a cometary fragment. To date, metallic iron, fragments of basement rock or fused glass, or tektites, have not yet been uncovered, except for a single fragment of deeply oxidized Sardis Meteorite. In general, however, we should not expect such evidence to be found in view of the porous soil and climactic conditions, unless extreme means were undertaken to deeply excavate under the bays.

Terrestrial origins for the bays have been considered and explored in great detail, however, they all fail to explain the parallel axis, size, overlapping structure, widespread distribution, southeastern thrust walls, elliptical shape, and magnetic residues common to nearly all of the bays. Since 1952, the majority of geologists who have dealt with the issue have generally accepted the meteor/comet collision theory. All the circumstantial evidence agrees with the hypothesis that a large celestial body entered the atmosphere from the northwest and exploded at a considerable height resulting in the formation of half a million craters we see today.

Present-day coast-line localities in the Carolinas show thousands of bays marked by erosion caused by oceanic flooding. In a few places there is direct evidence that the sea has *encroached seven miles inland upon some of the bays since their formation and has again withdrawn.* The bays are "filled to a considerable extent by the deposition of sand and silt, a process which doubtless occurred while the region was covered by the sea during the terrace-forming marine invasion of the Pleistocene [glacial] period. Dating of the Carolina Bays range from 800,000 years to 7,000 years, but studies by other geologists including Prouty suggest "the near-coast bays indicate that the bays are younger than the youngest

68 Ibid, page 173-174

69 Survey by Harvey Harlow Nininger 1923-1948

70 Ibid, page 175

Pleistocene terrace and some of the Coastal Plain drainage and older than the least of the more recent marine transgressions in the Carolina area," thus dating the cataclysm between 13,000 B.C. and 5,000 B.C.

Antonio Simora in a 2013 abstract, *Interpreting the Carolina Bays as glacier impacts*, writes *"The glacier ice impact hypothesis can be confirmed by finding evidence of an extraterrestrial impact in the Laurentide ice sheet. Conversely, proving that the Carolina Bays were created by impacts may be used as evidence that there was an extraterrestrial impact on the glacier ice sheet.* [71]

9,500 B.C. – Gobekli Tepe monument

c. 9500 B.C.; First building phase of the temple complex at Göbekli Tepe.

c. 9500 B.C.; There is evidence of harvesting, though not necessarily cultivation, of wild grasses in Asia Minor about this time.

c. 9300 B.C.; figs were apparently cultivated in the Jordan River valley.[3]

c. 9000 B.C.; Neolithic culture began in Ancient Near East.

c. 9000 B.C.: Near East: First stone structures at Jericho are built

c. 7640 B.C. – Tollman Impact. No ice records, but tree rings show a large Tunguska type of atmospheric impact.

Now within this book Knight and Lomas discuss the work of Edith & Alexander Tollman; geologists based at the Institute of Geology at Vienna University, who found evidence that the earth was hit by 7 major cometary fragments circa 7640 B.C. This was initially revealed by the study of tektites: small glassy objects found

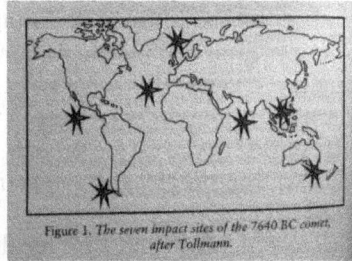

Figure 1. *The seven impact sites of the 7640 BC comet, after Tollmann.*

scattered in S Shaped patterns over large parts of the earth's surface. Irregular, rounded shaped rocks, tektites are formed by molten rock being ejected into the atmosphere, and then freezing into flattened and rounded spheres. Scientists now recognize that they are remnants of high energy cometary impacts, and that such rocks tend to be very old.

In 1970 Tektites were found embedded in fossilized wood in Australia and carbon dated to 9520 BP, or Before Present, + or − 200 years. Tektites were also found in Vietnam dated to about the same age. With yet further confirmation of this date coming from the study of core

---

71 Interpreting Carolina Bays as Glacier Ice impacts, Antonio Zamora, Abstract, Originally Posted on June 28, 2013 at 11:29:26 PM, http://www.scientificpsychic.com/etc/carolina-bays/carolina-bays.html

samples taken from the Indian Ocean. But still further, yet more evidence emerged to support this date, following other lines of inquiry.

Now in progressing still further in their studies concerning major earth changes as related in Uriel's Machine, Knight and Lomas uncovered yet more evidence establishing this date as marking intense global upheaval. However, in doing so they were also able to confirm a second date of disaster – and this using a technique known as Magnetostratigraphy, whereby the polarity of the earth's magnetic field is employed to date rock samples. Essentially, Knight and Lomas reasoned that friction from an incoming comet penetrating through the atmosphere would generate a powerful electric current, and that this would in turn generate an intense magnetic field at right angles to itself; one that would be far stronger than the earth's natural magnetic field. Consequently, remnant magnetization effects should exist embedded within rock samples, as would confirm a cometary impact, and allow one to fix its date. And this indeed proves to be the case.

Employing the technique of Magnetostratigraphy, Knight and Lomas found that in the last 10000 years there were 2 times when the direction of the earth's magnetic field suffered an abrupt change, suggesting an outside impulse. The first was a very close match to the date previously noted, of 7640 B.C. The second date uncovered, being more recent, at 3150 B.C. In other words, in addition to 7640 B.C., another cometary impact of great consequence must have occurred around the time of 3150 B.C.

Part III

Another Flood

...In one of my dreams I saw the image of white pyramids beaten by the ocean waves and surrounded by live bubbles. Unlike most of my paintings, I had no idea what those symbols meant, but I painted them anyway. Today the meaning still remains a riddle...

——— 7 ———

2,750 B.C: Noah's Deucalion's Flood

In 2,700 B.C *The Pyramid of Djoser (or Zoser), or the step pyramid* (kbhw-ntrw in Egyptian) was built in Saqqara necropolis, Egypt, northwest of the city of Memphis, for the burial of Pharaoh Djoser by Imhotep, his vizier.

In 2,470 B.C. **The Great Pyramid of Khufu** began construction... the pharaoh meant to be buried in the Great Pyramid, took power in 2470 B.C., according to Nur El-Din and colleagues. Egyptologists compared the modern calendar, the ancient Egyptian calendar, and the cycle of the star to find the exact day Sothis would have appeared that year. The team believes the ancient Egyptians observed the star from July 17 to 19, and the inundation period began 35 days later—on August 23.

Note: 3,300 B.C. Flood and pole shift was 10 degrees. The 10, 500 B.C flood and Pole Shift would have to be 180 degrees for the Sun to rise in the West.

The Flood of Deucalion (Ancient Greek Myth)

The story of Deucalion's flood is a folk memory of the flood waves that plagued the coasts of the Mediterranean following the tremendous eruption at Thera, or Stronghyli, as it is sometimes called.

"Zeus is said to have become angry with the rowdy and irreverent sons of Lycanon. When he paid them, a visit disguised as a poor traveler they served him a disgusting stew, a mixture of entrails of sheep and goats and one of their own brothers. Diving the horrid recipe, Zeus decided to forgo his meal and grimly returned to Olympus, whence he loosed a terrible flood, intending to drown mankind, so great was his disgust with the sons of Lycaon. The Titan Prometheus, however, warned his son, Deucalion, of impending disaster; the latter built an ark and rode out the storm with his wife, Pyrrha. A terrible south wind sprang up, "the rain fell, and the rivers roared down to the sea which, rising with astonishing speed, washed away every city of the coast and plain; until the entire world was flooded but for a few mountain peaks, and all mortal creatures seemed to have been lost, except Deucalion and Pyrrha. The ark floated about for nine days until, at last, the waters subsided." Deucalion, reassured by a dove sent out to find land, disembarked and went to the shrine to pray for the restoration of mankind. 'Veil your heads and throw the bones of your mother behind you,' came the answer. Deciding that this meant the bones of Mother Earth, they threw stones over their shoulders and where they landed men and women sprang up."

Thus, the flood of Deucalion, a natural event, was certainly caused by some massive tectonic commotion. This earth- and sea-shaking catastrophe has been at least roughly dated. From several sources, the Deucalion flood can be set approximately between 1529 and 1382 B.C. That these dates straddle those of the cataclysmic activity of Thera, as based on archaeological evidence, lends credence to the relation of the Deucalion flood to Thera.

The Greek myths of the Great Flood of Deucalion, describe the flood in three phases. It is also interesting that Plato mentions three floods.

However, accounts of the deluge of Deucalion, is also mirrored in the Old Testament flood of Noah recorded in Genesis, and the Babylonian inundation under Tunapishtim are remarkably similar, though they may not all be memories of the same event. (Not Likely!) While Deucalion's flood can be related to the Thera collapse, Utnapishtim's legend, according to archaeologists, is dated to the third millennium B.C., well before the Thera catastrophe, and indeed occurred in the Fertile Crescent, land of ancient Babylonia and Sumer, not in the Aegean. (Mayan Dating 3,300 B.C.)

Prior to the Deucalion / Noah Flood -An explosion of Knowledge Brought to Sumeria and Egypt

Around 3100 B.C., both Sumeria and Egypt appeared at the Zenith of their civilization. Suddenly (as 'suddenly' as five hundred years can be when compared to the preceding five and a half thousand), the

Sumerians exhibited a rapidly rising level of intelligence, the results of which were the appearance of mathematics, agriculture, architecture, astronomy/astrology and metallurgy. The same occurred in Egypt.

In Sumeria, Archaeologists have turned up evidence such as the 3000 B.C. smelter at Ezion Geber to the South, which had a modern style furnace system, complete with air ducting, and in which copper was undoubtedly refined. A Sumerian clay tablet has been found, dating back to 2000 B.C., on which was inscribed the solution to Euclid's classic triangle problem. This mathematical formula, appearing 1700 years before Euclid, was part of a schoolboy's textbook! Other researchers claim that the wheel and the horse appeared in Mesopotamia around 3000 B.C.

All the evidence, then, reveals that the Sumerians climbed to the peak of their civilization between 3000 and 2000 B.C., and when compared to the technological progress made in our own era— that is, over the last thousand years—you may find this in no way remarkable. But remember, one thousand years ago we already had the springboard of a number of long-established skills. No trace of these skills has been found relating to the Sumerians of 900 B.C., and it is assumed that they had to start from scratch. In Egypt, Narmer unified upper and Lower Egypt.....Need to fill in...

Noah's Flood

If the Hebrew new year began on March 21, and the deluge occurs 47 days after the new year, then the Deluge would have occurred on May 8th. If the Leonids caused the Deluge, we can extrapolate the exact year by using the formula of 33.33—years per day of regression through the calendar beginning from November 17, 1995--the current date of the Leonid shower and moving counterclockwise to May 8, thus discovering the possible year of the Deluge. The Leonid storm in 1833 occurred on Nov 12-13th.

We find that there are 190 or 191 days between November 17, and May 8. (using 30 day months) 33.33 multiplied by 190 days equals 6,333 years giving us the year 4334 B.C, just 202 years before the establishment of the age of Horakhte, and 105 years before the beginning of the age of Taurus in 4227 B.C. (Check the date of the 2nd month in the Bible.

Brazilian Record

The events prior to the Deluge itself are remembered in the Brazilian Flood legend, which tells us that Monan, the chief god, sent a great fire to burn up the world and its wicked inhabitants, and a magician caused so much rain to fall in extinguishing the flames that the earth became flooded. The Babylonian accounts continue by saying that during this period, "All the earth spirits leaped up with flaming torches, and with the brightness thereof they lit up the earth." (See Noah's Ark)

Mayan Record

At Palenque the Maya left a *record of a great flood that occurred near the end of their previous calendar cycle in 3300 B.C.* I interpreted this "mythical" event as a record of a real oceanic impact event that created enormous mega-tsunamis around the world. Evidence for such events has been found in both the sedimentary and ice core records. The Mayan eyewitness accounts of these events were recorded in their historical chronicles known as *Chilam Balam* and are consistent with an impact event with associated mega-tsunamis.

Why the Maya calendar starts in 3114 B.C.: a possible explanation. by M G L Baillie, School of Geography, Archaeology and Palaeoecology, The Queen's University, Belfast

The 13 baktun Mayan Long Count calendar spans 3114 B.C. to A.D. 2012; each baktun being 144,000 days. The calendar was probably devised by a gifted individual or group of astronomers (hereafter GIGA) in the 1st millennium B.C. who chose 3114 Aug 11 (proleptic Gregorian calendar equivalent) as the start date. How or why this choice was made has long been a mystery. Evidence from the chemistry of the American GISP2 ice core may provide a simple solution. GISP2 yielded records of ammonium (NH4) and nitrate (NO3). Plotting a combined index of NH4 plus NO3 against the original GISP2 timescale (Figure 1) is a revelation. The two highest NH4/NO3 indices in some six millennia – between 4558 B.C. and A.D. 1427 – occur within a year of two baktun transitions in the Mayan calendar: namely mean ice dates 2720.4 B.C. cf baktun 1.0.0.0.0 (2719 B.C.), and 1142.8 B.C. cf baktun 5.0.0.0.0 (1142 B.C.).

Figure 1. Index of (NH4%+NO3%) in GISP2 ice showing high values 1577 years apart at 2720.4 B.C. and 1142.8 B.C.; chemistry and dates from the ice record (Mayewski et al. 1997). Maya baktun changes and Tunguska (1908) are indicated.

Elsewhere it has been pointed out that NH4/NO3 occurred in the GISP ice at A.D. 1908.48 coincident with the Tunguska impact of 30 June 1908 (Baillie 2007; Melott et al. 2010). This makes it reasonable to suggest that some real physical atmospheric phenomena (almost

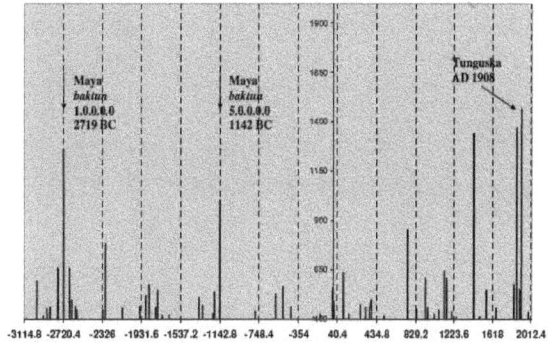

certainly cosmic) recorded 1577 years apart in the layer-counted ice record, may have been noticed by Maya/Olmec ancestors.

Perspective

Identification of a possible cosmic origin for the 2720 B.C. and 1142 B.C. NH4/NO3 spikes encourages investigation of the remainder of the Post Glacial GISP2 record. Figure 2 shows a continuation of the previous plot.

——— 8 ———

2,200 B.C.-Sodom & Gomorrah

~Working on the Sodom & Gomorrah chapter on Easter night, I asked God, 'I really need more research material for this chapter. I found more than I could ever imagine. Dr. Goodman Ph.D., who came to many of the celestial/comet impact scenarios and conclusions written in the Old Testament that I write about in the Armageddon Stones... He lives in Tuscon, AZ. He is author of The Genesis Mystery and the Comets of God-New Scientific for God: Recent archeological, geological and astronomical discoveries that shine new light on the Bible and its prophecies." Would he be a willing co-author? March 28, 2016. **~Chronicle 1020**

Sodom & Gomorrah

Lucas van Leden, Lot and his daughters, WGA12932.jpg
Article #1 –Sunday, April 1, 2012: Was Sodom and Gomorrah Wiped out by a Comet?
By Dr. Goodman, Ph.D.

An archeologist and a geologist, Dr. Goodman has devoted over twenty years to the study of the Bible. He holds a geological engineering degree from the Colorado School of Mines, an M.A. in anthropology from the University of Arizona, and a Ph.D. in anthropology from California Coast University. Dr. Goodman is the author of four archeological titles, including "American Genesis" and "The Genesis Mystery," which included accounts of his discovery of an early man site in the mountains outside of Flagstaff, Arizona. His current book is "THE COMETS OF GOD-New Scientific Evidence for God: Recent archeological, geological and astronomical discoveries that shine new light on the Bible and its prophecies." In part, "The COMETS OF GOD" tells of the linguistic and scientific discoveries Goodman made within the pages of the Bible. For more information on the book and links to over 50 articles that pertain to new scientific information that relates to the Bible go to **www.thecometsofgod.com**

"A number of astronomers believe the destruction of Sodom and Gomorrah involved comet activity. Dr. John S. Lewis, a retired professor of Planetary Sciences at the University of Arizona and Co-Director of the NASA Space Engineering Research Center at the University of Arizona, is one scientist who believes that the destruction of Sodom and Gomorrah was caused by cosmic bombardment (Rain of Fire and Ice: The Very Real Threat of Comet and Asteroid Bombardment). Genesis 19:24 says, "Then the Lord rained upon Sodom and upon Gomorrah brimstone and fire from the Lord out of heaven." "Brimstone (burning sulfur) and fire raining down from heaven" could be a description of the break-up and disintegration of a comet in the Earth's atmosphere above these ancient cities, since large chunks of rocky and icy material falling from heaven would be seen as fire raining down from heaven. In addition, cometary material is rich in sulfur. Even a small meteor fall can produce a smell of sulfur that is so strong that it is almost suffocating.

Few Bible commentators have grasped the true meaning of what is being described in this phrase about fire and brimstone falling from the sky. (Volcanic activity cannot be used to explain the fire and brimstone, because there are no volcanoes or volcanic deposits in the region.) In an effort to explain this phrase, some have said that the destruction of Sodom and Gomorrah was caused by an earthquake which somehow explosively ignited methane gas and sulfur found in the local tar deposits and shot it up into the sky. However, earthquake activity such as this is unprecedented and implausible in terms of geology. For those familiar with astronomy and ancient literature, the expression "fire falling from heaven" is not a cryptic expression nor a literary device that needs some sort of fanciful explanation; it is a simple descriptive phrase. "Fire from heaven" is an accurate description for cosmic material, either a meteor, asteroid, or comet, burning in the atmosphere ("fire") as it comes in to strike the Earth.

Further indication that a cometary bombardment took place during the destruction of Sodom and Gomorrah comes from Genesis 19:28. This verse tells how Abraham "looked toward Sodom and Gomorrah, and toward all the land of the plain, and beheld, and, lo, the smoke of the country went up as the smoke of a furnace." The "smoke of a furnace" speaks of the rising smoldering cloud that appeared after the explosive cometary impact. An earthquake opening a possible fissure would not produce a towering, smoldering cloud nor create the nodules of sulfur encased in ash found in the area. The impact of even a small cometary fragment over Sodom and Gomorrah could release energy equivalent to the explosion of many hydrogen bombs and raise a mushroom cloud like that in nuclear explosions. Note that the Bible also uses the expression "the smoke of a furnace" in Revelation 9:1-2, when telling how a star (a luminous heavenly body - a comet) from heaven falls to the Earth and opens a great pit (impact crater). It says smoke will arise out of the pit as the smoke of a great furnace and the air will be darkened and the sun will be darkened by the smoke. Without a doubt all of this describes an explosive cometary impact.

Knowing that a cometary impact is being described sheds a different light on what happened to Lot's wife. Genesis 19:23 says that Lot entered Zoar (the town of refuge agreed upon in verse 21) yet verse 26 says Lot's wife looked back. This is not an issue of Lot's wife simply turning her head to look back. It is an issue of her returning back in order to look. (In Luke 17:29-32 Jesus likens the day of his return to the day it rained fire and brimstone on Sodom and says "he that is in the field, let him likewise not return back. Remember Lot's wife.")

By returning back for a view of the impending destruction Lot's wife was close enough to be "consumed" (Genesis 19:17). Genesis 19:26 says "But his (Lot's) wife looked back from behind him, and she became a pillar of salt." The Hebrew word translated as "salt" is the word malach (#4417 in Strong's Concordance) which means "powder" as in something "pulverized" like salt or dust. So Genesis 19:26 can be retranslated to read "But his (Lot's) wife (returned and) looked back from behind him, and she became a pillar of dust." Now, this verse takes on new meaning because comet impacts raise and produce vast amounts of dust. After an impact the atmosphere rolls back, and then fierce winds containing superheated grains of dust shoot out from the impact site like the "base surge" which rolls away from the site of a nuclear explosion at ground level. Genesis 19:17 NIV warned of being "swept away." Genesis 19:25 says all "which grew upon the ground" was destroyed. Anyone caught in this surge at just the right distance could conceivably be suffocated and covered by hot dust and become a pillar of dust after the dust cooled and hardened. This is reminshful of those covered by ash and dust when a volcanic eruption rained down on the Italian city of Pompeii. This Biblical account of fire and destruction raining down from the sky to destroy Sodom and Gomorrah brings to mind the eyewitness accounts of the large comet fragment that broke up in the atmosphere over Tunguska, Siberia in 1908. Further, the concept of fire from heaven bringing destruction is also found in the literature of the ancient Sumerians, Akkadians, Babylonians, and Assyrians, the contemporaries of the peoples of the Old Testament. In this literature we read about their (cometary) gods throwing down fire and firebrands from heaven to Earth, burning brought by "hailstones and flames," and of the Queen of Heaven (Jeremiah 44:17-19 and 25) and her consorts "who rain flaming fire over the land" in contexts consistent with cosmic impacts.

More importantly, the Bible itself shows that the expression and concept of fire falling from heaven indeed pertains to cosmic material raining down from heaven because of the other times this expression or variations of it are used in the Scriptures in a context relating to cosmic impact. For example, see Exodus 9:23-25, Isaiah 30:3, Ezekiel 38:19-22, Revelation 8:7-10, and 9:1-2.)

The Bible reiterates the cometary nature of the destruction of Sodom and Gomorrah in Isaiah 13. Isaiah 13 describes what we recognize as cometary events, with comets coming in from "the end of heaven . . . to destroy the whole land," heaven and Earth being shaken, and the Sun and the Moon being darkened (Isaiah 13:5, 10, and 13). Then verse 19 says these events "shall be as when God overthrew Sodom and Gomorrah." In Luke 17:26-29, Jesus likens the day of his return to both the day of Noah and the day of the destruction of Sodom and Gomorrah. In terms of catastrophe, the common denominator between all three events will be comet activity. [See my March 19, 2012, CP blog entitled - Noah's Flood Was Really a Tsunami Caused by a Comet: A Retranslation of Genesis 7:11.]

Since natural phenomena can cause fire and brimstone to rain from heaven, the question is: Is there any scientific evidence to support the account of Sodom and Gomorrah's destruction or is it just a morality tale regarding the wages of sin? At the southern end of the Dead Sea in an area characterized by tar pits and oases (Genesis 14:10) archeologists have found the ruins of two ancient Bronze Age cities (Genesis 13:12 and 14:3). Burnt and reddened bricks have been found. Both cities were destroyed by fire. Abundant potsherds indicate a dense population dating to a period between 2500-2000 B.C. that ended abruptly around 2000 B.C.

It is also interesting to note that the surface of the Dead Sea suddenly dropped by several hundred feet around 2200 B.C., and some have speculated that the whole southern part of the Dead Sea may be a very shallow impact crater that was caused by a cosmic disaster. A very shallow impact crater would be consistent with a comet fragment exploding in the atmosphere high above the ground. For example, the 1908 atmospheric impact above Tunguska, Siberia left no discernible crater.

The most definitive evidence for the destruction of Sodom and Gomorrah would come from geological evidence associated with cosmic impact. Core samples from buried sediments dating to the time these cities were destroyed by fire should contain high concentrations of cosmic dust with very high concentrations of the elements iridium and nickel, and other materials created at impact. There could also be grains of shocked quartz, whose structure stems from the high pressures of impact or tiny spherules of fused glass-like material that stems from the high temperature of impact. While no formal scientific testing has been done yet, there is some geological evidence that indicates that a cosmic event took place. Dr. Benny Peiser, an expert on cosmic impact from Johns Moores University in England, reports that deposits of a form of calcite only found in meteorites has been discovered near the sites. Then there is the sulfur found in the area. In gypsum deposits, sulfur occurs in small marble to palm sized nodules or balls. The sulfur is tightly compacted and over 95% pure. Glassy ash encloses the sulfur nodules indicating burning and vitrification from great heat. Several different amateur groups have filed reports about this sulfur and posted pictures of these unusual sulfur nodules on the web. Petrographic study of sulfur could reveal its origins and how it came to form these unique nodules.

The bottom line is this: if the destruction of Sodom and Gomorrah was caused by cosmic bombardment as the Bible indicates, there should be more evidence waiting to be found. In a culture that generally believes that science and faith in the Bible are incompatible, what does it say if there is scientific evidence to support the Bible's account of what happened to Sodom and Gomorrah?

In the Bible story of Sodom and Gomorrah we have another incident like that of the Flood, where the God of the Bible said destruction was coming and behold, the destruction came in the form of cosmic impact. Since this happened on more than one occasion, it may not be a coincidence. We should give God's description of His "ministers of flaming fire" a closer look for what we can learn about comets and what this means for mankind's future.

Article #2-Evidence for God

Monday, April 23, 2012. Why Nuclear Weapons Are Not God's
Weapons of Wrath
By Dr. Goodman Hubble's Advanced Camera for Surveys took
images of the disintegration of Comet 73P/Schwassmann-Wachmann
3's fragment B. Click to EnlargeCredit: NASA, ESA, H. Weaver
(APL/JHU), M. Mutchler and Z. Levay. (STScI)
http://www.nasa.gov/mission_pages/hubble/Comet_73P.html

"While the Bible makes reference to the Weapons of God's Wrath
(Isaiah 13:5 NIV), few Bible scholars have addressed exactly what God's
weapons are. Yet dozens of books and scores of pastors have spoken
about the part nuclear weapons will play during the end times. Does God
need man's modern inventions to accomplish His will during the end
times? Based on the description of the disastrous events prophesied in
Revelation, the weapons of God's wrath must be able to:
Cause the stars or host of heaven to fall to the earth, and reign fire
and brimstone upon the land (Matthew 24:29, Isaiah 34:4 and Revelation
6:13, 8:7-10, 9:1-2, 15-19; also see Genesis 19:24).
Cause an event where a third part of men are killed by fire, smoke and
brimstone (Revelation 9:17-20).
Cause an earthquake large enough to make the cities of the nations
fall to the ground (Revelation 16:19). Cause planet earth to reel to and
from like a drunkard in space (Isaiah 24:18-21). Put up enough dust and
debris in the sky to blacken a third part of the sun, a third part of the
moon, and a third part of the stars (Revelation 8:12, 6:12).
Cause a tsunami with waves over a mile in height that destroys one-
third of the ships at sea (Revelation 8:9). Poison the waters in lakes and
rivers (Revelation 8:10-11). Bombard the earth with 100-pound
hailstones (Revelation 16:21, 11:19; also see Joshua 10:11, Exodus 9:18-
25 and Ezekiel 38:22). cause every mountain and island of the earth to
disappear. (Revelation 16:20). Cause the atmosphere to roll back
(Revelation 6:14 and Isaiah34:4). Scorch men with heat and even
consume eyes and tongues (Revelation 16:9 and Zechariah 14:12). Cause
grievous sores (Revelation 16:2). Engulf planet earth in fire so that the
elements melt with fervent heat (II Peter 3:12-13).
Nuclear weapons are not powerful enough to cause this type of global
destruction. Even if all the nuclear bombs of the world were stacked
together and simultaneously detonated, they couldn't produce events of
this magnitude. No fault triggered earthquakes of any magnitude or
volcano eruptions could cause these levels of destruction. On the other
hand, cosmic impacts, by comets or asteroids are powerful enough. The

impact of a several mile wide comet can release many thousands of times more energy than nuclear bombs. (Depending on where they hit [air, land or sea] comet impacts can have different types of catastrophic effects.) For example, in Time-Life's Comets, Asteroids, and Meteorites their scientific consultants calculated what would happen if a six-mile-wide meteorite (comet) crashed into the earth. They said, "the energy of the impact – equivalent to the explosion of five billion atom bombs – would transform cool blue earth into a flaming crucible."

The impact of an approximate 18-mile-wide comet produced the 312-mile in diameter Wilkes Land Crater in Antarctica 250 million years ago. The Wilkes Land crater is more than twice the size of the Chicxulub Crater in Mexico, the crater associated with the impact that killed the dinosaurs 65 million years ago. It seems that the Wilkes Land impactor penetrated the earth's crust with such force it caused a sudden outflow of molten material from the interior of the Earth through the opposite side of the Earth which produced the Siberian Traps. The Siberian Traps are a very large region of volcanic rock (basalt) thousands of feet in thickness that cover an area about the size of the lower 48 United States. A comet of this size penetrating the Earth's crust and exploding in the already hot interior of the Earth could produce enough heat to melt the Earth's crust and cause every mountain and island to disappear.

Since only comets can produce the catastrophic events of the magnitude called for in the Bible, what does the Bible say about comets being God's weapons of wrath? Few people are aware that comets are referred to in the Bible as "stars" and as the "host of heaven." During ancient times the word "star" was a generic term that meant any luminous body in the heavens. A "star" could be a hairy star (a comet), a shooting star (a meteorite), a wandering star (a planet), the sun, the moon, or a hot gaseous body. So, in scripture when we read about stars falling like figs, it is referring to meteorites or comets.

The words "host" or "host of heaven" in the Bible can refer to the objects of heaven such as the sun, the moon, and the stars which includes comets. This definition of "the host of heaven" is in Deuteronomy 4:19 which says, "And lest thou lift up thine eyes unto heaven, and when thou seest the sun, and the moon, and the stars, even all the host of heaven, shouldest be driven to worship them, and serve them, which the LORD thy God hath divided unto all nations under the whole heaven." The culture Abraham came out of mainly worshipped the host of heaven, a pantheon of sky gods led by the "queen of heaven" (Jeremiah 44:17-19, 25). These sky gods were mainly comets. Deuteronomy 17:3 warns about serving other gods "either the sun or moon, or any of the host of heaven ('stars of the sky' NIV); and II Kings 23:3 talks about "those who burned incense to the Sun, and to the Moon, and to the planets, and to all the host of heaven." (Also see Acts 7:42.)

According to ancient Near Eastern usage, when the Bible refers to the God of the Bible as the "Lord of Hosts" in a passage that pertains to the objects of heaven, it is in effect referring to the God of the Bible as the "Lord of Comets." Again and again, the God of the Bible represents Himself as being the one who is in charge of the heavens, including comets, which are part of the "host of heaven." Recognizing that the definition of the words "host" and "star" can refer to "comets" is the key to recognizing that comets are "the weapons of God's wrath."

God says in Isaiah 45:12 says: "I have made the Earth, and created man upon it: I, even my hands, have stretched out the heavens and all their host have I commanded." From this verse the God of the Bible, the "King of Heaven (Daniel 4:37) specifically says that it is He who created comets and He alone who commands the host of comets (Isaiah 40:26, 13:3-5, Daniel 4:35, Psalm 103:20-21, 104:4, and 148:8). Since comets can be referred to in the Bible as "snow" or "ice" or "hail" and comets can be surrounded by basketball sized hailstones, these words are also important to recognizing scriptures about God's use of comets as weapons. Job 38:22-23 (NIV) asks "Have you entered the storehouses of the snow or seen the storehouses of the hail, which I reserve for times of trouble, for days of war and battle?" In other words, God says that he has set aside comets in storehouses until the times that they are needed. This is scientifically consistent with the Hills Cloud and the Oort Cloud, vast reservoirs of comets that begin at the edge of the solar system and contain trillions of comets. In Daniel 4:35 the comets in these storehouses are referred to as God's "army of heaven."

So what else does the Bible says about comets?

God calls them His weapons of wrath, His mighty ones, His messengers and His ministers of wind and flaming fire (Isaiah 13:3, Isaiah 13:5 NIV and Psalm 104:4 NAS).

God calls the comets by name (Isaiah 40:26).

God commands the comets (Job 37:12 and Isaiah 45:12).

Comets respond in accordance to the physical laws and ordinances of heaven and Earth set by God (Job 38:33, Jeremiah 31:35, 33:25).

· Some comets were prepared for a specific destination, at a specific time (Revelation 9:15).

The following scriptures are just a few that speak of God's distinct connection to the heavens:

Concerning who created the heavens: Nehemiah 9:6 says "Thou, even thou, art Lord alone; thou hast made heaven (the solar system), the heaven of heavens (the Oort Cloud, a spherical reservoir of comets at the end of the solar system), with all their host (in this case comets)."

Isaiah 40:25-26 asks: "To whom then will ye liken me, or shall I be equal? saith the Holy One. Lift up your eyes on high, and behold who hath created these things, that bringeth out their host by number ('starry host one by one'- NIV, comets): he calleth them (the comets) all by names by the greatness of his might for that he is strong in power; not one (comet) faileth ('fails to appear' in Tanakh)."

Concerning God's control over comets: Isaiah 13:3-7, speaks of God's control of the comets to come during the end times, says: "I have commanded my sanctified ones (comets), I have also called my mighty ones (warriors' NAS - comets) for mine anger . . . the Lord of hosts mustereth the host of the battle (army of comets). They came from a far country (place), from the end of heaven (the Oort Cloud, the spherical reservoir of comets at the end of the solar system), even the Lord and the weapons of his indignation (wrath in NIV - comets) to destroy the whole land. Howl ye; for the day of the Lord is at hand; it shall come as a destruction from the Almighty. Therefore, shall all hands be faint, and every man's heart shall melt."

Comets are called by name: Isaiah 40:25 says that the God of the Bible calls the "host" or comets "all by names." Indeed, the name of a comet is given in the Bible. Revelation 9:11 says that the name of the "star" that falls to Earth opening a bottomless or very deep pit, is Abaddon in Hebrew and Apollyon in Greek. In both Hebrew and Greek this name appropriately means "destroyer."

The author is by no means the first scientist to recognize that the God of the Bible uses comets as His weapons. A growing number of astronomers and geoscientists have written about the connection between cometary impacts and the catastrophes recorded and prophesied in the Bible. It is interesting that the father of modern physics, Sir Isaac Newton (1642-1727), whose theory of gravitation permitted the calculation of the movement of the objects of heaven, also believed that comets represented one of the ways that God expressed His wrath.

In the book The Prophet and the Astronomer, Physics and Astronomy Professor Marcelo Gleiser of Dartmouth wrote that Newton believed God used comets as His tools or instruments; what the Bible calls "instruments of indignation" or "weapons of wrath" (Isaiah 13:5 NAS or NIV). Professor Gleiser wrote, "In Newton's scheme of the world, history was punctuated by catastrophes promoted by collisions with comets through the agency of God, in what might be called a causal theology." Dr. Gleiser also noted that, "In his (Sir Isaac Newton's) view, the scientist's search for a quantitative description of natural phenomena was part of a grander quest, that of deciphering God's plan, or mind: the scientist was a decoder of God's writing."

We have looked at how the scriptures identify comets as God's weapons of wrath. Added support for this thesis comes from the study of the phenomenal catastrophes of the Old Testament. These events can now be explained to be a consequence of comet activity. There is now physical evidence for events such as the Flood, the destruction of Sodom and Gomorrah, and the destruction of the Tower of Babel being caused by comets. For example, recently discovered impact craters such as Burckle Crater and the Amarah Crater are linked to the Flood and the destruction of the Tower. Astronomers and geophysicists are now theorizing what could happen if the earth took a very large impact. Eerily their descriptions match the level of destruction the Bible calls for in the end times.

When we know what we are looking for, we can see that the Bible contains a "textbook" of scientifically accurate information about comets, including details of their origins, composition, behavior, and impact effects. How did the writer of the Bible come to 'know' so much about comets, the great ice balls of space, before modern science? For example, in the fall of 2010 NASA was taken by surprise when they discovered that Comet Hartley 2 was surrounded by a huge cloud of basketball sized hailstones. Yet Revelation 16:21 refers to basketball sized hailstones, hailstones that weigh upwards of 100 pounds when it says "And there fell upon men a great hail out of heaven, every stone about the weight of a talent (70-100 pounds)

9

The Exodus
1800 B.C. or 1623 B.C. or
10,980 B.C.

Retelling of the Creation Myth

Wikepedia: The "Ipuwer Papyrus" is thought to have been written in
the Thirteenth dynasty of Egypt (18th century B.C.E), and certainly no
earlier than the 12th Dynasty.[48][49] Written in the form of a dialogue,
the sage Ipuwer accuses both the creator-god Ra and the king of having
neglected their roles, as a result of which the social order is overturned
and disasters fill the land.[50] Ipuwer has been put forward in popular
literature as an Egyptian confirmation of the exodus account, most
notably because of its statement that "the river is blood" and its frequent
references to servants running away, but these arguments ignore the
many points on which Ipuwer contradicts Exodus, such as the fact that
Ipuwer's Asiatics are arriving in Egypt rather than leaving, and the
likelihood that that the "river is blood" phrase refers to the red sediment
colouring the Nile during disastrous floods.[51] Scholars have identified
this and similar works (Ipuwer being the most ambitious) as examples
of a common Egyptian literary genre, with little or no basis in historical
events.

Tooke, the renowned mythologist, describes a comet called Typhon
from Greek mythology as it is personified in the forces of the cosmic
drama:

"Typhoeus, or Typhon the son of Juno, had no father. So vast was his magnitude that he touched the east with one hand and the west with the other and the heavens with the crown of his head. A hundred dragon heads grew from his shoulders; his body was covered with feathers, scales, rugged hair and adders; from the ends of his finger's snakes issued, and his two feet had the shape and fold of a serpent's body; his eyes sparkled with fire, and his mouth belched out flames."[72]

According to Napier and Clube, the Typhon catastrophe occurred in 1369 B.C. Pliny the Elder wrote in 23 A.D. about a terrible comet, "Seen by the people in Ethiopia and Egypt, which the King who reigned in that age named Typhon. It resembled fire, and was twisted like a wreath, hideous to the sight; and not to be counted a star, but truly a ball of fire." In mythology, Typhon was "a monster of the primitive world...described sometimes as a destructive hurricane and sometimes as a fire-breathing giant. He is described as a monster with a hundred heads, fearful eyes, and terrible voices; he wanted to acquire the sovereignty of gods and men, but, after a fearful struggle, was subdued by Zeus with a thunderbolt."

Hephaestion, in conjunction with the historian, Pliny, makes further mention of the comet Typhon.[73] He depicts the cosmic terror as an immense globe of fire, formed in the manner of a sickle. The comet appeared "blood red"[74] and moved slowly near the path of the sun. It caused destruction in its "rising and setting." Campster, a third or fourth century astrologer warned that if comet Typhon should return and meet the Earth, "a four-day encounter would suffice to destroy the world." The statement implies the comet brought the world near the brink of total destruction.

The 17th century scholar Rockenbach, who claimed to use only the earliest and most trustworthy writers, said of Typhon in his De cometis tractatus novus methodicus (1602);

"In the year of the world 2453...a comet appeared which Pliny also mentioned in his second book. It was fiery, of irregular circular form, with a wrapped head; it was in the shape of a globe and was of terrible aspect. It is said that King Typhon ruled at that time in Egypt...certain (authorities) assert that the comet was seen in Syria, Babylon and India, in the sign of Capricorn, in the form of a disc, at the time when the children of Israel advanced from Egypt to the Promised Land, led on

---

72 Olfield, The Encircled Serpent,

73 Johannis Laurentii, Lycli Liber de ostentis et cearia Braeca omnia (ed. by C. Wachsmuth, 1897, p. 171. cit.op, Velikovsky, Collision, pg. 85

74 During the superstitious Middle Ages, people believed that blood-red comets were a bad omen.

their way by the pillar of cloud during the day and by the pillar of fire at night."

Lydus, a 6th century Byzantine astrologer, in his treatise De Ostentis, also says that Typhon was a comet; "...the sixth comet is called Typhon after the name of the king Typhon, seeing that it was once seen in Egypt and which is said to be not of a fiery but a blood-red colour. Its globe is said to be modest and swollen and it is said that its `hair` appears with a thin light and is said to have been seen for some time in the north. The Ethiopians and Persians are said to have seen this and to have endured the necessities of all evils and famine."

If Typhon was a comet, it must have been enormous, perhaps of a magnitude exceeding any comet in recorded history. The comet Typhon, with an average nucleus about 20 kilometers in diameter, would at its nearest to Earth have attained a magnitude of −12, approaching that of the Moon. "It would have appeared as an intense yellow spot of light surrounded by a circular coma probably larger than the full Moon, with a tail stretching across a large part of the sky..."

Possibly its appearance originated the fear of comets which lasts even to the present day. But why should comets inspire fear? As Carl Sagan says, "Rarely have so many diverse cultures, all over the world, agreed so well...Everywhere on Earth, with only a few exceptions, comets were harbingers of unwanted change, ill-fortune and evil."

Perhaps the most well-known sighting and record of the effects of the appearance of Typhon and the explosion of Santorin can be found in the Hebrew story known as the Exodus, a story of the deliverance of the Israelite slaves from bondage in Egypt. Contained within the account are references to twelve plagues[75], which the prophet Moses brings down upon Egypt for refusal to free the Israelites. Various authors throughout the ages have suggested that the plagues of the Exodus were natural phenomenon associated with the near approach of comet Typhon which triggered the nearby volcanic eruption of Santorin.

The sighting of a comet or series of comets in the middle of the 2nd millennium is commonplace throughout the Near East. Abraham Rockenback, a Jewish author steeped in ancient traditional lore, describes in the Cometographia of Hevililus (1668) the worldwide spectacle of a massive comet named after Typhon, an Egyptian pharaoh, who at the same time drove the Israelites out of Egypt:

---

75 Traditionally there are ten plagues. Two listed in the Bible-- number one, serpents, and number eight,   pestilence-- are not usually listed in other traditions. The plagues in the Greek myth of Cephalus closely parallel those in the Biblical account

1496 B.C.

"In the year of the world two thousand four hundred fifty-three (1496 B.C.)[76], as many trustworthy authors, on the basis of many conjectures, have determined - a comet appeared which Pliny also mentioned in his second book. It was fiery, of irregular circular form, with a wrapped head; it was in the shape of a globe and was of terrible aspect. It is said that King Typhon ruled at that time in Egypt... Certain [authorities] assert that the comet was seen in Syria, Babylonia, India, in the sign of Capricorn, in the form of a disc, at the time when the children of Israel advanced from Egypt and into the Promised Land, led on their way by the pillar of cloud during the day and by the pillar of fire at night."[77]

Several historians, in an attempt to date the Exodus, have attempted to identify various kings of Egypt as the 'the pharaoh of the Exodus,' ranging from Amosis I of the Hyksos kings to Thutmose III, Amenhotep II, Thutmose IV, Merenptah and down to Ramesses II in 1270 B.C.; but the most prevalent choice is Ramses II who reigned about 1460 B.C. Immanuel Velikovsky believed Thaui Tom, (Typhon), to be the Pharaoh of the Exodus in 1490 B.C. The historian Josephus (c.37-100 A.D.), suggests that the Exodus occurred in 1628 B.C., a century before Hatchsepsut, at the end of the period of the Hyksos kings. This date perfectly coincides with the Santorini eruption, and thus inspires considerable confidence.

Comet Typhon Triggered the Exodus

First Plague: Comet sighting coincided with the violet explosion of Santorini

As far as the Mediterranean is concerned, the passage of the comet evidently coincided with the catastrophic explosion of the island mountain of Santorini, then called Calliste, meaning "the beautiful." This explosion is one of the largest identified in historical times, three times more powerful than the huge explosion of Krakatoa (Aug. 27th, 1883, A.D.). Before this explosion Calliste had resembled Mt. Fuji, in Japan; afterwards it was reduced to a ring of small islands, much as it appears today. The catastrophic eruption sent massive tidal waves crashing inland along the shores of the Mediterranean, which left behind folk memories of the disaster. Both the floods of Deucalion and Ogyges have been attributed to the Santorini explosion.

---

76 This date will later be show to be too recent

77 Hevelius, Cometographhia (1668), pp. 794f., cit op, Velikovsky, Collision, pg. 83

The association of Typhon with the volcanic eruption of Calliste fills the pages of Greek Legends. The allegorical descriptions of Apollodorus relates the traditions of the Thracians[78] in the following story: "Zeus pelted Typhon at a distance with thunderbolts, and at close quarters struck him down with an adamantine sickle....and in the fighting at Mt. Haemus he heaved whole mountains... at Mt. Haemus," a summit so named because of the "stream of blood which gushed out of the mountain." And when Typhon started to flee through the Sicilian Sea, Zeus cast Mount Etna upon him, a volcanic mountain which down to this day "blasts of fire issue from the thunderbolts that were thrown." After the skies darken with volcanic ash, it appeared that the heavenly god Zeus had thrown the fiery monster down to earth, imprisoning Typhon at the base of an erupting volcano.

Several absolute dates have been determined for large volcanic eruptions that have spewed large clouds of dust into the atmosphere and block out the sun over large areas of the earth's surface. The Santorini eruption produced large amounts of silicon dioxide, which readily formed aerosols with the water in the atmosphere. This in turn causes clouds, dimming of the sun, lower temperatures, which then affects the growth of trees and plants. The atmospheric effects of the explosion can be dated by frost damage and slowed growth found in ancient tree rings and by the high acidity in cores drilled in glacial ice. These methods in which individual rings and ice layers are counted are capable of being the most precise of any known dating methods and form the basis for carbon 14 chronologies. Many tree ring dates from America, Ireland, Germany, and England all confirm a particularly large volcanic eruption between 1630 and 1620 B.C. -- a remarkably consistent and precise result.

Confirmation of the tree ring dates have recently been found in ice cores brought up from deep below the Greenland ice cap. A New York Times report, "Santorin Volcano Ash, Traced Afar, gives a Date of 1623 B.C. describes the discovery: "Ash believed to be from a great explosive eruption that buried the Minoan colony on the island of Santorin 36 centuries ago has been extracted from deep in an ice core retrieved in 1993 from central Greenland. Its depth in the core indicated that the Aegean eruption, which may have given rise to the Atlantis legend, occurred in or about 1623 B.C." [79] According in a simultaneous report by G.A. Zielinski in Science, "High levels of Sulfuric Oxide residual found in the Ice Core layer at 1623 B.C. have been identified as particles unique

78 James W. Mavor, Jr., Voyage to Atlantis, Park Street Press, Rochester, Vermont, 1990, pg. 266., p. 127

79 Walter Sullivan, Santorini Volcano Ash, Traced Afar, Gives a Date of 1623 B.C., June 7, 1994, Section C8

to Santorin ash, an eruption thought to have occurred around 1626 to 1628 B.C."[80]

Ash from the Santorini explosion has already been identified deep in sediment layers on the floor of the Eastern Mediterranean, in Egypt's Nile delta and in parts of the Black Sea. There are also suspicions that the ash cloud persisted long enough to stunt the growth of oak trees in Irish bogs and of bristlecone pines in the White Mountains of California, producing tightly packed tree rings.[81] Atmospheric effects of the eruption would also be expected to have been carried by east winds to China, at the same latitude as Santorin. Author James. W. Mavor, Jr, presents evidence of the widespread effects of the Santorin explosion in Voyage to Atlantis. He reports that Chinese scholars, Kevin Pang, Santosh Srivastava, Robert Keston, and Hung-hsiang Chou "found written records of the atmospheric effects of a devastating volcanic eruption" which have been dated between 1630 and 1570 B.C. by Chinese royal genealogical records covering 37 generations.[82] The records were calibrated by "six astronomical events, which include lunar and solar eclipses and planetary conjunctions, from 1953 B.C. in the Hsia Dynasty to 841 B.C. in the Chou Dynasty."[83]

If, as seems likely, the Santorin eruption occurred about 1623 B.C., then the conventional Egyptian dating for the expulsion of the Hyksos / Israelites from Egypt is correct; the Santorin eruption and the Exodus all occurred at the end of the Second Intermediate, or Hyksos period in Egypt, one of the many dark ages in history. In Egyptian and Hebrew traditions, it is described as a series of plagues. In 1828 the Museum of Leiden in the Netherlands acquired the papyrus containing the writings of an Egyptian scribe, Ipuwer, who appears to have recorded the catastrophe. It is only by a careful examination of the writings of the Egyptian scribe, when compared and examined beside the writings of the Hebrew writer of Exodus, that one can see if indeed both writings describe the same event, the same cataclysmic disaster.

We should note the plagues of the Exodus preceded the massive flooding of the Mediterranean. In the Bible, twelve Plagues are mentioned, so to avoid confusion I have numbered them in chronological order for easy reference:

80 G.A. Zielinski, Science, Vol. 264, May 13, 1994

81 Ibid, Walter Sullivan, Santorini Volcano Ash, Traced Afar, Gives a Date of 1623 B.C., June 7, 1994,   Section C8

82 James W. Mavor, Jr., Voyage to Atlantis, Park Street Press, Rochester, Vermont, 1990, pg. 266.

83 Ibid, pg. 266

1. Serpents (Firefall trails)	7. Blains and Boils (leprosy/heat/radiation impact sym)
2. Water turned to blood	8. Pestilence
3. Frogs	9. Hail & fire
4. Lice	10. Locusts/famine
5. Flies	11. Darkness
6. Murrain of animals	12. Death of the firstborn (cannibalism)

It is important to realize that the plagues of the Exodus were not limited in scope to Egypt. The Exodus Plagues are a cataloguing of general phenomena associated with the appearance and disintegration of a comet, accompanied by a nearby volcanic eruption. From a scientific point of view, the catastrophes follow a logical sequence one might expect from the disintegration of a comet including; the appearance of serpents (meteor storms, falling Stars); falls of ferruginous meteoric dust turning the skies, rivers and oceans a bloody red color; the iron oxide poisons the waters, triggering massive "red tides" suffocating the fish; the entire ecosystem is radically affected; disease spreads rapidly with some species multiplying out of control, while others teeter near extinction; burning meteorites slam into the earth; fires spread throughout the land; crops destroyed, food supplies dwindle, and famine begins; the disturbed atmosphere spawns severe thunderstorms and hurricanes; volcanoes erupt darkening the skies; and children born during this period perish.

Plague of Serpents:

The first plague of serpents mentioned in Exodus is not described by the Ipuwer papyrus. Strictly speaking it was not a plague at all, but an allegorical reference to a frightful meteor storm which preceded the appearance of comet Typhon. A meteor storm occurs when the earth passes through tail of a comet or a mass of cometary debris, triggering several hundred thousand brilliant streaks of light as they collide with the atmosphere. We will find throughout this Epilogue, scriptural and mythological references to meteor storms as "warning signs" which precede the "Apocalypse."

A contemporary Chinese sighting of the meteor storm in 1623 B.C. given by King Chieh confirms this hypothesis: "In the night, the stars fell like rain" Author Joseph Goodavache writes that during the reign of Emperor Yao, Chinese astronomers talk of the "Valley of Obscurity" and the "Somber Residence" resulting from the aftermath of a cosmic catastrophe. Buddhist scholars preserved records which declared that the entire world was filled with meteors and smoke, "There is no distinction of day and night. The gloom is caused by a world-destroying great cloud of stones of cosmic origin and dimensions."[84]

In the manuscripts of Avila and Molina, who collected the traditions of the Indians of the New World, it is related that "a cosmic collision of stars preceded the cataclysm."[85]

Ancient literature often refers to falling stars as "arrows" or "serpents" because of their long, luminous tails. In the Bible, Deut: 32: 23-36, the Hebrew author writes: "I will spend mine arrows upon them. They shall be burnt with hunger, and devoured with burning heat, and with bitter destruction; I will also send the teeth of beast upon them, and the poison of serpents of dust."

The meteoric "serpents of dust" descended upon Egypt and the entire world, precipitating the next plague which turned the waters of the world blood red.

Plague of Blood

Heat causes algae blooms.... Dying of fish, flies, pestilence.

The initial stage of comet crash, in scores of legends and prophecies, begins with a massive fall of toxic, ferruginous dust, turning the skies, earth and seas into blood. Red skies coinciding with comet and meteorites is a common theme throughout history. A curious anomaly discovered at the Quaternary / Tertiary boundary containing the fossils of dinosaurs was the excavation of large quantities of red clay -- a possible indication that 65 million years ago, a similar comet infused the atmosphere with enormous quantities of iron oxides and meteorites.

Likewise, the first visible effect of the approaching disaster in Egypt was the reddening of the earth's surface by massive quantities of a reddish dust known as hematite, composed primarily of ferric oxide. In sea, lake and river this pigment gave a bloody coloring to the water. It is a phenomenon which appears to be associated with the fall of ferruginous particles or other soluble pigments from meteorites and meteorite dust turning the waters red.

Shedding Blood & Remission (See also Personification chapter)

Adam and Eve: One of the oldest theories of the origin of life was, as the Bible puts it, "the blood is the life." (Deut. 12:23) Even the name

84 The Comet Kohoutek, Joseph Goodavage, 1974

85 Brasseur, Sources de rhistoire primitive du Mexique, p. 40, cit. op, Velikovsky. p.61

of Adam gave away the primal secret. it meant literally "man of red earth or man, made of blood." "Without the shedding of blood there is no remission" (Heb.9:22)

Egypt: Osiris and the Mutilation of the Penis....

Egypt's sun father Ra castrated himself to bring forth a new race from his genital blood. The Hindu "Great God" had his penis removed, chopped up, and buried to bring forth a new generation. A similar Babylonian god, Bel, mutilated his penis for blood that he could mingle with clay to produce people and animals, imitation of the older Goddess Ninhursag, maker of life forms from the mixture of clay and menstrual blood. A Mexican savior-deity, Quetzalcoatl, likewise gave blood from his sliced penis to the Earth Goddess's clay vessel, to repopulate the earth after the Deluge.

Jesus Christ and the Crucifixion

Jewish Sacrifices

Mayan Sacrifices

The second plague is the plague of blood. Ipuwer turns the waters to blood, is recorded in both Exodus and the papyrus Ipuwer, who was an Egyptian eyewitness of the Exodus catastrophe, and wrote his lament on papyrus:[86]

Ipuwer: "The river is blood."

Exodus (7:20): "All the waters that were in the river were turned to blood."

Ipuwer: "[The] plague is throughout the land. Blood is everywhere."

Exodus (7:21): "There was blood throughout all of the land of Egypt."

As the water absorbed the meteoric dust and volcanic ash, the oxygen became depleted, suffocating the fish, which in the desert heat, was followed by rapid decomposition and stench.

Ipuwer: "And the river stank.... "Men shrink from tasting; human being's thirst after water," and "That is our water! That is our happiness! What shall we do in respect thereof? All is ruin."

Exodus (7:21): "And the fish that was in the river died; and the river stank, and the Egyptians could not drink of the water of the river.

---

86 A.H. Gardiner, Admonitions of an Egyptian Sage from a hieratic papyrus in Leiden (1909).

While some scholars believe volcanic ash from Santorin can account for the bloody hue of the water -- a fact which certainly compounded the disaster, it is a hypothesis that cannot explain the worldwide plague of blood which is seen as far away as South America.

The Manuscript Quiché of the Mayas tells us that in the Western Hemisphere, in the days of a great cataclysm, when the earth quaked and the sun's motion was interrupted, the water in the rivers turned to blood.[87] The Babylonian myth from Persia says the world was colored red by the blood of the slain Tiamat, the heavenly monster, the comet.

The phenomenon of "blood" raining from the sky has also been observed in limited areas on a smaller scale in recent times. We know this to be true of recent meteorite falls associated with the appearance of comets recorded in the latter half of the 19th century. (See Chapter 6) The red, ferruginous dust, soluble in water, descending from the heavens, does not originate in clouds, but must come from volcanic eruptions or from the ferric oxide of iron meteorites. The fall of meteorite dust, though in smaller quantities, is a phenomenon generally known to be associated with meteorites as they collide with Earth's atmosphere. This dust, called Brownlee particles, collects daily in Arctic and mountains snow all over the world.

The Plagues of Frogs, Lice, flies, Murrain, Blains, Boils, Pestilence, Locusts & Famine

Seven plagues in the Exodus account that follow the fall of red meteoric dust are related and belong in a group together. The tremendous fish kills that contaminated the waters of the nearby Nile, Red Sea, and Mediterranean, became fodder for frogs and other aquatic reptiles, which quickly multiplied from the dramatic increase in the food supply. Subsequently, a fall of "small dust" like "ashes of the furnace" fell "in all the land of Egypt,"[88] which is magically transformed by Moses into a "plague of lice." The reference to "ashes of a furnace" in all likelihood describes the initial fallout of ash from the erupting volcano of Santorin, several hundred miles away. Volcanic ash is a very fine, grey-white composite of ROCK, MINERAL, AND GLASS PARTICLES. Its ubiquitous, pervasive nature is known by anyone who has experienced ash fallout from a volcanic eruption. Its heavy, fine particles find their way into the smallest cracks and crevices, covering bedding and cookware, and are a menace to the human body. Volcanic dust is a veritable "plague" to the human body; it can't be stopped from entering

87 The Seven Tablets of Creation, ed. L. W. King (1902)

88 Exodus 9:8

the eyes and ears or breathed through the nose and mouth. It is no wonder the Hebrew scribes described this "little dust" as lice -- they couldn't get rid of it! Another reference to the fallout of volcanic ash is found in the Finnish epic of Kalevala, several thousand miles from the Santorin eruption, which describes how, in the days of the cosmic upheaval, the world was sprinkled with red milk.[89]

The Exodus account continues with the dying out of the frogs, possibly due to the fall of ash. The death and decomposition of millions of frogs triggered a subsequent plague of flies. Volcanic ash continued to fall throughout the land. The caustic nature of the ash, which contained traces of sulfuric acid, triggered boils on humans, and murrain in animals. The death of millions of insects and animals, the poisoning of the water, and the spreading of disease, naturally led to pestilence and famine.

The Plague of Hail

If a comet were of sufficient size and mass, the main body would not have to impact the Earth to wreak havoc. As Typhon came disastrously close to the Sun, its strong gravitational field may have caused the comet to break up under the stress, and the pieces slowly spread apart. The phenomenon is not without precedent. Over the past 150 years, astronomers have watched more than 20 comets break apart, including Comet-Shoemaker-Levy 9 which impacted Jupiter.

The eighth plague of Hail recorded in Exodus probably resulted from a mass of fragmented comet debris, and not the large central core of comet itself. While the main body of Comet Typhon continued to orbit the sun, several fragments broke off, including the Carolina Meteorite, and crashed into the Earth. The Exodus account describes the comet fall as "fire mingled with the hail, very grievous, such as there was none like it in all the land of Egypt since it became a nation." [90] The English translation "hail,"[91] is taken from Hebrew, "barad," and wherever it is mentioned in the Bible, is the term used to describe the fall of meteorites.

In the Biblical Book of Joshua, God Himself threw "Hailstones [stones of barad] of about a hundred pounds" on the heads of the Israelite enemies, destroying 31 kingdoms, killing every man, woman, children, and animal, leaving nothing but dust. This could not be referring to a severe thunderstorm with large hailstones of ice. Hail (ice) has killed a few hundred people on single occasions [China, 1845?], but never entire

89 Kalevala, Rune 9., cit. op. Velikovsky, p.

90 Exodus 9:24

91 A. Mcalister, "Hail," in Hastings, Dictionary of the Bible (1901-1904)

kingdoms. It is destruction on an order of magnitude greater than a terrestrial hailstorm.

The Kalevala of the Finns, tell of a time when hailstones of iron fell from the sky, followed by a period of darkness.[92] Here again, we find iron meteors [stones of barad], not ice. Misrashic and Talmudic sources say the stones which fell in Egypt were "hot." But this fact by itself does not prove the stones of barad were meteorites. While meteorites can be hot enough to spark a fire, very often when they are recovered immediately after impact, they have been found to be frost covered and icy-cold to the touch. This is because the core of the meteorite, cooled to near absolute zero in space, still retains a temperature far below freezing despite its fiery passage through the atmosphere.

Let's research the matter further. The Exodus plague included "thunder and hail [barad,]" and the fire ran along upon the ground." [93] The fall of large meteorites or bolides are usually accompanied by crashes or explosion-like noises. In a similar manner, the fall of the stones of barad were accompanied by "loud noises," a description in Hebrew rendered _____, but interpreted as "thunderings." (see collision text pg. 52 for Hebrew word) It is a translation which is only figurative, and not literally correct because the word for "thunder" is "raam," which is not used here. According to the Exodus narrative, the stones of barad made such a roar that the people in the palace were terrified as much by the din of the falling stones as by the destruction they caused.[94]

Ipuwer writes that one day the Egyptian fields are turned into a wasteland:

Ipuwer: "Trees are destroyed.... No fruits, no herbs are found... Grain has perished on every side... That has perished which yesterday was seen... The land is left to its weariness like the cutting of flax."[95]

The Exodus version attributes the destruction to the stones of barad:

Exodus 9:25 - "And the hail [stones of barad] smote every herb of the field and brake every tree of the field."

It is written that fire "ran along the ground" when the hailstones fell. The Papyrus Ipuwer describes this consuming fire: "Gates, columns, and walls are consumed by fire. The sky is in confusion."[96] Later the papyrus says that this fire almost "exterminated mankind." The fall of fire and

---

92 Kaleva, (transl. J. M. Crawford, 1888), p. xiii., cit. op., Velikovsky, p. 61

93 Exodus 9:23

94 Exodus 9: 28

95 Papyrus Ipuwer 4:14-6: 1-6, cit. op. Velikovsky, Collision, pg. 51

96 Papyrus Ipuwer 2:10: 7:1; 11 : 11; 12 : 6., cit. op Velikovsky, pg 54

stones from the heavens is one of the most common motifs associated with the appearance of the comet.

The Washo Indians of California "tell of a great terrestrial revolution which caused the mountains to blaze up, the flames rising so high that the stars of heaven melted and fell upon the Earth. Then the sierras rose up from the plains, while the other parts of the country were inundated."[97] In the Mexican codex Chimalpopoca, we are told that "in the third aeon-Kiauhtonatiuh, that of the 'fire-rain sun'-the god of fire descended upon the Earth in a rain of fire which burnt everything, while a hail of stones destroyed whatever was still left. Then the rocks rose in uproar and the red mountains grew."[98] We will read later how the mountains were violently uplifted at this time as well.

In far-away South America, the Popol Vuh, sacred book of the Quiché Indians says, Hurakan,[99] God of terror, rained fire and stone from the sky: "Masses of sticky material [burning mud and stony debris] fell.... The face of the Earth was obscured, and a heavy darkening rain began. It rained by day, and it rained by night...There was heard a great noise above, as if by fire."[100] The tradition of the Cahinaua of western Brazil witnessed a similar event, "The lightnings flashed and the thunders roared terribly, and all were afraid. Then the heavens burst, and fragments fell down and killed everything and everybody. Heaven and earth changed places. Nothing that had life was left upon the earth."

In the Aztec codex we find another massive fall of meteorites associated with a lowering of the sky: "... a rain of fire came following the sun of rain... all was burned.... a rain of rocks came, and the sky drew near the waters and the earth"[101] The Ute Indians of California say that 'When the magical arrow of Ta-wats struck the sun-god full in the face, the sun was shivered into a thousand fragments, which fell to the Earth causing a general conflagration.' [102]

On the other side of the world, in Siberia, the Voguls carried down through the centuries and millennia this memory: "God sent a sea of fire upon the earth. . .. The cause of the fire they call 'the firewater.' " [103]

---

97 H. S. Bellamy, Moons, Myths and Man, Fagen and Fagen, London, p. 95

98 Ibid, p. 96

99 Hurakan, a name signifying 'the furiously hurrying one' is also contained in the word 'hurricane,' a word we used today to describe tropical cyclone. Hurakan is compared with the name of the raging storm-giantess of Norse mythology, Hyrrockin, 'demon of the abyss'. Hurikan, is was also known as the Heavenly heart, or the location of the pole star. In 1,600 B.C. no star occupied the coveted position, but was a "dark place" surround by stars; thus the name, "heavenly heart," but also a place intimately connected with the serpent.

100 Popul Vuh, (See author)

101 Roy Norville, The Road in the Sky, pg 125

102 H. S. Bellamy, Moons, Myths and Man, Fagen and Fagen, London, p. 99

103 Holmberg, Finno-Ugric, Siberian Mythology, p. 368, cit.op. Velikovsky, pg 55

Where does the fire come from? Comet fragments are known today to be a conglomerate of mostly water ice, methane-ice, and various sizes of stone and iron boulders. Yet, repeatedly throughout history we find legends of an all-consuming fire associated with comets. If the elements of water are separated into hydrogen and oxygen, you have one of the most flammable chemicals available for explosive combustion. Can the kinetic energy of a colliding comet fragment separate these elements, creating one of the hottest fires known to man-- hot enough to melt stone and cause iron to burn like wax? The question will be tackled and answered in the following chapter.

Notes: Impact Ecology: Heat, Acid Rain, Tsunami, Volcanic Eruptions, Earthquakes. (get article I wrote for the SS at FIL)

The Plague of Darkness

Legends of darkness literally fill the pages of history during the middle of the second millennium. Volcanic ash from the violent eruption of Santorini saturated the atmosphere[104], creating total darkness in many areas of the world from five to ten days. Hebrew tradition passed down in the Midrashim gives further indication of the volcanic origin of the plague: "An exceedingly strong wind endured seven days. All the time the land was shrouded in darkness."
On the fourth, fifth, and sixth days, the darkness was so dense that they [the people of Egypt] could not stir from their place." "Nothing could be discerned... None was able to speak or to hear, nor could anyone venture to take food, but they lay themselves down their outward senses in a trance. Thus, they remained, overwhelmed by the affliction."[105]
In the Finnish Kalevala, when, in the days of a cosmic upheaval, the world was sprinkled with red milk Then came the darkness, which lasted several days.
Exodus: "And there was a thick darkness in all the land of Egypt three days." According to Caius Julius Solinus, "Following the deluge which is reported to have occurred in the days of Ogyges, a heavy night spread over the globe for nine consecutive days." Avilia and Molina reported the traditions of New World Indians that after a "cosmic collision of stars", the Sun did not appear for five days. The Iranian book Bundahis describes "a war between the stars and the planets" which resulted in the world being dark at midday as though it were in deepest

---

104 Dr. Galanopoulos, who spent years researching the Aegean Sea and Minoan civilization, says the Santorin eruption may have lasted for as long as 25 years.

105 Ginzberg, Legends, II, 360., cit. op. Velikovsky, p.

night. According to Ovid, after Phaethon`s disastrous ride across the sky, "one day passed without the appearance of the Sun."

In Egypt, the darkness was so dense, "Their eyes were blinded by it and their breath choked";[106] —describing volcanic ash. According to the rabbinical tradition, contradicting the Scriptural narrative, during the plague of darkness the "vast majority of the Israelites perished and that only a small fraction of the original Israelite population of Egypt was spared to leave Egypt. Forty-nine out of every fifty Israelites are said to have perished in this plague."[107]

Evidence of the Exodus?

At El-Arish on the border of Egypt and Palestine is a shrine of black granite which bears a long inscription in hieroglyphics. It reads: "The land was in great affliction. Evil fell on this earth... There was a great upheaval in the residence....

Nobody could leave the palace during nine days, and during these nine days of upheaval there was such a tempest that neither men nor gods could see the faces of those beside them."[108]

That the eruption was massive enough to cause total darkness is found in the traditions of the Indians of the New world: "During the five days that the cataclysm lasted, the sun did not show its face and the earth remained in darkness."[109]

A remarkable account of this period of darkness occurred several hundred miles to the Northeast of Egypt, in Persia. The eleventh tablet of the Epic of Gilgamesh describes the horrific event. "From out of the horizon rose a dark cloud and it rushed against the earth; the land was shriveled by the heat of the flames..." Desolation... stretched to heaven; all that was bright was turned into darkness... Nor could a brother distinguish his brother... Six days ... the hurricane, deluge, and tempest continued sweeping the land... and all humans back to its clay was returned."[110]

Death of the First Born:

The death of the first born is a plague not well understood by mythologists and historians. Actually, the truth of it is quite simple.

106 Josephus, Jewish Antiquities (transl. H. St. J. Thackeray, 1930), Bk. II, xiv. 5.

107 E. Velikovsky, Worlds in Collision, pg. 59., Targum Yerushalmi, Exodus 10:23 Mekhilta d'rabbi Simon ben Jokhai (1905), p. 38.

108 Velikovsky reference

109 Brasseur, Sourced de rhistoire primitive du Mexique, p. 40, cit. op. Velikovsky, p. 61

110 The Epic of Gilgamish (trans. R.C. Thompson, 1928), cit op. Velikovsky, p. 61

Throughout the ancient world, the first born, usually the son, inherited the family holdings, the family name, and the family blessing. The plague of the death of the first born describes an earthquake which destroyed much of upper Egypt. So many people died in the catastrophe, that the majority of firstborn sons died, leaving no successor to inherit the family name:

Exodus 12:30- "And Pharaoh rose up in the night, he and all his servants, and all the Egyptians; and there was a great cry in Egypt; for there was not a house where there was not one dead."

From the account of the Egyptian scribe, Ipuwer, we can be certain that the deaths were caused by the collapse of stone masonry by a massive earthquake:

Ipuwer: "The towns are destroyed. Upper Egypt has become a waste... All is ruin." "The residence is overturned in a minute.' [111]

The meaning of the Egyptian word 'overturn' implies 'to overthrow a wall.'[112] And, as it is written, the walls were thrown down in a minute, a catastrophe which could only be caused by an earthquake. Corroborating evidence for the "earthquake" theory can be found from a passage of Artapanus in which he describes the last night before the Exodus, and which is quoted by Eusebius: "There was "hail and earthquake by night, so that those who fled from the earthquake were killed by the hail, and those who sought shelter from the hail were destroyed by the earthquake. And at that time all the houses fell in, and most of the temples."[113]

It appears, however, that only the Egyptian residences were destroyed. "The [angel of the Lord] passed over the houses of the children of Israel in Egypt, when he smote the Egyptians and delivered our houses." [114] The Hebrew word nogaf translated as "smote" describes a very violent blow, as in this case, an earthquake.

But why were the homes of the Israelites delivered from destruction? The answer can be found in the type of materials used to build their dwellings. During the earthquake, Egyptian buildings made of stone and brick, came crashing down on their sleeping occupants. Israelite dwellings on the other hand, were constructed of reeds and red clay, and were more resilient than brick or stone. It is an interesting fact that the red clay used in the construction of Hebrew dwellings is described in the Exodus account as painted with "Lamb's blood," which allowed the

111 Papyrus Ipuwer 2:22:3 :13, cit. op, Velikovsky, p.63

112 Gardiner's commentary to Papyrus Ipuwer. cit. op. Velikovsky, p.63

113 Zohar ii, 38a-38b, cit. op. Velikovsky, p.64

114 Exodus 12:27

"Destroyer," the earthquake, to pass by their houses, saving them from certain death.

As buildings crumbled in Egypt, earthquakes and tidal waves wreaked havoc overseas. American Indian legends describe people and animals seeking shelter from the catastrophe in mountain caves. "Scarcely had they reached there when the sea, breaking out of bounds following a terrifying shock, began to rise on the Pacific coast. But as the sea rose, filling the valleys and the plains around, the mountain of Ancasmarca rose, too, like a ship on the waves."[115]

The Popol Vuh, sacred book of the Quiché Indians says, "Now men were seen running, pushing each other, filled with despair. They wished to climb upon their houses, but the houses, tumbling down, fell to the ground. They wished to climb upon trees, but the trees shook them off. They wished to hide in caves, but the caves caved in before them. Water and fire contributed to the universal ruin at the time of the last great cataclysm which preceded the fourth Creation.'

In the Aztec codex darkness covered the earth... men went to the caves but they were sealed in by falling rocks... men climbed into trees, but they fell... there was no sun and for five days' blackness was everywhere... earthquakes shook the land... flames came from the earth, and flaming stones dropped from the heavens...."[116]

Indeed, the whole world appeared to shake, causing mountains to crumble and others to rise to great heights. High up in the Andes mountains of South America, the megalithic[117] city of Tiahuanacu, appears to have been suddenly abandoned around 1,600 B.C.[118] On the surface, the reason for the abandonment is obvious. "Ancient agricultural terraces there rise to a height of 15,000 feet, twenty-five hundred feet above Tiahuanacu, and still higher, up to 18,400 feet above sea level, or to the present line of eternal snow on Illimani. The ancient city and terraces, belonging to a mythological race of white giants known as Virachocca, occur at altitudes far too high to support the growth of crops for which they were originally built. Some rise to 15,000 feet above

115 Brasseur, Sourced de rhistoire primitive du Mexique, p. 40, cit. op. Velikovsky, p. 61

116 Roy Norville, The Road in the Sky, pg 125

117 "The term "megalithic" fits the dead city only in regard to the great size of the stones in its walls, some of which are flattened and joined with precision. It is situated on the Altiplano, the elevated plain between the Western and Eastern Cordireras, not far from Lake Titicaca, the largest lake in South America and the highest navigable lake in the world, on the border of Bolivia and Peru."(Upheaval)The ruins of Tianhuanaco are not much a comparison with the Inca fortresses, for it is apparent that the City of the Sun was built by a race who possessed a great deal of artistry as well as architectural skill. And yet, the same huge stones, some weighing two hundred tons, were a feature of its original construction, quarried and transported from the volcanic Kiappa region, some forty miles away.

118 Radiocarbon datings of materials from Tiahuanaco indicate the site is much younger than expected. The early classic style there is dated to about the fifth century B.C., although Arthur Posnansky believed it to have been built and occupied much earlier, and the following cultural period to about the time of Christ. (Ref. Radiocarbon, Vol. IV, 1962, p. 91) The city continued to be occupied as late as the eighth century A.D. (Ref. Radiocarbon, Vol. 1, 1961, pp. 54-57).

sea level, or about 2,500 feet above the ruins of Tiahuanacu, and on Mt. Illimani they occur up to 18,400 feet above sea level; that is, above the line of eternal snow where corn will not grow today. (46:39)." (Path of Pole)

Charles Darwin, on his travels in South America in 1834-35, noted with incredulity that the beaches at Valparaiso, Chile, at the foot of the Andes, had been thrust upward to an altitude of 1300 feet.

He was impressed even more by the fact that the sea shells found at this altitude were still undecayed, to him a clear indication that the land had risen 1300 feet from the Pacific Ocean in a very recent period, "within the period during which upraised shells remained undecayed on the surface."[119] Darwin noted that only a few intermediary surf lines could be detected down to the coastline, therefore, the ocean could not have receded little by little.

Geological evidence for the recent catastrophic uplift of the Andes Mountains abounds. Standing some 12,500 feet above sea level, there is a white tide mark to be seen all along that portion of the Andes. Composed of calcified marine plants, it is the sea level marking the days before the cataclysm, when Tiahuanacu overlooked the Pacific Ocean by a mere 1000 feet. "Titicaca and Poopo, lake and salt bed of Coilaga, salt beds of Uyuni--several of these lakes and salt beds have chemical compositions similar to those of the ocean.' "[120] As long ago as 1875 Alexander Agassiz demonstrated the existence of a marine crustaceous fauna in Lake Titicaca.' At a higher elevation the sediment of an enormous dried-up lake, whose waters were almost potable, "is full of characteristic mollusks, such as Paludestrina and Ancylus, which shows that it is, geologically speaking, of relatively modern origin."[121]

"With the great upheaval of the mountains, the ground on which the city stood was jacked up to a height of 13,000 feet! The conservative view among evolutionists and geologists is that mountain making is a slow process, observable in minute changes, and that because it is a continuous process there never could have been spontaneous uplifting on a large scale. In the case of Tiahuanacu, however, the change in altitude apparently occurred after the city was built, and this could not have been the result of a slow process that required hundreds of thousands of years to produce a visible alteration."

119 Charles Darwin, Geological Observations on the Volcanic islands and Parts of South America, Pt. II, Chap. 15., cit ops, Immanuel Velikovsky, Upheaval, pg. 85

120 H. P. Moon, "The Geology and Physiography of the Altiplano of Peru and Bolivia," The Transactions of the Linnean Society of London, 3rd Series, Vol. 1, Pt. 1 (1939), p. 32., cit ops, I. Velikovsky, Earth in Upheaval, pg. 82

121 Posnansky, Tiahuanacu, p. 23., cit op, The Path of the Pole, Charles Hapgood,

(Upheaval) Geologists must shake their heads with incredulity, but the evidence for a recent upthrusting of the Andes at this most recent period is so overwhelming no other explanation will do.

What strange force could thrust mountains several thousand feet above sea level almost overnight, or trigger planet-wide earthquakes and devastating oceanic floods? According to the account of Exodus and sacred texts gathered from many nations, it is the most feared event of the appearance of the world ending meteor storm -- a phenomenon known as Pole Shift-- a sudden movement of the axis of the Earth, accompanied by a radical shift of the crust. But did a pole shift occur during the appearance of comet Typhon in 1623 B.C. or during its previous visit in the 3rd Millennium B.C.?

Book of Exodus describes a Comet & Meteor Storm

Remarkably, however, the obvious "dimming of the sun" and the disastrous effects on the weather coincided with the sighting of three "suns" in the sky, possibly indicating that comet Typhon had disintegrated into three large fragments:

"At the time of King Chieh the sun was dimmed, and three suns appeared. Winter and summer came irregularly, Frosts [occurred] in the sixth month. Ice formed in the morning. There was heavy rainfall and communities were destroyed."[122]

The day after the great earthquake, the Israelites fled Egypt, led on by a portent which was described as "a mighty band with great terribleness, and with signs, and with wonders," [123] but further delineated as 'the Lord who went before them by day in a pillar of a cloud, to lead the way; and by night in a pillar of fire, to give them light; to go by day and night."[124] From the evidence given here, the description could be interpreted as a volcanic pillar of smoke and pillar of fire ejected by Santorin, some six hundred miles distant. But was it? A volcanic pillar of fire and smoke, no matter how far it was ejected into the atmosphere, could not have been distinguished as such, due to the curvature of the earth blocking the line of sight. More likely, it was the dragon-like appendage of comet Typhon burning low over the horizon in the heavens!

It should be noted that as comets approach the Sun, they become hotter, dustier, and consequently redder because of selective absorption of sunlight by the dust in the tail. In this manner, the deep red hue of

---

122 Ibid, pg. 266.

123 W. Max Miller, Egyptian Mythology (1918), p. 126, cit. op. Velikovsky, p.65

124 Exodus 13:21

Typhon, would have been seen in the daylight as a pillar of "smoke" and in the evening as a pillar of "fire."

Further evidence supporting the comet hypothesis is later seen when the pillar of fire and smoke, [the angel of God] was "removed" and "went behind them." This perfectly describes the behavior of a comet! The cosmic "pillar" first appeared over the eastern horizon. But as Typhon orbited the sun and [was removed] from sight, it then reappeared from behind the sun and moved closer and closer towards the western horizon, until it disappeared a few months later. In essence, the comet moved from before the Israelites in the East, to behind them in the West.

During the comet's movement toward the western, evening horizon, the waters of the Red Sea receded so that "dry ground" appeared, allowing the Israelites to cross over. Pharaoh's army attempted to follow but were drowned when the sea violently returned and crashed down upon them. Moses and several thousand Israelites then spent the next forty years wandering the Sinai Peninsula, looking for the Promised Land. Moses died at the ripe old age of 120 and passed the mantle of responsibility to his lifelong apprentice, Joshua.

52 Years Later: Venus Cycle in Genesis

Approximately fifty-two years[125] after the Exodus, during a battle with King Azekiah, the "Lord cast down great stones [stones of barad] from heaven upon them unto Azekiah, and they died: they were more which died from the hail stones."[126] Shortly thereafter, "the sun stood still, and the moon stayed.... So, the sun stood still in the midst of heaven, and hasted not to go down about a whole day."[127]

Religious scholars have noted that the description of the motion of the luminaries implies that the sun was in the forenoon position.[128] Joshua indicates the sun stood in the "midst" of the sky. The biblical narrative describes the sun as remaining in the sky for an additional day. The Midrashim, the books of ancient traditions not embodied in the Scriptures, relate that the sun and the moon stood still for thirty-six or

---

125 Period after the Exodus range from 52 to 60 years.

126 Joshua 10: 11

127 Joshua 10: 13

128 H. Holzinger, Josua (1901), p.40, in "hand-comentar sum Alten Testament," ed. K. Marti. R. Eisler, "Joshua and the Sun," American Journal of Semitic Languages and Literature, XLII (1926), 83: "It would have had no sense in early in the morning of a battle, with a whole day ahead, to have prayed for the lengthening of the sunlight even into the night time."

eighteen hours,[129] and thus from sunrise to sunset the sunlight lasted about thirty hours.

A Shifting of the Poles or Cessation of Earth's Rotation?

Tradition told to Herodotus by the Egyptians, that "on four several occasions, (the Sun) moved from his wonted course, twice rising where he now sets, and twice setting where he now rises."

Is it simply a metaphor, a fairy tale, a poetic image of God suspending the motion of the universe in answer to the prayers of a righteous man? Let's assume for a moment the sun actually stood motionless in the ancient Hebraic sky. What was happening, then, on the other side of the planet? Because half the world is bathed in sunlight, we must assume the Western Hemisphere experienced an extended period of darkness, or dusk. Searching the records, we find repeated statements in Andean legends that describe a frightening darkness long ago. Completely ignored by scholars, are clues describing the nonappearance of the sun when it was due, both spoken in the tale of Teotihuacan and its pyramids as well as the Mexican Annals of Cuauhtitlan. For if there had been such a phenomenon, that the sun failed to rise extending the night, then it would have been preserved throughout the Americas.

In the Mexican annals of Cuauhtitlan it is stated that "the world was deprived of light and the sun did not appear for a fourfold night."[130] In the Andean legends, Montesinos and other chroniclers write: "The most unusual event took place in the reign of Titu Yupanqui Pachcuti II, the fifteenth monarch in Ancient Empire times. It was in the third year of his reign, when 'good customs were forgotten and people were given to all manner of vice,' that 'there was no dawn for twenty hours."[131] We must marvel at this statement, "there was no dawn for twenty hours." The sun did not rise! Can you imagine the fear that must have gone through the minds of the ancient Andean people so long ago? Apparently, after a great outcry, confession of sins, and sacrifices, the sun finally rose.

It could not have been an eclipse; their astronomical experience knew that the dark shadow of the moon covering the sun never lasts more than seven minutes. The story does not say that the sun disappeared. It says the sun "did not rise" for twenty hours.

129 Sefer Ha-Yashar, ed. L. Goldschmidt (1923): Pirkei Rabbi Elieser (Hebrew sources differ as to how long the sun stood still); the Babylonian Talmud, Tractate Aboda Zara 25a; Targum Habakkuk3:11. cit. op. Velikovsky, p.
130 Velikovsky, Worlds in Collision, pg 46
131 Zecharia Sitchen, The Lost Realms, Avon Books, New York, 1990, pg. 151

We find a similar story of the cosmic event recorded by a Spanish savant named Sahagun, who came to America a generation after Columbus and gathered the stories indigenous to the land. He wrote that "at the time of one cosmic catastrophe the sun rose only a little way over the horizon and remained there without moving; the moon also stood still."[132]

While some may suggest that the Sahagun tale was influenced by Catholic missionaries, the only resemblance to the Canaanite version is the fact that the sun and moon stood still. Important differences, however, remain. In the Sahagun tale we find no war, and the position of the sun in the sky is quite different, just above the horizon, instead of in the midst of the sky. This innocent account may actually serve as a clock to identify the timing of the Joshua event. Columbus first made contact with the natives of the West Indies, nearly two thousand miles distant from Mexico or the Andes, where in fact, stories record total darkness. If the Sahagun account is accurate, the appearance of the sun "just over the horizon" in the West Indies indicates it was dawn, and the sun was just rising, let's say about 6:00 am. Sunrise in the land of Canaan occurs eight hours earlier, therefore, it must have been 2 or 3 o'clock in the afternoon when the diurnal motion of the sun stopped. In Mexico, or the Andes, it would have been 3:00 or 4:00 am, and shrouded in total darkness!

Not only is the length of total darkness or sunshine corroborating, but also three separate legends serve to verify the timing and accuracy of the cosmic event - a temporary cessation of the diurnal motion of the planet.

Scholars have struggled for generations to explain the marvelous tale of two hemispheres. Some discount it as a fairy tale, a myth, an allegory. Once thought to be unique, the Joshua story is described by other nations halfway around the world. Do not the three tales, then, describe the same event, and by coming from different sides of the Earth attest to its accuracy, its veritable truth?

How do we explain the association of the fall of meteorites and the sun standing still? The author of the Book of Joshua was surely ignorant of any connection between the two phenomena.

The meteorites must have fallen in torrents, for they struck down more warriors than the swords of their adversaries. Such a cataract of stones would mean that the earth had passed through a meteor swarm, possibly associated with the fragmented mass of comet Typhon still orbiting the sun, some fifty-two years later -- or was it fifty -two weeks?

---

132 Ibid, pg 46.

The mystery deserves exhaustive study, and therefore, for the moment we will leave it until later.

In the Biblical narrative, it appears Typhon appeared fifty-two years later. Once more there was another "Sun" in the sky, as Joshua besieged Jericho. Perhaps there were other passages; Typhon appearing at intervals of about 52 years, not passing close enough to cause catastrophe, but with its enormous coma and tail terrifying at even a great distance it was a source of comet-fear. The early Mexican scholar Fernando de Alva Ixtlilxochitl (c. 1568 – 1648) wrote that according to ancient tradition, the multiple of 52-year periods played an important role in the recurrence of cosmic catastrophes. He said that once only 52 years elapsed between two such catastrophes. The natives of Pre-Columbian Mexico expected a new catastrophe at the end of every period of 52 years.

The array of catastrophes of the Exodus and the rest of the known ancient world during this period, taken literally, may seem hardly believable. But we will discover that the Exodus legend represents only one catastrophic comet and meteor event.

What appeared as "Judgment Day" in Exodus, was in the eyes of prophets and mystics, but one event in a millennia-long cycle of Earth-Creation and destruction caused by a regular, periodic, and predictable Meteor Storm. Ancient memories of the apocalypse, passed down through the ages in sacred traditions and mythology, eventually became future predictions of Apocalypse, Armageddon, and Judgment Day; a Day of the Lord associated with a rain of blood, falling stars, meteorites, consuming fire, earthquakes, floods, plagues, famine, and darkness

Historical Validity of the Exodus

Wikepedia: "The archaeological data do not accord with what could be expected from the Bible's exodus story: there is no evidence that the Israelites ever lived in Ancient Egypt, the Sinai Peninsula shows almost no sign of any occupation at all for the entire 2nd millennium B.C.E, and even Kadesh-Barnea, where the Israelites are said to have spent 38 years, was uninhabited prior to the establishment of the Israelite monarchy.[15]

Scholars generally agree that while the exodus narrative contains late 2nd millennium elements, it has not been demonstrated that these elements could not belong to any other period and they are consistent with "knowledge that a 1st millennium B.C.E writer trying to set an old story in Egypt could have known."[16] A few scholars, notably Kenneth Kitchen and James Hoffmeier, continue to discuss the historicity, or at least plausibility, of the story, although historians of ancient Israel rarely respond.[17] They advance a range of arguments to explain the lack of evidence: possibly the Egyptian records of the presence of the Israelites and their escape have been lost or suppressed; possibly (or probably) the fleeing Israelites left no archaeological trace in the desert; possibly the huge numbers reported in the story are mistranslated.[18]

Numbers and logistics

According to Exodus 12:37–38, the Israelites numbered "about six hundred thousand men on foot, besides women and children," plus many non-Israelites and livestock.[19] Numbers 1:46 gives a more precise total of 603,550 men aged 20 and up.[20] It is difficult to reconcile the idea of 600,000 Israelite fighting men with the information that the Israelites were afraid of the Philistines and Egyptians.[21] The 600,000, plus wives, children, the elderly, and the "mixed multitude" of non-Israelites would have numbered some 2 million people.[22] Marching ten abreast, and without accounting for livestock, they would have formed a line 150 miles long.[23]

Te entire Egyptian population in 1250 B.C.E is estimated to have been around 3 to 3.5 million,[24][22] and no evidence has been found that Egypt ever suffered the demographic and economic catastrophe such a loss of population would represent, nor that the Sinai desert ever hosted (or could have hosted) these millions of people and their herds.[25] Some have rationalised the numbers into smaller figures, for example reading the Hebrew as "600 families" rather than 600,000 men, but all such solutions have their own set of problems.[26] The most probable explanation is that 600,000 symbolises the total destruction of the generation of Israel which left Egypt, none of whom lived to see the Promised Land,[27] while the 603,550 is a gematria (a code in which numbers represent letters or words) for bnei yisra'el kol rosh, "the children of Israel, every individual".[28]

Archaeology

A century of research by archaeologists and Egyptologists has found no evidence which can be directly related to the Exodus captivity and the escape and travels through the wilderness,[29] and archaeologist generally agree that the Israelites had Canaanite origins.[30] The culture of the earliest Israelite settlements is Canaanite, their cult-objects are those of the Canaanite god El, the pottery remains are in the Canaanite tradition, and the alphabet used is early Canaanite.[31] Almost the sole marker distinguishing the "Israelite" villages from Canaanite sites is an absence of pig bones, although whether even this is an ethnic marker or is due to other factors remains a matter of dispute.[31] Despite the Bible's internal dating of the Exodus to the 2nd millennium B.C.E, details point to a 1st millennium date for the composition of the Book of Exodus: Ezion-Geber, (one of the Stations of the Exodus), for example, dates to a period between the 8th and 6th centuries B.C.E with possible further occupation into the 4th century B.C.E,[32] and those place-names on the Exodus route which have been identified – Goshen, Pithom, Succoth, Ramesses and Kadesh Barnea – point to the geography of the 1st millennium rather than the 2nd.[33]

Similarly, the Pharaoh's fear that the Israelites might ally themselves with foreign invaders seems unlikely in the context of the late 2nd millennium, when Canaan was part of an Egyptian empire and Egypt faced no enemies in that direction, but does make sense in a 1st millennium context, when Egypt was considerably weaker and faced invasion first from the Achaemenid Empire and later from the Seleucid Empire.[34]The mention of the dromedary in Exodus 9:3 also suggests a later date of composition – the widespread domestication of the camel as a herd animal is thought not to have taken place before the late 2nd millennium, after the Israelites had already emerged in Canaan,[35] and they did not become widespread in Egypt until c.200–100 B.C.E.[36]

——— 10 ———

Elijah & Elisha

Extra-biblical sources verify Elisha died at the age of 60. He was struck three times with illness and on the third time he died. Costa postulates, "Elisha's illnesses tie in with dendrochronological records which show 3 major periods of trauma 854, 826, and 811 B.C.E. Likewise, at his age of death, we can arrive at the year of his birth 870 B.C.E."[133]

    c. Elijah: 892-854 B.C.E
    c. Elisha: 870 - 810/811 B.C.E

Who was Elijah?

Elijah figures predominantly in Judaism, Christianity, and Islam. When his Mission was over, God took him up to Heaven. All three religions believe Elijah will return as a Harbinger at the End Times. Elijah's name means, 'God is God.'

Legends ascribed to Elijah

Ginzberg's Legends of the Jews (IV p.201): "Elijah's miraculous deeds will be better understood if we remember that he had been an angel from the very first, even before the end of his earthly career…"

---

133 Nicola Costa 2013. Adam to Apophis, Asteroids, Millenarianism and Climate Change. D' Aleman Publishing, Lemona 8545, Cypress. p. 63.

St. Epiphanius of Cyprus, "When Elijah was born, his father Soback saw in a vision, angels of God around him. They swaddled him with fire and fed him flames."

Lives of the Prophets, 23 "Elijah, a Thesbite from the land of the Arabs of Aaron's tribe…When he was to be born, his father Sobacha saw that men of shining white appearance were greeting him and wrapping him in fire, and they gave him flames of fire to eat."

Who was Elisha?

Comet Halley appearance 911 B.C. "42 years later in 870 B.C., Elijah was "taught' by Michael, when a meteor swarm/storm blasted Persia. It explains a hitherto unexplained preponderance of the number 42 in the Elijah/cycle." [134]

Interestingly, according to recent calculations done by Japanese meteor scientists Mikiya Sato and Jun-ichi Watanabe cite, "the [elevated activity of 2006 of the Orionid meteor shower] was caused by the dust trails ejected from 1P/Halley in 1266 B.C., 1198 B.C., and 911 B.C." [authors note: the debris that hit Earth in 2006 was noted for being rich in large fireball-producing meteoroids]

---

134 Nicola Costa 2013. Adam to Apophis, Asteroids, Millenarianism and Climate Change. D' Aleman Publishing, Lemona 8545, Cypress. p. 63.

Part IV

History Becomes Myth

11

Myths of Ancient Apocalypse

During the last 13,000 years, man witnessed on several occasions, the ancient horrors of heavenly Firefall, caused by the disintegration of a world-destroyed comet, the rain of fire, ironstones, boulders, gravel, and iron dust; the world-consuming fires, earthquakes, oceanic-floods, earthquakes, and volcanic eruptions that followed.

Scientists today naturally ask: if so many mega-collisions have occurred in recent Earth history, where are they depicted in the artifacts or petroglyphs carved by Neolithic man or perhaps civilized man?

Myths and legendary motifs recorded in the sacred texts of ancient cultures throughout the world chronicle such extraordinary events, but often go unrecognized because they are lost in mythological allegory. In this book, I shall endeavor to show that humankind has been an intelligent audience, watching the development of the cosmic drama, its rising to a climax, and the solution of its crisis, from safe seats on certain mountain heights, and from other places of refuge.

History becomes myth, but as we see in Iliad, and the myth of Troy, it can become reality.

Troy Mythology

The tale of high-walled Troy and its destruction was long held to be a myth. The poems describing the city, Homer's Iliad and Odyssey, (*2's note: an hour after editing this chapter, I turned on the 100, a science fiction show about future survivors of a nuclear war, and turn on to a scene where the Iliad, one of their rare books is handed to the actor. March 17, 2016, ~Chronicle 1009*) are ancient; the great Greek poet created them before 700 B.C. Although classical Greeks read Homer as history, later scholars consigned him to the ranks of literature, conceived in an age of fantasy.

It took Heinrich Schliemann, a nineteenth-century millionaire, amateur archeologist, and dreamer, to prove the scholars wrong. Stubborn and romantic, the German-born businessman became convinced that Homer had told the truth about Troy. Armed with maps tracing Greek locations gleaned from the fabled stories of Greek heroes and wars, and despite the ridicule and contempt for his educational ignorance, he set out to find the lost city.

In the late 1860s, Schliemann decided that the Turkish town of Hissarlik, known for its fortress like earth mounds, best matched the scene of the Iliad. There he hired help from the natives for the purpose of excavation. Again, following the Homeric story, he studied the ground, compared it with the directions he had gained from the printed legends, nearly 3000 years old, and in 1871, he began to dig.

In 1873, he found that a city did indeed lie beneath Hissarlik's earthworks. In fact, several stages of an ancient city were buried there, one atop the other. And one of the layers, scorched by fire, looked very much like Homer's Troy.

When this news broke across Europe, taking the banner headlines of the papers, one of the scientists who had laughed the loudest and longest was so mortified that he committed suicide. Subsequent archeologists have confirmed that the city he unearthed is very probably Troy, albeit a Troy that underwent drastic change through the centuries. The German businessman's conversion of a myth into reality clearly shows, that while mythology may be couched in some rather fantastic allegory and metaphors, they are often based on historical fact.

Biblical metaphors of Firefall are considered 'cosmological'. Up until now the term 'cosmological myths' has been used with no thought that they were reports of real events in the distant past, dramas of mankind with a vast cosmic background. We must recognize that these myths are by no means the wild conjectures of an ignorant age, about the 'beginning of things'; rather we must regard them as the finished, though much worn, much overgrown, outcome of close observation.

Myths are not immensely exaggerated tales of local happenings, but matter-of-fact reports of universal events-which may have become rounded off, interpreted, and idealized in the course of time. The realm of mythology, therefore, is not fable-land. Myths are primeval lore, sacred lore, the 'science' of unknown, unsuspected forefathers living in the dark days far beyond our earliest history. Myths, to stress it once more, have a real, material background and describe ancient historical happenings of which only geology has up till now been able to give some account.

Hebrew Myths

This brings to mind Isaiah 24:1 KJV "Behold, the Lord maketh the earth empty, and maketh it waste, and turneth it upside down, and scattereth abroad the inhabitants thereof."

Isaiah 51:9-10 recalls the Apocalypse: "the primeval days" when the might of the Lord "carved the Haughty One, made spin the watery monster, drained off the waters of the mighty Tehom."

*"Awake, awake! Clothe yourself with strength, O arm of the Lord; awake, as in days gone by, as in generations of old. Was it not you who cut **Rahab*** to pieces, who pierced that monster through? Was it not you who dried up the sea, the waters of the great deep, who made a road in the depths of the sea so that the redeemed might cross over?* *Rahab the Harlot, Ref. Revelation, The Whore of Babylon (Astronomy/ Stars)

The Kalevala of the Finns

The Kalevala tell of a time when *hailstones of iron* fell from the sky, followed by a period of darkness. [135] Here again, we find iron meteors, not ice. Misrashic and Talmudic sources say the stones of Barad, which fell in during the Exodus in Egypt, were hot. However, this fact by itself does not prove the stones of barad were meteorites. While meteorites can be hot enough to spark a fire, very often when they are recovered immediately after impact, they have been found to be frost covered and icy-cold to the touch. This is because the core of the meteorite, cooled to near absolute zero in space, still retains a temperature far below freezing despite its fiery passage through the atmosphere.

Chinese Apocalypse

According to Sir William Jones, other undisputed sources of authenticity suggest the Chinese considered the Flood a world-wide event, not a local flood. It so devastated the land that after several generations some descendants of the survivors regressed to living like beasts, eating raw flesh and knowing their mothers but not their fathers, a trait not exclusive to the ancients or the Chinese, remarks Durant. [136]

---

135 Kaleva, (transl. J. M. Crawford, 1888), p. xiii., cit. op., Velikovsky, p. 61

136 Will Durant, Our Oriental Heritage (New York: Simon & Shuster, 1954)

And again, from the dynasty of King Chieh,"In the 10th year the 5 planets went out of their courses. *In the night, stars fell like rain.* The earth shook. Summer and winter came irregularly."

Corroboration of the Chinese astronomical sightings also come from another Chinese Taoist text of *Yinsee:*

"When the sky hostile to living beings wishes to destroy them, it burns them. The sun and the moon lose their form and are eclipsed, the five planets leave their path, the four seasons encroach one upon another. Daylight is obscured, growing mountains collapse, rivers dry up, it thunders in winter. Hoarfrost falls in summer. The atmosphere is thick and human beings are choked. The state perishes. The aspect and order of the sky are altered. The customs of the age are thrown into disorder. All living beings harass one another."

And in the Sing-li-ta-tsirn-cho, an ancient Chinese Encyclopedia, we find: *"In a general convulsion of nature, the sea is carried out of its bed, mountains spring out of the ground, rivers change their course, human beings and everything are ruined, and the ancient traces effaced."*

Washo Indians of California

"A great terrestrial revolution which caused the mountains to blaze up, the flames rising so high that the stars of heaven melted and fell upon the Earth. Then the sierras rose up from the plains, while the other parts of the country were inundated."[137]

Mexican codex Chimalpopoca

In the third aeon-Kiauhtonatiuh, that of the 'fire-rain sun'-the god of fire descended upon the Earth in a rain of fire which burnt everything, *while a hail of stones destroyed whatever was still left.* Then the rocks rose in uproar and the red mountains grew."[138]

South American Quiché Indians

The Quiché Indians say in their sacred book, *Popul Vuh,* "Hurakan,"[139] the god of terror, rained fire and stone from the sky:

137 H. S. Bellamy, Moons, Myths and Man, Fagen and Fagen, London, p. 95

138 Ibid, p. 96

139 Hurakan, a name signifying 'the furiously hurrying one' is also contained in the word 'hurricane,' a word we used today to describe tropical cyclone. Hurakan is compared with the name of the raging storm-giantess of Norse mythology, Hyrrockin, 'demon of the abyss'. Hurikan, is was also known as the Heavenly heart, or the location of the pole star. In 1,600 B.C. no star occupied the coveted position, but was a "dark place" surround by stars; thus the name, "heavenly heart," but also a place intimately connected with the serpent .

"Masses of sticky material [burning mud and stony debris] fell.... The face of the Earth was obscured, and a heavy darkening rain began. It rained by day, and it rained by night...There was heard a great noise above, as if by fire."[140]

Cahinaua of Western Brazil

The Cahinaua witnessed a similar event, "The lightning flashed and the thunders roared terribly and all were afraid. Then the heavens burst, and fragments fell down and killed everything and everybody. *Heaven and earth changed places. Nothing that had life was left upon the earth.*"

The Aztec codex

The Aztecs witnessed a massive fall of meteorites associated with a lowering of the sky: "... a rain of fire came following the sun of rain... all was burned.... a rain of rocks came, and the sky drew near the waters and the earth"[141]

The Ute Indians of California

Their legends recall the splintering of a comet fragment, "When the magical arrow of Ta-wats struck the sun-god full in the face, the sun was shivered into a thousand fragments, which fell to the Earth causing a general conflagration."[142]

The Siberian Voguls

On the other side of the world, in Siberian Voguls carried down through the centuries and millennia this memory: "*God sent a sea of fire upon the earth... The cause of the fire they call 'the fire-water.'*" [143]

Get more from When the SKY Fell

140 Popul Vuh, (See author)

141 Roy Norville, The Road in the Sky, pg 125

142 H. S. Bellamy, Moons, Myths and Man, Fagen and Fagen, London, p. 99

143 Holmberg, Finno-Ugric, Siberian Mythology, p. 368, cit.op. Velikovsky, pg 55

12

Personification of the Myth

Michael & Gabriel

"When we look at Jewish sources for record of the parents is of the angel Gabriel you find it astonishing synchronization between the calculated apparitions of Haleys comet and recorded apparitions of the angel Gabriel, not only that but we also find the presence of Hitherto unknown celestial body known to choose as Michael which is placed repeatedly 42 years after the appearances of the angel Gabriel. The big difference between the two is that when Gabriel appears he is invariably on on his own, but Michael is many times described as appearing with his host-indicating something in the form of a meteor swan that followed in the wake of Halley's comet.

And even though they are not so named in Persian or Egyptian sources find the same pattern of disruption throughout the entire history of Persian empire as well as Egypt particularly evident during the era of the Ptolomies. Ref. p. 151. Adam to Apophis.

Michael

The name Michael translates directly as a question, "Who is like God?" Or "Who is strong and powerful?" With underlying meaning in the last part deriving from an unused root meaning to twist.

Michael was represented in the Haggadah as the most prominent of the archangels. And Daniel (Dan. xii 1) he is described as "the great prince," later Jewish writings go to the great pains to describe his greatness. Although both Gabriel and Michael are described as "great princes"; see (Ber. 4b; Yoma 37a). He was God's viceroy who ruled over the world (Enoch, 1xix. 14 et seq.), and wherever he appeared the Shekinah was also seen (Ex. R.ii. 8).

Notably when we look at the history of Michael if I am identified with Angel known as Metatron, and the memory perhaps of a larger celestial body.

G. Scholem has deduced, from a statement in Perek Re 'Kyoto Yehezkel (Wertheimer, Battei Midrashot, 2 (1955), 132-3) and from other sources, that at first Michael and Metatron were identical. Matatron was depicted as the guardian of the interior and the highest figure in the domain of the angels in the Merkaba literature and in the Kabbalah which succeeded it.

Metatron is not mentioned in the Bible but is remembered in later Jewish sources have the angel intimately associated with Noah's flood, with his last major appearing during the time of Moses' his death (authors note; 1623 B. C 1511/1510 B.C. N need to fix the state to match up with the exodus in chapter. And page 151 Adam to Apophis.

Begin page 152. There was no unanimity as to the origin of the name Metatron, and many interpret it as a title rather than a name. It could derive from the Hebrew Matara, which means "keeper of the watch," or from the Hebrew Kabbalistic term metator, which means "guide or messenger," or from two Greek words, meta thronos, which means "one who serves behind the throne."

The Zohar tells us that, he is the highest archangel, esteem more than any other of gods hosts. The letters [of his name] are a great mystery. Period. He rules over all, the living things below and the living things above."

Even the princess of the heavens, when they see Metatron, Trumbull before him, and prostrate themselves; Is magnificence and majesty, the Splendora and beauty radiating from him overwhelm them. (Ginz 1 p. 159)

Metatron is mostly detected as the twin of Sandaphon (with Elijah was equated) or even as an identical to Sandolphin.

He is huge. In Hagiga 13b Rabbi Eliezar is quoted as saying, one Angel was standing on the ground and his hand reaches into the living creatures. Authors note letting creatures equals the zodiac. In a Matnina [small Midrashot, or sayings of the rabbis] we learned that he is Sandolphon. This is his name. He is taller than his fellows [A height that would take] 500 years to walk. He stands behind the chariots and fashions crowns for his maker. [Authors note: 500 years is the same period of the Phoenix in Egypt. Also related to the Benben Stone that sits on top of the pylon as a representative of the Naval of the earth and the symbolic representation of the serpent and Osiris which was cut up into 14 pieces.

Metatron and Sandolphon is represented by the colors green, white and red, and is remembered as a flaming princely angel (sar) and and ofan, wheel-shaped angel that stretches from heaven to earth, the distance of "500" (Beit ha-Midrashot 1:58-61).

He is said to be the wheel with the eyes described (Ezekiel, 1: 15) that help propel the divine chariot. Some rabbis regard to be the source of having my thunder. In this aspect he was also referred to him by the name Ophan (Hebrew for "wheel").

Daniel 12:1-13:
And at that time shall Michael stand up, the great prince which standeth for the children of thy people: and there shall be a time of trouble, such as never was since there was a nation even to that same time: and at that time thy people shall be delivered, every one that shall be found written in the book. 2And many of them that sleep in the dust of the earth shall awake, some to everlasting life, and some to shame and everlasting contempt.

3And they that be wise shall shine as the brightness of the firmament; and they that turn many to righteousness as the stars for ever and ever. 4But thou, O Daniel, shut up the words, and seal the book, even to the time of the end: many shall run to and fro, and knowledge shall be increased. 5Then I Daniel looked, and behold, there stood other two, the one on this side of the bank of the river, and the other on that side of the bank of the river. 6And one said to the man clothed in linen, which was upon the waters of the river, how long shall it be to the end of these wonders? 7And I heard the man clothed in linen, which was upon the waters of the river, when he held up his right hand and his left hand unto heaven, and sware by him that liveth for ever that it shall be for a time, times, and an half; and when he shall have accomplished to scatter the power of the holy people, all these things shall be finished. 8And I heard, but I understood not: then said I, O my Lord, what shall be the end of these things? 9And he said, Go thy way, Daniel: for the words are closed up and sealed till the time of the end. 10Many shall be purified, and made white, and tried; but the wicked shall do wickedly: and none of the wicked shall understand; but the wise shall understand. 11And from the time that the daily sacrifice shall be taken away, and the abomination that maketh desolate set up, there shall be a thousand two hundred and ninety days. 12Blessed is he that waiteth, and cometh to the thousand three hundred and five and thirty days. 13But go thou thy way till the end be: for thou shalt rest and stand in thy lot at the end of the days.

Personification in Mythology

Apocalyptic literature was often deliberately written in obscure terminology, employing the mystical arts of astrology, mathematics, geometry, numerology, temple measures, and mythical symbolism, to hide the primal secret. In some cases, as with Nostradamus, this was done to protect life and family from religious persecution, but more often was to conceal the inner meaning of the prophetic tradition. Why? The knowledge was bluntly considered to be dangerous. What if you knew the precise day and hour of your death? How would it affect the living of your life? The cyclic nature of the Apocalypse was only revealed to the initiates of the secret mystery school traditions of the Ancient World.

The most famous of these were the Greek schools of Pythagoras, the Egyptian Schools of Temple Building, including the Masonic Order and Freemasons, and perhaps the most ancient, the Chaldean Schools of Astronomy, where the patriarch Abraham, learned of the secret traditions from the descendants of Noah, the survivors of the World Deluge. According to the biblical narrative, Noah departed from the Ark in the mountains of Aararat, (mountains in northern Turkey), and settled in the nearby Valley of the Fertile Crescent, the cradle of the ancient Sumerian civilization (c. 3800 B.C.). Following the opinion of Josephus, Origen said that Noah brought with him the knowledge of astronomy passed on from Enoch, Seth, and Adam, where it is written the constellations were "already named and divided" into the Dodecatemory (12) divisions, a very ancient history indeed. It is no wonder Chaldea became preeminent in their observation and understanding of the heavens as early as 4,000 B.C.!

Abraham's father, Terah, learned of these same traditions because he was a temple builder and part of the royal priesthood. In the days of Abraham, sons followed in their father's footsteps, therefore, it is reasonable to assume Abraham became a temple builder, which means he was an important and powerful "wise man," a member of the Magi who understood the secret, apocryphal language of the ancient oral traditions. Apocryphal mythology often combines literal, historical events to convey physical, mental and spiritual meaning. References to the deluge, a fact I will prove to be an historical event (s), is also the genetrix of allegorical expressions of ideas or concepts expressed by the term 'deluge," for the term referring to "a flooding" is not only an overflow of water.

Common Firefall Comet Symbols

Evil Star / Comet
Even words associated with destruction and disaster today have a cosmological relationship with the apocalypse. The meaning of the word 'disaster' comes from the Greek language, 'dis' meaning evil, and 'aster' meaning star— thus meaning evil star. The reason for the choice of this term is that disaster is caused by evil stars, or comets implicating an astronomical relationship to the apocalypse. The word cataclysm has an astronomical relationship to the position of the stars at the time of the apocalypse. Censorinus wrote: "There is a period called 'the supreme' by Aristotle, at the end of which the sun, moon, and all the planets returned to their original position. This 'supreme year' has a great winter, called by the Greeks kataclysmos, which means deluge, and a great summer, called by the Greeks ekpyrosis, or combustion of the world. The world, actually, seems to be inundated and burned alternately in each of these epochs."[144]

CELESTIAL MESSIAH
JESUS / MA.D.HI / ANTI-CHRIST /DAJJAL
SON OF MAN / SON OF GOD / KNOWLEDGE OF GOOD & EVIL SERPENT / TREE OF LIFE SERPENT
John 3:13-14 "And no man hath ascended up to heaven, but he that came down from heaven, even the Son of man which is in heaven. And as Moses lifted up the serpent in the wilderness, even so must the Son of man be lifted up."

Islamic, Christian, and Jewish SON OF MAN Apocalyptic figures are identical. The Celestial Messiah that descends from Heaven is NOT A MAN.

"And then shall appear the sign of the Son of man in heaven; and then shall all the tribes of the earth mourn, and they shall see the Son

---

144 Joseph Goodavage, The Comet Kohoutek, 1974 .

of man coming in the clouds of heaven with power and great glory." Matt: 24:30.

Jesus denies that he is the Son of man. Peter calls him the Son of the living God which translates to Jesus as the sun of the living God. Now when Jesus came into the district of Caesarea Philippi, he asked his disciples, "Who do men say that the Son of man is?"

And they said, "Some say John the Baptist, others say Elijah, and others Jeremiah or one of the prophets."

He said to them, "But who do you say that I am?"

Simon Peter replied, "You are the Christ, the Son of the living God." (Matt. 16:13-16)

Mutilation of the Phallus (Symbol of the Serpent)

One of the oldest theories of the origin of life was, as the Bible puts it, "the blood is the life." (Deut. 12:23) Egypt's sun father Ra castrated himself to bring forth a new race from his genital blood. The Hindu "Great God" had his penis removed, chopped up, and buried to bring forth a new generation. A similar Babylonian god, Bel, mutilated his penis for blood that he could mingle with clay to produce people and animals, imitation of the older Goddess Ninhursag, maker of life forms from the mixture of clay and menstrual blood. Menstrual blood shed by Pagan mothers evolved to Patriarchal ritual killing of bulls, stags, rams, he-goats or stallions, where it was traditional to cut off the animals' genitalia.

Apocalypse Blood

A Mexican savior-deity, Quetzalcoatl, likewise gave blood from his sliced penis to the Earth Goddess's clay vessel, to repopulate the earth after the Deluge.145 Even the name of Adam gave away the primal secret. It meant literally "man of red earth or man-made of blood."

The underlying belief was that followers of the gods would be forgiven by drinking or shedding of their own blood. The ritual of the castration of the male for religious reasons was gradually abandoned,

145 Joseph Campbell 1974. The Mythic Image. Princeton, NJ. The Princeton University Press. p.156.

except for a few notable exceptions such as the priests of Attis and the early Christian castrati; also, Jesus who mentioned as having made themselves eunuchs for the sake of the Kingdom of heaven. Matthew (19:12). Castration later evolved into "ceremonial" circumcision and was often performed on a "blood moon."146 The shedding of blood was essential in the remission of sin. See (Heb.9:22)

Human and animal sacrifice; castration and circumcision—the shedding of blood all points an ancient massacre of blood where millions and millions perished for what was believed to be a judgment from God for sin

—an abhorrent practice that Jesus Christ put an end to, with this Crucifixion and the shedding of his blood once for all time. Because of His sacrifice, Apocalypse blood is no longer required.

However, in all apocalypse legends, the fall of blood is the first plague. It personifies the massacre of humans, but not only humans, it is the massacre of plants and wildlife by fire, burning heat and drowning floods; a river of blood reddening Earth's rivers, oceans and sweet waters drenched with the deaths of countless millions and billions of beings, a story retold in the book of Exodus later in the book.

Archangels: Gabriel & Michael & Halley's Comet

Afterward you will come to the hill of God where the Philistine garrison is; and it shall be as soon as you have come there to the city, that you will meet a group of prophets coming down from the high place with harp, tambourine, flute, and a lyre before them, and they will be prophesying. "Then the Spirit of the LORD will come upon you mightily, and you shall prophesy with them and be changed into another man." 1 Samuel 10:5-6

6000 years ago, knowledge of the sciences sprang up fully developed in Sumeria and Egypt. Writing, mathematics, geometry, architecture, government, and religion. Stories were passed down in the Old Testament through Tera, the Father of Abram, who met the High Priest Melchizedek in the desert.

The Levite Priestly Order & the Order of Melchizedek

Jesus is portrayed as "a priest forever in the order of Melchizedek" (Ps. 110:4). He plays the role of the king-priest once and for all. According to the Apostle Paul, Hebrews (7:13-17), Jesus is considered a priest in the order of Melchizedek because, like Melchizedek, Jesus was not a

---

146 Barbara G. Walker 1985. The Crone: Woman of Age, Wisdom, and Power. New York: Harper and Row. p.48.

descendant of Aaron, and thus would not qualify for the Jewish priesthood under the Law of Moses. Melchizedek is referred to again in Hebrews 5:6-10; Hebrews 6:20; Hebrews 7:1-21: "Thou art a priest forever after the order of Melchizedek"; and Hebrews 8:1.

And verily they that are of the sons of Levi, who receive the office of the priesthood, have a commandment to take tithes of the people according to the law, that is, of their brethren, though they come out of the loins of Abraham: But he whose descent is not counted from them received tithes of Abraham, and blessed him that had the promises" (Hebrews 7:5-6).

If therefore perfection were by the Levitical priesthood, (for under it the people received the law,) what further need was there that another priest should rise after the order of Melchizedek, and not be called after the order of Aaron? For the priesthood being changed, there is made of necessity a change also of the law" (Hebrews 7:11-12). The author of the Epistle to the Hebrews in the New Testament discussed this subject considerably, listing the following reasons for why the priesthood of Melchizedek is superior to the Aaronic priesthood:

Abraham paid tithes to Melchizedek; later, the Levites would receive tithes from their countrymen. Since Aaron was in Abraham's loins then, it was as if the Aaronic priesthood were paying tithes to Melchizedek. (Heb. 7:4-10) The one who blesses is always greater than the one being blessed. Thus, Melchizedek was greater than Abraham. As Levi was yet in the loins of Abraham, it follows that Melchizedek is greater than Levi. (Heb. 7:7-10) If the priesthood of Aaron were effective, God would not have called a new priest in a different order in Psalm 110. (Heb. 7:11)

The basis of the Aaronic priesthood was ancestry; the basis of the priesthood of Melchizedek is everlasting life. That is, there is no interruption due to a priest's death. (Heb. 7:8,15-16,23-25)

Christ, being sinless, does not need a sacrifice for his own sins. (Heb. 7:26-27)

The priesthood of Melchizedek is more effective because it required a single sacrifice once and for all (Jesus), while the Levitical priesthood made endless sacrifices. (Heb. 7:27)

The Aaronic priests serve (or, rather, served) in an earthly copy and shadow of the heavenly Temple, which Jesus serves in. (Heb. 8:5)

The epistle goes on to say that the covenant of Jesus is superior to the covenant the Levitical priesthood is under. Some Christians hold that Melchizedek was a type of Christ, and some other Christians hold that Melchizedek indeed was Christ. Reasons provided include that Melchizedek's name means "king of righteousness" according to the author of Hebrews, and that being king of Salem makes Melchizedek the "king of peace." Heb. 7:3 states, "Without father or mother, without genealogy, without beginning of days or end of life, like the Son of God he (Melchizedek) remains a priest forever."

Melchizedek gave Abraham bread and wine, which some Christians consider symbols of the body and blood of Jesus Christ, the sacrifice to confirm a covenant. Catholics find the roots of their priesthood in the tradition of Melchizedek.

In Genesis 14:18, Melchizedek offers a sacrifice of bread and wine. Christ therefore fulfilled the prophecy of Ps 110:4, that he would be a priest "after the order of Melchizedek," at the Last Supper, when he broke and shared bread with his disciples. Catholics take seriously Christ's command that the Apostles should "do this in memory of Me". As such, the Catholic Church continues to offer sacrifices of bread and wine at Mass, as part of the sacrament of the Eucharist.

Abraham carried the knowledge of Persian astronomy and astrological sciences until a regional cataclysm destroyed Sodom and Gomorrah with "fire and brimstone." From their perspective, they believed the entire world had been destroyed, and that Abraham and his family were the only living people left alive on the planet.

Dates of Abraham vary widely from 2150 B.C. to 1850 B.C.

Survive he and his family did. He bore sons who became the 12 tribes of Israel. During the great famine and drought that occurred after the firestorm impact of Sodom & Gomorrah, (see Vol. III, the Armageddon Stones), Jacob was forced to flee to Egypt and join his Son Joseph, who became Vizier, second only to Pharaoh of Egypt.

400 years later, the Hebrew saviors of Egypt became slaves of Egypt, requiring a deliverer from their bondage. Moses. Their deliverance is preceded by 12 plagues. Blood, fire, and brimstone. See Ch. 9, The Exodus, Vol. III, The Armageddon Stones.

Passing through the Red Sea, and carrying the Ark of the Covenant, the Children of Israel wandered the Sinai Wilderness, and eventually gathered at the foot of Mt. Sinai, and waited for Moses to receive the handwritten commandments on stone written by the finger of God. When Moses finally came down from the mountaintop.

Aaron of the Levites kept the priestly secrets of God.

The Personification of the Prophets

Serpents & Dragons
The cosmic origins of the benu bird, or phoenix, is discussed at length in Robert Bauval and Adrian Gilbert's excellent book, The Orion Mystery. They record that before the pyramids were built, in the ancient

city of Heliopolis, there stood an important sacred hill or mound arising out of the primordial sea upon which the First Sunrise had taken place and set upon the mound was a pillar symbolizing Atum, the Creator of the Universe.

The Egyptians called iron the 'bones of Typhon', or the 'gift of Set'; both of these names belong to spirits of darkness and of evil, The Jews call iron ore nechoshet, which literally means the 'droppings of the [cosmic] serpent, a nonsensical term unless our interpretation of it is allowed.

BenBen Stone On top of the pillar was placed the Benben, a mysterious conical stone, which was credited with cosmic origins.

Primodial Hill: The first land after the flood

Benben Stone place on top

The authors write: "The Benben Stone was housed in the Temple of the Phoenix and was symbolic of this legendary cosmic bird of regeneration, rebirth and calendrical cycles. In ancient Egyptian art the phoenix was usually depicted as a grey heron, perhaps because of the heron's migratory habits; it was believed that the phoenix came to Heliopolis to mark important cycles and the birth of a new age.

Its first coming seems to have produced the cult of the Benben Stone, probably considered the divine `seed' of the prodigal cosmic bird. This idea is evident from the root word ben or benben which can mean human sperm, human ejaculation or the seeding of a womb."Since Osiris was often identified with the phoenix, it is likely that the Benben Stone symbolised, among other things, his seed. (bolides, meteorites) Egyptian legends of the dismemberment and scattering of the fourteen pieces of the body of Osiris by his evil brother Seth, (all except his penis which was never found. The Sword. The Penis. Of course, it was never found. Depicted the Comet. The Osirion body parts are most likely meteorites which were found and later worshipped as the Osirion Seed.

Osiris Djed Pillar

Consider the mystery play of the Djed column, the magical resurrection of Osiris, the birth of Horus. Note the 4 disks that separate the Djed column like or circles at the top. If Osiris represented a form of the Phoenix, then he most likely represented the "Sun" God of the Age, who then later was killed by his brother Seth, the Comet, which exploded into 14 pieces. Set represented the dark force, or Lucifer.

Venus. The Bright and Morning Star. Venus Lucifer / Venus Jesus Christ. Legend of Quetzecotl. Christ of the Americas

.

104 Amulets of Osiris

It is also interesting to note that in the chamber in the Temple of Denderah were found the "104 Amulets of Osiris" made of gold and precious stones which had been placed in the sarcophagus called "the House of God" in order to protect him.[147] The 13th amulet is called "Uraeus wearing the crown of the North."

The numbers 104 and 13 specifically relate to the planet Venus. The Asp relates directly to the Lion, tail, serpent, and represent the dangerous, violent celestial phenomena associated with comets and meteors. Venus came to symbolize a comet.

Like Venus, comets are seen as a morning and evening `star` as they pass round the Sun, and thus associated the remnant of the comet with Venus, the brightest object in the sky.

Marcus Varro wrote, "There was seen...a surprising prodigy in the heavens, with regard to the brilliant star Venus. Castor affirms that Venus changed its colour, size, figure and track...Adastrus of Cyzicus and Dion the Neapolitan refer this great prodigy to the reign of Ogyges." (Ogyges was the most ancient of Greek mythical kings. During his reign, supposedly in about 1764 B.C., there was a deluge which left Attica waste for 200 years). Hyginus relates how Phaethon, that caused the conflagration of the world, was placed by the Sun among the stars. "It was the general belief that Phaethon changed into the Morning Star."

A Chinese astronomical text from Soochow says that in the past, "Venus was visible in full daylight and, while moving across the sky, rivalled the Sun in brightness." The pre-Columbian Mexicans called Venus "the star that smoked", a phrase they also used to describe comets.

Leo the Lion

Images of Lucifer riding Leo's back. In Egypt, the pharaohs and Queens wore the Uraeus, a Greek name for the asp-crown meaning "of the Lion's Tail."

"Uraeus" is probably derived from "ouros" meaning mountain. But "ura" means "tail of the lion"[148] and since a lion expresses its anger with

---

147 Ibid.

148 R. Graves, The Greek Myths, (Penguin, Middlesex, 1955)

its tail, the word hints at violence. Moreover, the asp was sacred to the Goddess, Isis. The Lion tail and the asp are both implements able to inflict pain, one by lashing, the other by biting.

13

The Orion Mystery, The Pleiades & Halley's Comet

Halley's Comet & 12 B.C. The Birth of Christ

The Dead Saints Chronicles: Halley's Comet (Sign of the Lord) (Acceptable Year of the Lord) 2029 (See Apophis & 2! Signs...) 69th week – Year of Jubilee (50th Year) – Year 2029.

12 B.C to 2029 = 2041 years.....

Other Celestial signs signifying important events

Comets are frequently described as 'swords' in ancient literature because of their upward tails (in a direction away from the sun).

And the Meaning of Comets.... Pliny (Natural History 11, 23) records that Augustus (63 B.C. to A.D. 14) dedicated a temple to a comet that appeared during athletic games he sponsored in 44 B.C., just after the assassination of Julius Caesar. The common people assumed that the comet was taking the soul of Caesar to the heavens where the gods lived. The emblem of a comet was added to a bust of Caesar that was dedicated in the forum. Augustus then used an emblem of a comet on some of his own coins, presumably as a symbol of his own greatness and possibly as a symbol of his assumed deity.

It seems clear therefore that at the time of Christ comets were associated with great kings and with important events. The ancient root of the word comet is "dis" "aster" --- meaning a "bad star," an omen foretelling disaster.

The Star prophecy

The "Star Prophecy" (or Star and Scepter prophecy) is a Messianic reading applied by radical Jews and early Christians to Numbers 24:17:

"I shall see him, but not now: I shall behold him, but not nigh: there shall come a Star out of Jacob, and a Sceptre shall rise out of Israel, and shall smite the corners of Moab, and destroy all the children of Sheth."

The Star Prophecy was often employed during the troubled years that led up to the Jewish Revolt, the destruction of the Second Temple in Jerusalem (70 CE) and the suicidal last stand of the Sicarii at Masada in 73 CE. The Star Prophecy appears in the Qumran texts called the Dead Sea scrolls. "This was the prophecy that was of such importance to all resistance groups in this period, including those responsible for the documents at Qumran and the revolutionaries who triggered the war against Rome, not to mention the early Christians"149

The Star Prophecy was applied to the coming Messiah himself in contemporary radical Jewish documents, such as the apocalyptic War Scroll found at Qumran. In a pesher applied to the text from Numbers, the War Scroll's writer gives the following exegesis:

...by the hand of the Poor whom you have redeemed by Your Power and the peace of Your Mighty Wonders... by the hand of the Poor and those bent in the dust, You will deliver the enemies of all the lands and humble the mighty of the peoples to bring upon their heads the reward of the Wicked and justify the Judgment of Your Truth on all the sons of men.

Matthew: The Star of Bethlehem, the Birth of Christ & the Magi

"Many scholars, seeing the Gospel Nativity story told in Matthew as a later apologetic account created to establish the Messianic status of Jesus, regard the Star of Bethlehem as a pious fiction. Aspects of Matthew's account which have raised questions of the historical event include Matthew is the only one of the four gospels which mentions either the Star of Bethlehem or the magi. The author of the Gospel of Mark, considered by modern text scholars to be the oldest of the Gospels, does not appear to be aware of the Bethlehem nativity story. A character in the Gospel of John states that Jesus is from Galilee, and not

---

149 Eisenman, Robert, 1997. James the Brother of Jesus: The Key to Unlocking the Secrets of Early Christianity and the Dead Sea Scrolls (VikingPenguin), pg 23.

Bethlehem. The Gospels often described Jesus as "of Nazareth," but never as "of Bethlehem." Scholars suggest Jesus was born in Nazareth and the Bethlehem nativity narratives reflect a desire by the Gospel writers to present his birth as the fulfillment of prophecy. The Matthew account conflicts with that given in the Gospel of Luke, in which the family of Jesus already living in Nazareth, travel to Bethlehem for the census, and return home almost immediately.

The Church of the Nativity, a basilica located in Bethlehem, Palestine, sits over the site that is still traditionally considered to be located over the cave that marks the birthplace of Jesus of Nazareth, symbolized by a 14-petal silver star embedded in marble on its floor. The church was originally commissioned in 327 A.D. by Constantine and his mother Helena.

The earliest known account of the star of Bethlehem is in Matthew 2:1- 12. Most scholars believe that the final text of this gospel was composed in about A.D. 80 (some scholars give an earlier date, e.g. ref. 5) from sources written in earlier times. Presumably one of these sources recorded the star of Bethlehem and the visit of the Magi. The account in Matthew describes how the Magi saw a star which they believed heralded the birth of the Messiah-king of the Jews. They travelled to Jerusalem and informed King Herod of the time when the star appeared, which indicates that the star was not a customary sight. The advisers of Herod told the Magi that, according to the prophecy of Micah, the Messiah should be born in Bethlehem, so the Magi journeyed there. The star moved before them and 'stood over' Bethlehem. The Magi found the place where the child was and presented him with gifts.

The identification of a comet with the star of Bethlehem goes back to Origen in the third century, and this is the earliest known theory for the star. Origen (Contra Celsum 1, 58) stated 'The star that was seen in the East we consider to be a new star ... partaking of the nature of those celestial bodies which appear at times such as comets ... If then at the commencement of new dynasties or on the occasion of other important events there arises a comet... why should it be a matter of wonder that at the birth of Him who was to introduce a new doctrine ... a star should have arisen?

"Babylon (in Mesopotamia) was the world center of astronomy and astrology at that time and Magi were important members of the Babylonian royal court. In about 586 B.C. the Babylonians sacked Jerusalem and took the Jews into Exile. From the time of the Exile onwards Babylon contained a strong Jewish colony, and the knowledge of the Jewish prophecies of a Saviour-King, the Messiah, may have been well-known to the Babylonians and to the Magi.

"In the Hellenistic age some of the Magi left Babylon and travelled to neighboring countries to teach and practice astronomy/astrology, which was a core educational subject in the ancient world (e.g. Plato, The Republic, 529). Thus, the first century A.D. Jewish scholar Philo of Alexandria stated that the student of astronomy perceives 'timely signs of coming events' since 'the stars were made for signs' (De Opificio Mundi, 22). There is a strong tradition that the Magi who visited Jesus came from Arabia (now Saudi Arabia), which lies between Mesopotamia and Palestine. Thus, in about A.D. 160 Justin Martyr wrote 'Magi from Arabia came to him [Herod]' and in about A.D. 96 Clement of Rome6, associated frankincense and myrrh, two of the gifts of the Magi, with 'the East, that is the districts near Arabia'. We conclude that the Magi who saw the star of Bethlehem were astronomers/astrologers, who may have been familiar with the Jewish prophecies of a Saviour-King, and who probably came from Arabia or Mesopotamia, countries to the east of Palestine. Matthew 2:1 simply states, 'Magi from the East arrived in Jerusalem'. It is important to realize that there are many references in ancient literature to Magi visiting kings and emperors in other countries. For example, Tiridates, the King of Armenia, led a procession of Magi to pay homage to Nero in Rome in A.D. 66 (Suetonius, Nero 13 and 30; Tacitus, Annals, 16:23: Dio Cassius, Roman History, 63:1), and Josephus records that Magi also visited Herod in about 10 B.C.

Thus, a visit by the Magi to pay homage to Jesus, the new King of the Jews, would not have appeared as particularly unusual to readers of Matthew's gospel. However, the Magi must have had an unmistakably clear astronomical/astrological message to start them on their journey.

Prophecies of a Redeemer/Messiah who will save fallen humankind predate the birth of Christ by thousands of years. The Patriarch Job expected his coming. "I know my Redeemer lives. For in my flesh, I shall see God." It is of New Testament record that certain wise men among the Gentle people of Persia looked to the stars as signs for a coming Redeemer, who would suffer on the cross, even the Lord Jesus himself? "A scepter shall come out of Israel.

And so it is in 12 B.C., a scepter star arose in the East.

The Star Prophecy and the Star of Bethlehem.

Jeshua (Hebrew) means Salvation. Christ (Logos) means Messiah. There is convincing, but controversial evidence Jesus was born in 12 B.C. Halley's Comet appeared in August – October 10, 12 B.C. A census was performed by Quinarus of Syria in 12 B.C. Only shepherds were present. The arrival of the Magi was recorded later by Josephus in 10 B.C.

The Persian, Babylonian Magi knew a King would be born to the East in Israel. The Birth of a REDEEMER. Why were astrologer's (greek—magios) called wise men? Doesn't the Bible teach against astrology? Why does Matthew mention the "kings of the East?" Of course, assigning a 12 B.C. birthdate to Jesus would require a slight reshuffling of the Chronology of the birth and death of Jesus of Nazareth.

Joseph R. Seiss, a nineteen century Christian author, groundbreaking book, "The Gospel in the Stars" describes the story of the Redeemer/Messiah written in ancient stories that have of the constellations that make up the 12 signs of the Zodiac. He points to Roberts, in his Letters to Volney, "that the emblems of the stars refer to the primeval promise of the Messiah and His work of conquering the Serpent through His sufferings..."150 He references Dupuis, in L' Origines des Cultus, a collection of vast number of traditions prevalent in all nations of a Divine Person, born of a virgin woman, suffering in conflict with a Serpent, and eventually triumphing over him at last, as a story told in the ancient constellations."151

The Patriarch Job declared, "By His Spirit He hath garnished the heavens; His hand hath formed the crooked Serpent." Job 26:13 and the Psalmist agreed, "The heavens do declare the glory of God."

The Constellations that garnish the heavens are ancient. According to Drummond, "Origen tells us that was asserted in the book of Enoch (quoted by the apostle Jude) that in the time of that patriarch the constellations were **already named and divided**." Seiss states that "Josephus and the Jewish Rabbi's affirm that the starry lore had its origin with the antediluvian Patriarchs, Seth and Enoch."152

He references Smith and Sayce in the Chaldean account of Genesis, who write, "It is evident, from the opening of the inscription of the first tablet of the great Chaldean work on astrology and astronomy, that the function of the stars were, according to the Babylonians, to act not only as regulators of the year, but to also used as signs; for in those ages it was generally believed that the heavenly bodies gave, by their appearance and positions, signs of events which were coming to the earth. "

"The Hope of the Nations," beginning with Virgo, and the virgin birth of the SON who becomes the "Hope of the Nations" His story of Crucifixion and REDEMPTION.

---

150 Gospel in the Stars, p. 11

151 Gospel in the Stars, pp. 11-12

152 Gospel in the Stars, p. 22

The Peculiar Movements of the Star of Bethlehem

"(i) It was a star which had newly appeared. Matthew 2:7 states 'Then Herod summoned the Magi secretly and ascertained from them the exact time when the star had appeared'.

(ii) It travelled slowly through the sky against the star background. The Magi 'saw his star in the east' (Matthew 2:2) then they came to Jerusalem where Herod sent them to Bethlehem, then 'they went on their way and the star they had seen in the east went ahead of them' (Matthew 2:9). Since Bethlehem is to the south of Jerusalem the clear implication is that the star of Bethlehem moved slowly through the sky from the east to the south in the time taken for the Magi to travel from their country to Jerusalem, probably about one or two months (see later).

(iii) The star 'stood over' Bethlehem. Matthew 2:9 records that the star 'went ahead of them and stood over the place where the child was'.

Popular tradition has the star pointing out the very stable in which Christ was born, but Matthew neither states nor implies this: according to Matthew, viewed from Jerusalem the star stood over the place where the child was born, i.e. Bethlehem.

The first suggestion that the star of Bethlehem was a nova was made by Foucquet in 1729, and possibly earlier by Kepler in 1614 (see also Sachs and Walkerg9) and it has received considerable recent support.10 A supernova has also been suggested.11 A nova or supernova satisfies the requirement that the star of Bethlehem was a single star which appeared at a specific time, but cannot account for the star moving through the sky. Similarly, all other suggestions for the star of Bethlehem (e.g. that it was Venus, etc) can be ruled out except one: a comet."

Was the Star of Bethlehem a Comet?

(copied from an article: Colin Humphreys From Science and Christian Belief, Vol 5, (October 1995): 83-101. **http://www.asa3.org/ASA/topics/AstronomyCosmology/S&CB%20 10-93Humphreys.html**) Needs to be re-written....

"Comets probably have the greatest dramatic appearance of all astronomical phenomena. They can be extremely bright and easily visible to the naked eye for weeks or even months. Spectacular comets typically appear only a few times each century. They can move slowly or rapidly across the sky against the backdrop of stars, but visible comets usually move through the star background at about 1 or 2 degrees per day relative to the Earth. They can sometimes be seen twice, once on their way in towards perihelion (the point in their orbit which is closest to the sun) and again on their way out. However, from a given point on the Earth's surface, a comet is often only seen once, either on its way in or its way out, because of its orbit relative to the Earth. Since a comet usually peaks in brightness on its way out, about one week after perihelion, the most visible comets are seen on their way out from perihelion.

If the star described in Matthew was a comet, was it seen twice, first in the east on its way in towards perihelion and again in the south on its way out, or was it seen continuously moving from east to south (and then to west) on its way out? Matthew 2:9 states 'the star they had seen in the east went ahead of them [to Bethlehem in the south]'. It was not generally recognized 2000 years ago that a comet seen twice, once on its way in towards perihelion (where it would disappear in the glare of the sun) and again on its way out was one and the same comet. It was normally regarded as two separate comets.

Since Matthew 2:9 clearly implies that the star seen in the south was the same star as that originally seen in the east we deduce that the star was continuously visible and suggest that it was a comet on its way out from perihelion travelling east to south (to west). In particular it is suggested that the Magi originally saw the comet in the east in the morning sky (see later). They travelled to Jerusalem, a journey time of 1- 2 months (see section 7), and in this time the comet had moved through about 90's, from the east to the south. In one month, the star background would move through 30', and in two months through 60 degrees. For the comet to have moved through 90' an additional 60' (in one month) or 30' (in two months) motion is required, which is broadly consistent with the 1 or 2 degrees per day typical motion of a comet. In Jerusalem, Herod's advisers suggested the Magi go to Bethlehem, six miles to the south and a journey time of one or two hours. The Magi set off next morning and saw the comet ahead of them in the south in the morning sky. Hence it appeared that the comet 'went ahead of' the Magi on this last lap of their journey.

Two comets are candidates for the "Star of Bethlehem." A comet in 5 B.C described by ancient Chinese records and Halley's Comet in 12 B.C.

5 B.C Theory

The earliest possible date for the birth of Christ can be deduced from Luke 3:2 3, which states that he was 'about 3 0' when he started his ministry, which commenced with his baptism by John the Baptist. Luke 3:1-2 carefully states that the ministry of John the Baptist started in the fifteenth year of Tiberius Caesar. Depending on whether Luke used the Julian calendar or the Roman regnal year calendar, the fifteenth year of

Tiberius was Jan. 1-Dec. 31, A.D. 29 or autumn A.D. 28-29, respectively. The Lucan term 'about 30' is a broad term covering any actual age ranging from 26 to 34,21 thus the earliest possible year for the birth of Christ is obtained by subtracting 34 years from A.D. 28, giving 7 B.C. Hence, we can rule out as being too early for the star of Bethlehem the comet of 12 B.C. (Halley's comet) in Table 1, although the 12 B.C. comet has recently been revived as the star of Bethlehem.

The Chinese Catalogue 63, Ho Ping-Yoke 20, records a tailed comet (sei) March-6 April 5 B.C., that was observed for 70 + days with no movement recorded.153

The latest possible year for the birth of Christ is given by the date of the death of king Herod the Great, since Matthew 2:1 state that Herod was king when the star was seen by the Magi. The generally accepted date for the death of Herod the Great is the spring of 4 B.C. although other dates have also been suggested (e.g. 5 B.C.,23 1 B.C.24,25,26,27 and 1 A.D.28). The evidence that Herod died in 4 B.C. is strong, and the accounts in Josephus of the reigns of his three sons, Archelaus, Antipas and Philip, all correlate perfectly with a 4 B.C. date .21,29 Josephus (Antiquities 17:167) records that Herod died between an eclipse of the moon (usually taken to be that of 12/13 March 4 BQ and the following Passover (on 11 April 4 BQ. Josephus also describes that following the death of Herod his funeral occurred, then a seven-day mourning period, then demonstrations against his son Archelaus and then the Passover. Thus, the latest date for the death of Herod is the end of March 4 B.C. and hence the comet that appeared in April 4 B.C. is too late to be the star of Bethlehem. In addition, the Chinese records give no details of the 4 B.C. comet (e.g. its duration) hence it was probably short-lived and insignificant.

Colin Humphreys believes evidence from the Bible and astronomy suggests that the Star of Bethlehem was a comet which was visible in 5 B.C. and described in ancient Chinese records. He feels the evidence points to Jesus being born in the period 9 March-4 May, 5 B.C., probably around Passover time: 13-27 April, 5 B.C. Birth in the spring is consistent with the account in Luke that there were shepherds living out in the fields nearby keeping watch over their flock by night. Birth in 5 B.C. also throws light upon the problem of the census of Caesar Augustus. A new chronology of the life of Christ is given which is consistent with the available evidence. This chronology suggests that Christ died close to his 37th birthday."154

The Magi therefore set off and went to Jerusalem to King Herod and asked 'where is the one who has been born king of the Jews'

---

153 Mark, Kidger. "Chinese and Babylonian Observations". Retrieved 2008-06-05.

154 Colin Humphreys: The Star of Bethlehem, From Science and Christian Belief, Vol 5, (October 1995): 83-101 Used by permission.

(Matthew 2:2) The legend that the star guided them to Jerusalem is not required (and Matthew neither states nor implies this): it is suggested that the Magi went to Jerusalem because their interpretation of the 7 B.C. conjunction and the 6 B.C. planetary massing was that a Messiah-king would be born in Israel and the appearance of the 5 B.C. comet told them this had happened. The advisers of Herod told them where: in Bethlehem, according to the prophet Micah.

Again, they did not need guidance from the star, but were, 'overjoyed' when, on this last lap of their journey, the star 'went before them' in the morning sky and when it 'stood over' Bethlehem, where Jesus was born. Bethlehem was a small town, and a few enquiries may have quickly revealed the location of the child recently visited by the shepherds.

Montefiore, Finegan and Hughes have previously suggested that the 'star' may have involved both the 7 B.C. conjunction and the 5 B.C. or 4 B.C. comets, with the Magi setting out in 7 B.C. and arriving in Jerusalem in 5/4 B.C. Hughes has rightly criticized this theory (and withdrawn his own earlier theory, Hughes) as having the 'almost insurmountable difficulty' that the star the Magi saw when they set out, and the star they saw in Jerusalem, should have been one and the same star not a conjunction and a star, since Matthew 2:9 states 'the star, which they saw in the East, went before them' on the final leg of their journey to Bethlehem. We agree: the proposal here is that two events, the conjunction of 7 B.C. and the planetary massing of 6 B.C., alerted the Magi to the coming birth, but they did not set out until the 5 B.C. comet appeared in the east, indicating that the birth was imminent.

How long did their journey take? Hughes has noted that Lawrence of Arabia in The Seven Pillars of Wisdom states that in 24 hours a fully loaded camel can cover 100 miles if hard pressed and 50 miles comfortably. The furthest the Magi are likely to have travelled is from Babylon to Jerusalem, a distance of about 550 miles going directly across the Arabian desert and about 900 miles travelling via the Fertile Crescent. Wiseman38 has shown that as crown-prince, Nebuchadrezzar took 23 days to travel from north of Jerusalem to Babylon in a rapid return to take up the throne in Babylon in 605 B.C. Hence allowing one or two months for the journey, including preparation time, seems not unreasonable, so that if the Magi commenced their journey soon after the comet appeared in March/April 5 B.C., they would have arrived in Jerusalem in April 4-June 5 B.C. As noted in Section 5, this journey time for the Magi is consistent with the probable time for the comet to move from east to south and with the 70+ days visibility of the 5 B.C. comet noted in the Chinese records.

The theory proposed here fits well with Herod giving orders to kill all the boys in Bethlehem who were two years old and under 'according with the time he had learned from the Magi' (Matthew 2:16). Earlier Herod had asked the Magi 'the exact time the star had appeared' (Matthew 2:7).

It is suggested that the Magi spoke with Herod when they arrived in Jerusalem in April/June 5 B.C. and recounted not only the appearance of the comet about one month previously but also described the significance of the planetary massing in 6 B.C. and the triple conjunction of Jupiter and Saturn in May, October and December 7 B.C. Herod, leaving nothing to chance, decided to kill all boys born since the first stage of the triple conjunction in May 7 B.C., two years previously.

The 5 B.C. date for the star of Bethlehem also fits well with the textual evidence for the length of stay of Jesus and his family in Egypt. According to Matthew 2:13-15, after the Magi had left Bethlehem, Joseph was warned that Herod planned to kill Jesus, so the family left for Egypt (a classic refuge for those trying to flee the tyranny of Palestine) and returned after Herod died. Both Origen and Eusebius state that Jesus and his family were in Egypt for two years, and they returned in the first year of the reign of Archelaus. Archelaus, one of Herod's sons, started his reign when Herod died. Thus, if Herod died at the end of March, 4 B.C., the first year of the reign of Archelaus was from April 4 B.C. to April 3 B.C. Jesus and his family probably left for Egypt shortly after the Magi left Bethlehem, in about April- June 5 B.C. If they stayed in Egypt a reasonable time after the death of Herod, to be absolutely sure of the news, they could have returned to Israel in, say, March 3 B.C., when travelling conditions would be good, in the first year of Archelaus and having spent about two years in Egypt. Thus the 5 B.C. comet is consistent chronologically with both Herod's massacre of the infants and the two year stay in Egypt.

The Problem of the Census

According to Luke 2:1-5, a census was taken by Emperor Caesar Augustus around the time of the birth of Christ, and Joseph travelled with Mary from Nazareth to his hometown of Bethlehem in order to register. This census is one of the thorny problems of the New Testament about which much has been written. There are three well-documented censuses conducted by Augustus: in 28 B.C., 8 B.C. and A.D. 14, but these were apparently only for Roman citizens.

In addition, there are various records of provincial censuses under Augustus for non-citizens for purposes of taxation, for example in A.D. 6, a decade after the death of Herod the Great, Josephus refers to a census in Judea administered by Quirinius the governor of Syria, and Luke also refers to this census (Acts 5:37).

So how do we resolve this problem? There is no record of a census for taxation purposes in Judea around the time of the birth of Christ, hence the problem in interpreting Luke 2:1-5.

The 5 B.C comet theory resolves the problem by suggesting the census was not for taxation purposes but was instead a census of allegiance to Caesar Augustus (some translations of Luke 2:1-5 refer to taxation, but this is not implied in the Greek text). The fifth century historian Orosius (Adv. Pag. VI.22.7, VII.2.16) states '[Augustus] ordered that a census be taken of each province everywhere and that all men be enrolled. So at that time, Christ was born and was entered on the Roman census list as soon as he was born. In this one name of Caesar all the peoples of the great nations took oath, and through the participation in the census, were made part of one society'.

Josephus (Ant. XVII, ii, 4) appears to refer to the same event: 'when all the people of the Jews gave assurance of their goodwill to Caesar, and to the king's government, these very men [the Pharisees] did not swear, being above six thousand.' From the context of these words in Josephus, this census of allegiance to Caesar Augustus occurred about one year before the death of Herod the Great.

There is one further problem with this census. Luke 2:2 is usually translated 'This census was first made when Quirinius was governor of Syria', but Quirinius did not become governor of Syria until 6 B.C.

However, the Greek sentence construction of Luke 2:2 is unusual, and an alternative translation is 1,39: 'This census took place BEFORE the one when Quirinius was governor of Syria'.155 As noted above, from Josephus this latter census can be dated to A.D. 6, and Luke (Acts 5:37) was well aware of it. Thus the earlier census reference by Luke in Luke 2:1-5 provides a chronological clue to the birth of Christ, and from the context in Josephus (Ant. XVII, ii, 4) this census of allegiance occurred about 1 year before e death of Herod the Great, which is consistent with our placing the birth of Christ in the spring of 5 B.C.

5 B.C Revised Chronology of the Birth and Death of Jesus Christ

Date in Julian Calendar	Event
9 March-4 May, 5 B.C.	
	Birth of Jesus in Bethlehem. Matthew 2:1, Luke 2:1-7 (13-27 April, 5 B.C.)

155 Bruce, F. F., 1969, New Testament History, Nelson, London

9 March-4 May, 5 B.C.
Visit of the Shepherds Luke 2:8-20

13-27 April, 5 B.C.
Circumcision on the eighth day (counting inclusively). Luke 2:21.

16 March-1 I May, 5 B.C.
Presentation of Jesus at the Temple in Jerusalem after 40 days from birth. Luke 2:22-24.

20 April-4 May, 5 B.C.
Visit of the Magi. Matthew 2:2-12.

24 May-8 June, 5 B.C.
Late April/mid June, 5 B.C. - Flight to Egypt from Bethlehem. Matthew 2:13-15.

End March, 4 B.C.
Death of Herod. Matthew 2:20-23

March, 3 B.C.
Return from Egypt to Nazareth. Matthew 2:20-2

Autumn A.D. 29
Baptism of Jesus when he was 33.

Friday, 3 April, A.D. 33
Crucifixion on 14 Nisan when Jesus was near his 37th birthday. All four gospels.

If Jesus was born in the Spring of 5 B.C. then he would have been 33 when he commenced his ministry. This is consistent with Luke 3:23 that he was 'about thirty' at this time. (As noted above, the Greek translated 'about thirty' means any age between 26 and 34). If the crucifixion was on 3 April A.D.33, 27,29,40,41 then Jesus was around his 37th birthday when he died.

12 B.C Theory

In July, 12 B.C, a new star appeared after sunset in the Eastern sky over Jerusalem. As the days passed, the star grew brighter and developed a long, feathery tail that spanned much of the heavens. The new heavenly visitor was Halley's Comet, last seen by the Chinese in 86 B.C., returning from its usual 75-year journey around the Sun.

The curious terminology in Matthew 2:9 that the star 'stood over' Bethlehem will now be considered. Phrases such as 'stood over' and 'hung over' appear to be uniquely applied in ancient literature to describe a comet, and no other astronomical object. The historians Dio Cassius and Josephus were broadly contemporary with the author of Matthew's gospel. Dio Cassius (Roman History, 54, 29) describes the comet of 12 B.C. (Halley's comet) which appeared before the death of Marcus Agrippa: "The star called comet hung for several days over the city [Rome] and was finally dissolved in flashes resembling torches."

The apparition of 12 B.C. was recorded in the Book of Han by Chinese astronomers of the Han Dynasty who tracked it from August through October.156 The Chinese Catalogue 63, Ho Ping-Yoke 20, records Halley's Comet (Po) on 26 August – 20 October, 12 B.C. It was observed for 56 days, traversing through the constellations of Gemini to Scorpio. An imperial edict following its appearance read, "Lately, reproaches in the form of solar eclipses and meteors have been in the sky. These great strange signs were repeated and yet those in official positions remained silent; rarely has there been loyal advice. Now a bushy star has been seen in Tung-chin. We are very dismayed. The ministers, grandees, doctors and advisors are each to think solemnly as to the meaning of these changes and compare them clearly with the Classical texts: nothing is to be concealed..."

The 12 B.C flyby of Halley's Comet was very close. It passed within 0.16 AU of Earth, just slightly more than the 1909-1911 spectacular display, when Comet Halley reached a total apparent brightness of about magnitude -1 and passed within about 0.15 AU [23 million km: 14 million miles] of Earth.) If estimates of Halley's brightness during past appearances are a guideline, a .16 AU approach in 12 B.C., assuming the comet behaved as it did in 1909-1911, would have been spectacular.

Luke 2:8 states that at the time of the birth of Jesus 'there were shepherds living out in the fields nearby, keeping watch over their flock by night'. Bethlehem is cold and very wet during December, January and February and flocks of sheep were not normally kept in the fields in these months. Sheep were usually put out to grass between March and November, the shepherds being with the flocks at night particularly tending their sheep before the fields were plowed, and winter rains began soaking the parched ground.

In about A.D. 1303, Giotto painted a comet above the head of the infant Jesus in a fresco in the Arena Chapel in Padua, presumably using

---

156 G. W. Kronk. "1P/Halley". cometography.com. Retrieved 13 October 2008

as a model the A.D. 1301 appearance of Halley's comet (the European Space Agency mission to Halley's comet in 1985/6 was called 'Giotto' in commemoration of the artist's nativity comet). Adoration of the Magi by Florentine painter **Giotto di Bondone** (1267–1337). The Star of Bethlehem is shown as a **comet** above the child. Giotto witnessed an appearance of **Halley's Comet** in 1301.

Comet Standing over Bethlehem

Hence we interpret Matthew's description of a star, standing over' the place where Jesus was born as meaning that when the Magi left Herod and headed towards Bethlehem, as he had suggested, they looked up and saw the comet in front of them, with a near vertical tail, the head of the comet appearing to stand over Bethlehem.

Census problem resolved. New evidence may overturn the requirement of retranslating Luke. In 1985, Biblical scholar, Jim Fleming, dean of the Jerusalem Center for Biblical studies, announced new evidence of a Census done in 12 B.C during a lecture at Hebrew University in Jerusalem. His thesis was based on an unpublished work by Jerry Vardaman, professor of archaeology at the Cobb Institute of Archaeology at Mississippi State University concerning a recently deciphered Tablet known as the Aemilius Secundus inscription, discovered 300 years ago in 1685 in Beirut, Lebanon. The document, which now resides in Italy, at the Venice Museum, found a census was ordered by Quirinius, the Governor of Syria in 12 B.C., the same year Halley's Comet lit up the skies.

Notes: Jack Finagen points out that as well.

Jesus was not a baby when the Magi visited Herod the Great in 10 B.C

While tradition depicts three wise men bearing gifts, gold, frankincense, and Myrrh, it is likely they were not present at the Nativity scene. The wise men (greek Magios) were astrologers from the east in Persia (notice the Greek to English translation considers these astrologers wise), but not royalty or kings. The astrologers did not journey to the side of a newborn in a manger but arrived when Jesus was a child and was living in a house. Look closely at the account of Jesus' birth in Luke 2:8-16: "There were shepherds living out of doors and keeping watches in the night over their flocks. And suddenly God's angel stood by them, and . . . said to them: '. . . You will find an infant bound in cloth bands and lying in a manger.' . . . And they went with haste and found Mary as well as Joseph, and the infant lying in the manger. Only

Joseph, Mary, and the shepherds were present with baby Jesus. No one else is listed in Luke's report.

Mathew 2:1-11 (KJV) indicates the visit by the "wise men" not "three wise men" happened later when Jesus was a "young child."

"Now when Jesus was born in Bethlehem of Judea in the days of Herod the king, behold, there came wise men from the east to Jerusalem . . . And when they were come into the house, they saw the young child with Mary his mother." They first traveled from the east to Jerusalem, not to the birth city of Jesus, Bethlehem. By the time they finally reached Bethlehem, Jesus was a "young child"—no longer a baby—and no longer in a stable but in a house "And being warned of God in a dream that they should not return to Herod, they departed into their own country another way. And when they were departed, behold, the angel of the Lord appeareth to Joseph in a dream, saying, Arise, and take the young child and his mother, and flee into Egypt, and be thou there until I bring thee word: for Herod will seek the young child to destroy him." (Matthew 2:12-13)

But here's the critical, astonishing part. Luke indicates that Joseph and Mary went directly into Jerusalem just after Jesus was born! Luke says, "And when the days of her purification according to the law of Moses were accomplished, they brought him to Jerusalem, to present him to the Lord; ... And to offer a sacrifice." (Luke 2:22-24). They did this because the wise men had not yet paid a visit to Herod and felt safe to do so.

If the wise men visited Herod the Great in 10 B.C, two years after the birth of Jesus in 12 B.C, to inquire about the birth of "a king" foretold by the dazzling comet, it would make sense that Matthew called Jesus a "young child," a toddler who may have begun speaking.

When Herod realized the wise men were not coming back to him to tell him where Jesus was, he was enraged: "Then Herod, when he saw that he was mocked of the wise men, was exceeding wroth, and sent forth, and slew all the children that were in Bethlehem, and in all the coasts thereof, from two years old and under, according to the time which he had diligently enquired of the wise men." (Matthew 2:16)

"The Bible goes out of its way to indicate Herod clearly believed that Jesus was not a newborn babe by the time the wise men had arrived." He slaughtered all the young boys up to two years old. If this meeting between Herod and the Magi recorded by historian Josephus is accurate, it is another arrow pointing to Halley's Comet and the birth of Jesus in 12. B.C.

Redating Jesus' Age: Jesus was 42 or 45

If Jesus was born in 12 B.C., and died on the Cross on the 14th of Nisan, 33 A.D., he would have been 45 years old. Is this hinted at in the Gospels? John 8:56-57: "Before Abraham Was, I Am.....Your father Abraham rejoiced to see My day, and he saw it and was glad." So the Jews said to Him, "You are not yet fifty years old, and have You seen Abraham?" Jesus said to them, "Truly, truly, I say to you, before Abraham was born, I am."

John 2:18-21: "Then answered the Jews and said unto him, What sign shewest thou unto us, seeing that thou doest these things? Jesus answered and said unto them, Destroy this temple, and in three days I will raise it up. Then said the Jews, Forty and six years was this temple in building, and wilt thou rear it up in three days? But he (Jesus) spake of the temple of his body."

We're all aware our body is a temple of God (1 Peter 2:5, 1 Corinthians 3:16-17, 1 Corinthians 6:19, 2 Corinthians 6:16, Ephesians 2:21-22) - but here we have Jesus comparing His own body to the Second Temple Herod the Great who had begun repairing or rebuilding the Second Temple in the eighteenth year of his reign 37 B.C. – 4 B.C., in 20-19 B.C, seven or eight years (using 12 B.C birth date) before the birth of Christ. Josephus reports (Josephus, Antiq. 15:38-425; Wars: 184-247) some 10,000 (to 18,000) workmen were employed, plus 1000 priests, since only priests were allowed to work on the sanctuary proper. The major work occurred in the first three years and finished by 16 B.C., although the workers continued improvements well beyond Herod's death in 4 B.C.E. to 64 C.E., just four years before it was destroyed by the Roman Emperor Titus.

The beginning of Christ's ministry began either in 27 A.D or 30 A.D and continued for three and a half years.

19 B.C. – 27 A.D is 46 years. (Reference to Temple construction in John: 2:22 of 46 years) A 12 B.C. birth makes Jesus 39 years old in 27 A.D., with the crucifixion occurring on the 14th of Nisan in 30 A.D. when he was 42 years old.

19 B.C. – 30 A.D.. Is 49 years. (Reference to Temple construction in John: 2:22 of 46 years) A 12 B.C birth makes Jesus 42 years old, with the crucifixion occurring on the 14th of Nisan in 30 A.D. when he was 45 years old.

The common explanation is Herod began to rebuild the temple sixteen years before the birth of Jesus 1 A.D, and what is here mentioned happened in the thirtieth year of the age of Jesus, so the time which the Temple had been occupied was "forty-six years," which is not accurate.

12 B.C Revised Chronology of the Birth and Death of Jesus Christ

Julian Calendar	Event

20-19 B.C.

Herod begins reconstructing the Temple (second Temple)

26 August – 25 October, 12 B.C.

Birth of Jesus in Bethlehem. Matthew 2:1, Luke 2:1-7

26 August – 25 October 25, 12 B.C.

Visit of the Shepherds Luke 2:8-20

26 August – 25 October 25, 12 B.C.

Circumcision on the eighth day (counting inclusively). Luke 2:21.

6 October – December 4,12 B.C.

Presentation of Jesus at the Temple in Jerusalem after 40 days from birth. Luke 2:22-24.

1 January - 31 December, 10 B.C.

Visit of the Magi. Matthew 2:2-12. Recorded by Josephus 10 B.C.

1 January – 31 December, 10 B.C.

Flight to Egypt from Bethlehem. Matthew 2:13-15.

March - 4 B.C.

Death of Herod. Matthew 2:20-23.

March - 3 B.C.

Return from Egypt to Nazareth. Matthew 2:20-2

1 A.D. ?

Jesus teaches on the steps at the Temple at age 12.

Autumn - 27 A.D.

Baptism of Jesus. Ministry begins. Jesus is 39.

Or

Autumn - 30 A.D.

Baptism of Jesus. Ministry begins. Jesus is 42.

Friday, 3 April, 33 A.D.

Crucifixion on 14 Nisan when Jesus was 45, 6 months from his 46th birthday.

If Jesus was born in the fall of 12 B.C. then he would have been 39 in 27 A.D., when he commenced his ministry in the summer/fall of 27 A.D. While this age appears inconsistent with Luke 3:23 that he was 'about thirty' at this time. Furthermore, Rabbis were not typically considered qualified to teach until 40 years of age. (As noted above, the Greek translated 'about thirty' means any age between 30 and 39). If the crucifixion was on 3 April 33 A.D. then Jesus was 45, (which is remarkable implication Jesus spoke of the Temple of his body.)

Wise Men knew Halley's Comet would return in 66 A.D.?

Since Halley's visit in 12 B.C, historians have recorded every perihelion. These occur every 74-80 years, its irregularity due to perbutations by Jupiter and Saturn. The next appearance was January, 66 A.D., recorded by Josephus (Jewish War 6,5,3) who states 'a star, resembling a sword, stood over the city Jerusalem]'. Halley's Comet is referenced by in 66 A.D by Tacitus (Annals, 15,47), when Rome was visit to Rome was visited by a delegation of magi, about the time the Gospel of Matthew was being composed. This delegation was led by King Tiridates of Armenia, came seeking confirmation of his title from Emperor Nero. Ancient historian Dio Cassius wrote that, "The King did not return by the route he had followed in coming," a line echoed in Matthew's account. The same year Jewish historian Flavius Jesephus reported, 'A comet of the kind called Xiphias, because their tails appear to represent the blade of a sword' was seen above Jerusalem before its fall."

The 66 A.D. apparition of Halley's Comet presaged the destruction of Jerusalem and the Temple in 70 A.D. While some believe the destruction of Jerusalem fulfilled Isaiah's and Revelation's prophecies, there is good evidence John the Evangelist fuses the Jerusalem destruction, with a much later future catastrophe caused by a "mountain of stone" falling from the heavens, impacting the Earth destroying 1/3 of the planet; the entire event coincides with earthquakes, a darkening of the sun and moon; events which obviously did not occur in A.D. 70.

What is even more interesting is that the Book of Revelation itself that gives strong if not convincing evidence that St. John (and other learned men) knew the brilliant comet which stood over Bethlehem in 12 B.C. at the birth of Christ, would return and shine over Jerusalem within his lifetime.

In Rev. 2:16 St. John warns: "Repent; or else I will come quickly [Greek tachu] and will fight against them with the sword (comet) of my mouth" And again in Rev. 22.7: "Behold, I come quickly [Greek tachu]: blessed [is] he that keepeth the sayings of the prophecy of this book" (Rev.22:7).

These words, in their various tenses, are translated as "shortly," and "quickly." The words do not mean "soon," in the sense of "sometime," but rather "swift," "now," "immediately," "hastily," and "suddenly." The word meanings here are critical to understanding the "imminency" that is being communicated in the vision of the book! Parts of his vision is NOT something that would be expected to take place two thousand, or more years into the future!

John uses other Greek words, eggus, Rev. 1:13 and Rev. 22.10, mello, and mellie, in Rev. 1:19 and Rev. 3:10, meaning "is about to come." When these words are used with the aorist infinitive the preponderance of use and preferred meaning is "be on the point of, be about to," portraying an expectation of soon or quick future occurrence.

This kind of language should lead us to conclude that the prophecy in the vision was something that was to take place very close to its being revealed to John. I see this as being fulfilled by the appearance of Halley's Comet in 66 A.D., and his warning of the destruction of Jerusalem, which occurred 3.5 years later in 70 A.D.

Is there historical evidence the ancients knew about the cycle of Halley's Comet?

An interesting tale comes to us from the 1st century CE that Rabbi & Jewish Scholar Yehoshua Ben Hananiah may have known something of "a star that appears every 70 years." The tale, as told in the Horayoth (rulings) of the Talmud and described in Mr. Livio's blog is intriguing:

"Rabbi Gamliel and Rabbi Yehoshua went together on a voyage at sea. Rabbi Gamliel carried a supply of bread. Rabbi Yehoshua carried a similar amount of bread and in addition a reserve of flour. At sea, they used up the entire supply of bread and had to utilize Rabbi Yehoshua's flour reserve. Rabbi Gamliel then asked Rabbi Yehoshua: "Did you know that this trip would be longer than usual, when you decided to carry this flour reserve?" Rabbi Yehoshua answered: "There is a star that appears every 70 years and induces navigation errors. I thought it might appear and cause us to go astray."

The Rabbi's assertion is a fascinating one. There aren't a whole lot of astronomical phenomena on 70-year cycles that would have been noticeable to ancient astronomers. With an orbital period of 75.3 years, Halley's Comet seems to fit the bill the best.

Certainly, the appearance of Halley's Comet at the birth of Jesus would have been common knowledge recognized by the Rabbi's in Jerusalem, knowledge passed by Joseph and Mary, and later to the apostles and St. John.

This thesis is intriguing because it requires St. John to have penned the Book of Revelation before 66 A.D. In this fresco by Giotto di Bondone, the Star of Bethlehem is shown as a comet - likely the most famous of all, Halley' Comet. This painting by Florentine artist Giotto di Bondone, who lived from 1267 to 1337, is titled Adoration of the Magi and was painted around 1305. It's one of his many frescoes depicting the lives and Mary and Christ in the building today known as the **Arena Chapel** in Padua, Italy. What's special about this particular one, is the "star" painted above the manger.

According to the Bible, the Wise Men followed a star which stopped above the place where Jesus was born. But if you look closely, you'll see Giotto painted a comet, not a traditional star. And it's a good resemblance, too. A large, ball-like head or coma with a pointed, upward-slanting tail. Looking back through my own comet sketches, more than a few resemble Giotto's, though mine lack his artistic sensibility and color.

I suppose you could argue that it also resembles a meteor, but it's unlikely that's what Giotto had in mind. A meteor, however bright, flashes by in a second or two and is gone. Assuming the artist wanted to hold true to the Biblical narrative, the "star" would need to hover and stay lit for some time. That's just what comets do – they look meteoric but move slowly across the sky orbit like the planets do.

Scholars think it likely Giotto was inspired by none other than Halley's Comet. It appeared low in the northwestern sky over Italy only a few years earlier in the fall of 1301. Compare Giotto's painting with a photograph taken under similar circumstances during its swing by Earth in 1901. There are good similarities between the two appearances. Who wouldn't have been inspired by the appearance of a comet in the evening sky? With light pollution far off in the future, Italian skies would have been as dark as the best rural skies are today. You couldn't miss it at dusk. Halley's Comet would have been the talk of the town that fall. While our ancestors generally gave comets a bad rap, blaming them for every **pestilence and ill fortune**, they were also seen as signs of change, sometimes for the good as when a new king ascended the throne. It's not far-fetched that Giotto might choose a comet, especially one with which he was familiar.

As a symbol of change for the Star of Bethlehem in his painting. Interestingly, no one at the time knew they were seeing Halley's Comet. People had no idea comets orbited the sun and reappeared after a span of years. They were unpredictable, fiery, one-off phenomena thought to be part of the atmosphere, not frozen dust balls obediently following the laws of celestial mechanics. It was Englishman Edmund Halley, using Newton's newly formulated laws of gravity, who found that the comets of 1531, 1607 and 1682 were different appearances of the same comet. In his honor, it was named **Halley's Comet**.

When Giotto painted Halley's Comet, he didn't know that Halley had also appeared in the year 12 B.C., within a half dozen years of the best estimate of Jesus' birth. The other comes to us from American writer and humorist **Mark Twain**, who was born when Halley passed by Earth in 1835 and died in 1910, the year of its next return. In his own words:

Mark Twain born and died when Halley's during the flyby of Halley's comet. This is a somewhat uncommon occurrence given that Halley's comet only passes by the Earth approximately every 76 years. What makes it even more remarkable is that Clemens predicted the year of his death. In 1909, Clemens—known by his pen name, Mark Twain—said, "I came in with Halley's Comet in 1835. It is coming again next year, and I expect to go out with it. It will be the greatest disappointment of my life if I don't go out with Halley's Comet. The Almighty has said, no doubt: 'Now here are these two unaccountable freaks; they came in together, they must go out together." Clemens was indeed born just after Halley's Comet appeared in 1835, and he died of a heart attack one day after it appeared at its brightest in 1910.[157]

~Chronicle 994, Wednesday, March 2, 2016:

9:30 am. 5 minutes before Delynn walked in, I... finally... discovered the meaning of the number 82. In Chronicle 555, December 20, 2014, I wrote: "I had been pondering again **the meaning of the 82 days**...Both A.D.C dreams of Mr. Takanohashi (August 8 & October 29, 2014) and Paul Solomon visitation and dream (August 10 & October 31, 2014) were separated by 82 days. Lead is #82 on the periodic table. I hate being too symbolic. But were the ghosts of my teachers trying to say: convert lead of my cancer into a gift of gold? Well, my friends, I have news for you. You give it a whirl! December 20, 2014. Halley's comet's approach in 466 B.C., then looked for ancient texts that matched its probable appearance. The previous earliest recorded appearance - or, as it's more technically known, apparition - was by Chinese scholars noting the passage of the comet in 240 B.C., which is three apparitions later than the 466 B.C. event. The model suggests the comet would have been visible for about **82 days in 466 B.C.**, from the ancient equivalent of June 4 to August 25.

Amazingly, the ancient Greeks recorded an event that pretty much perfectly matches that description. Authors from that year described a wagon-sized meteor that struck northern Greece one day, which quite understandably terrified the neighboring population and created one of the ancient world's most popular tourist destinations. Significantly, the authors note the presence of a comet in the sky at the time the meteor struck, and the comet remained in the sky for about 75 days.

Considering atmospheric conditions might have slightly reduced **the 82-day presence of Halley's Comet**, that's an excellent match. It gets better. The records say the comet appeared in the western sky, and it came at a time of high winds and shooting stars. Strong winds are

---

157 http://www.todayifoundout.com/index.php/2013/10/samuel-clemens-born-died-year-halleys-comet/

common in Greece during the month of July, and Halley's Comet moved into the western sky on July 18.

At that time, Earth would have been directly under the comet's debris field, which explains the shooting stars. As for that wagon-sized meteorite, it's possible the comet pushed an asteroid off course and made it hit Earth, but as researcher Eric Hintz explains, it's probably more likely that "it was just a really cool coincidence."

So that's the earliest known record of Halley's Comet, at least for now. The comet has probably been in its current orbit for anything from 16,000 to 200,000 years. Indeed, if it's the latter, then Halley's Comet has been visiting Earth since the first emergence of anatomically modern humans, which would be another really cool coincidence. Either way, Halley's Comet has been visiting humanity for a long time, and writing predates the 466 B.C.E apparition by at least three millennia, so it's possible an even earlier sighting will turn up in the records of ancient Sumer or Egypt.

"And then shall appear the sign of the Son of man in heaven; and then shall all the tribes of the earth mourn, and they shall see the Son of man coming in the clouds of heaven with power and great glory." Matt: 24:30 ~The Son of Man is the Celestial Messiah.

Truth: Mt 23:34; 24:34 speaks about how the Herodian temple in Jerusalem will be destroyed within one generation. This indeed came to pass with the destruction of Jerusalem in 70 A.D. by Titus. Notice the parallel: "But when you see Jerusalem surrounded by armies, then recognize that her desolation is near. " (Luke 21:20). ~**Chronicle 555**

My discovery occurred when I read the following verse in Chapter 21: Halley's Comet: Messiah Messenger Heralded by Gabriel:

Babylon and the Bible

After these first recorded appearances in 466 and 240 B.C.E, Halley's Comet starts appearing with regularity every 75 or 76 years. Babylonian tablets mark its apparitions in both 164 and 87 B.C.E, and that second appearance may actually have been recorded in the local currency. A coin featuring the Armenian king Tigranes the Great features a star with a curved tail on his crown, and there's some thought that this is meant to be Halley's Comet.

Tigranes could have seen the comet during its closest approach to the Sun on August 6, 87 B.C.E, in the eighth year of his reign, and uses its brilliant appearance as a sign that his rule marked the beginning of a new era, a time of the king of kings.

Speaking of which, there's some thought that the comet's apparition in 12 B.C.E provided the basis for the biblical tale of the Star of

Bethlehem. Although it's not impossible, there's nothing about the New Testament's descriptions that clearly indicate it was Halley's Comet, and there were other comets that passed Earth closer in time to the assumed date of Jesus's birth. Still, even if it didn't inspire the original Star of Bethlehem, its 1301 appearance probably did influence Giotto di Bondone's depiction of the star in his 1305 depiction of the Nativity.

The Battle of Hastings

Halley's Comet made its closest approach in 837 CE, coming within 3.2 million miles of Earth. That's only about ten times the distance between the Earth and the Moon, and its tail would have stretched across about a third of the night sky. The comet made the ancient equivalent of international news, showing up in records from Japan, China, the Middle East, and northern Europe.

But the comet's most celebrated appearance probably came in 1066. Although not quite as close as the 837 apparitions, it was a remarkable sight, four times bigger than Venus and about a fourth as bright as the Moon. English astrologers took its appearance as omen for the upcoming battle between the English and invading Normans.

Of course, it was only after the Battle of Hastings that the English learned what type of omen it was - a good one for the new king William the Conqueror, but a very bad one for the now dead King Harold II. William the Conqueror certainly liked the look of it - he's said to have called it a "a wonderful sign from heaven" when he sighted it off the French coast and took it as final proof that his invasion was destined to succeed.

How do we know it was actually Halley's Comet that these people saw? The iconic Bayeaux Tapestry features the comet, attracting fearful attention from both commoners and King Harold himself, who seems to cower in its presence. (Although one could be forgiven for thinking the comet looks more like a strange alien craft come to tip the battle in the Normans' favor, but that's just how celestial phenomena tended to be drawn back then.)

As a final bit of proof, the long-lived monk and astrologer Eilmer of Malmesbury - who is rather awesomely best known for trying to fly using a pair of artificial wings - wrote of Halley's Comet as the destroyer of Anglo-Saxon Europe in 1066 while also recognizing it as the same comet that had appeared 75 years before in 989:

"You've come, have you? ... You've come, you source of tears to many mothers, you are evil. I hate you! It is a long time since I saw you; but as I see you now you are much more terrible, for I see you brandishing the downfall of my country. I hate you!"

Genghis Khan's inspiration

But spurring on the Norman invasion of the British Isles wasn't enough for Halley's Comet - its encore in 1222 was even more world-changing. Genghis Khan is said to have considered the comet his own personal star, and its westward trajectory inspired him to head west himself, launching his invasion of southeastern Europe that would leave millions dead. In fairness to the comet, invading Europe is the sort of idea Genghis Khan probably would have thought up without any celestial intervention.

Beginning in the sixteenth century, the comet moved from the domain of astrology to that of astronomy. In 1705, British astronomer Edmond Halley published Synopsis Astronomia Cometicae, demonstrating the comets of 1456, 1531, 1607, and 1682 were all one and the same object, and it would return again in 1758. He didn't live to see that he was correct, but he at least gained the immortality of having the comet named after him.

The death of Mark Twain

Anyone who lives a decently long life should have the opportunity to see the comet for themselves, although very few people actually plan their lives around its apparitions. Mark Twain, as with so many other things, was an exception. The legendary humorist was born two weeks after it appeared in 1835, and he said it would be the greatest disappointment if he didn't die when it returned in 1910. Thankfully for him, he kept his promise, dying a day after its return.

Unfortunately, practically no one alive today has firsthand experience with the full brilliance of Halley's Comet - its 1986 appearance was the dimmest ever recorded. Of course, the people of 1986 had one thing going for them that their predecessors didn't - they could send up spaceships to get a look up close at the comet. Tragically, the ill-fated Challenger mission was slated to investigate the comet for NASA, so it fell to the rest of the so-called Halley Armada, made up of Soviet, European, and Japanese probes, to investigate the comet. You can see one of the images snapped in 1986 above.

Tracking Halley's Comet

These days, Halley's Comet can be found in the outer solar system, its eccentric orbit taking it far from the orbital plane on which the planets revolve. You can track its current progress through the solar system here. Those waiting to see Halley's Comet again—or those who weren't around in 1986 - have got another 51 years to wait, as it won't be back until July 18, 2061.

Thankfully, the 2061 apparition will be quite a bit more impressive than the 1986 vintage, and those planning on living another 124 years should be in for a real treat on May 7, 2134, when the comet comes within 14 million miles of Earth. Who knows what future invasions it'll predict, future generals it will inspire or terrify, and authors it will kill off. Whatever happens, Halley's Comet should definitely be worth waiting for.

Part V

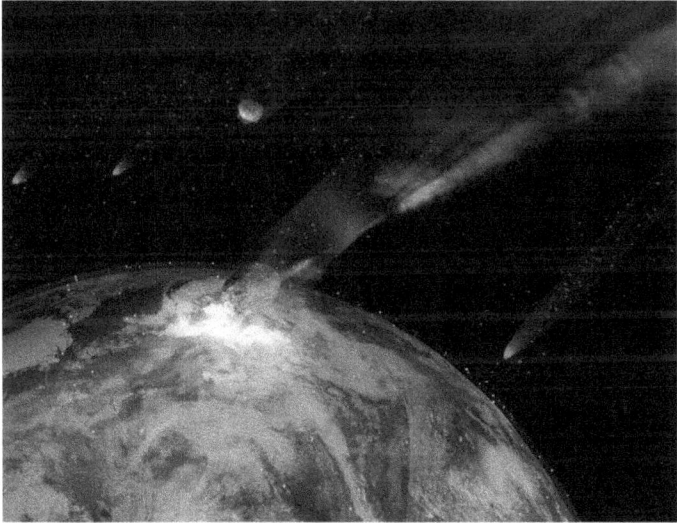

Impact Science

—— 14 ——

Houston We Have a Problem

Time Magazine's dramatic cover story of *Comet Crash* in May 1994 featured the predicted collision of comet Shoemaker-Levy 9 with Jupiter. Two months later, the comet, snagged by the immense gravitational pull of the giant planet, disintegrated into 23 mountain-sized chunks of ice and rock fragments, nearly 3 kilometers in diameter, and headed ominously for the Jovian South Polar region. Just before the fragments hit, some scientists were predicting the celestial show would fizzle. The comet debris, they said, was simply "weakly cohesive water-ice," incapable of producing a major explosive event. But, in fact, the impacts defied comprehension. Over a five-day period, legions of stargazers watched in amazement as telescopes observed brilliant nuclear-sized explosions with a combined energy reaching 21 million megatons, nearly 500,000 times larger than the Soviet Union's 58-megaton blast in 1963.[158] The collision of comet Shoemaker-Levy 9 served Earth a warning: *it could happen here.*

Many scientists believe it was a large comet such as the one that collided with Jupiter that killed off the dinosaurs and many other terrestrial species 65 million years ago. Evidence corroborating cometary impacts and their catastrophic effects on Earth recently was provided by Dr. Walter Alvarez of the University of California and his Nobel prize-winning father, physicist Luis Alvarez. Based on the periodicity of extinctions noted in the Quaternary / Tertiary rock layers containing the fossil remains of dinosaurs, their 1981 study identified deposits of stishovite (a dense form of silica formed by extreme pressures, only found

---

[158] James Reston, Jr., Collision Course, Time Magazine, May 23, 1994, p. 56

at impact sites), iron-rich H-chondrites (stones of meteoric origin), and a sudden increase in the abundance of the rare earth elements, iridium and osmium, seldom found on the earth's surface, but relatively abundant in meteorites. Walter and Luis Alvarez concluded that the cosmic collision spewed up enough dust and water vapor to block out the sun, triggering worldwide changes in weather, lowering temperatures to near freezing in summer, dimming sunlight and preventing photosynthesis, vital to food production. Scientists now believe two-thirds of the species then living both, plant and animal died, terminating the Cretaceous period and beginning the Tertiary geologic age.

The Alvarez discovery sent a shockwave through the geological and astronomical community crumbling nearly overnight the popular doctrine of gradualism, a belief that all rock formations visible today is the result of gradual, uniform processes we can still observe at work, such as erosion and the accumulation of sediment, an evolution that occurs at steady rates over millions of years. Achieving dominance in the early nineteenth century, uniformitarianism effectively excluded any theory of global catastrophe, not because it was supported by overwhelming evidence, but because certain shapers of scientific opinion wished to sever geology from the biblical tradition of the Great Flood. The latest generation of geologists cautiously employ a diluted form of "neo-catastrophism" to explain certain otherwise mysterious phenomena, such as the extinction of the dinosaurs and the onset of the Ice Ages.

With the floodgates of new thought wide open, astronomers and geologists have begun reexamining Earth's topography, geological strata, and paleontological records, in an attempt to uncover what may have been overlooked. Unlike the moon, where the lack of weathering has left craters undisturbed for millennia, Earth craters are difficult to identify, having been filled with sediment, eroded by wind and rain, or covered by the ocean. Years of painstaking exploration, however, have paid off. Fifteen years ago, only thirty craters had been identified by astronomers.

Today over 140 craters have been catalogued with a discovery rate of five per month. Some have been identified by great circular indentations in the surface which later, filled with water or covered with vegetation, formed depressions more easily recognized from the air than from the ground. Several craters have been pinpointed by the discovery of shocked quartz created by pressure and heat caused by the impacting cosmic fragment. Using this method, three craters on the ocean floor have been located 100 to 200 miles off the U.S. Eastern Seaboard, and nearly all date to a catastrophic meteoric bombardment occurring 33-35 million years ago.

We have proof of some of these huge meteor strikes; the well-known Barringer Crater in Arizona, the Sudbury Igneous Complex which yields the world's largest single source of nickel, believed now by astronomers to be the remnant of an immense iron-nickel meteorite. The huge depression of the Hudson Bay, including the Gulf of St. Lawrence. Canada's Deep Bay Crater, the large Ries Kessel crater in Central Europe, Clearwater Lake, and the declivity of the Western Sahara. In Africa a series of meteor craters litter the continent, the most famous being the Vredefort Dome in South Africa. (Australia and Siberia). A research team from the US geological Survey link a 50-mile-wide crater near Cape Charles, Virginia, to a one-mile-wide meteor which gouged out the Chesapeake Bay 35 million years ago.

Early in 1991, researchers located one of the largest craters ever found a 360-kilometer-wide depression near the Yucatan Peninsula in Mexico. The giant impact site, known today as the *Chicxulub* crater—is now thought to be one of several craters formed from a bombardment of comets which triggered the mass extinction of the dinosaurs. Another crater connected with the extinction, measuring 32 kilometers, has been recently located in Iowa.[159]

It is commonly believed that primordial debris capable of causing such catastrophic collisions to have dissipated to such an extent that the Earth today is relatively safe. But this, we shall find, is a false sense of security.

Impact Danger Underestimated

Five seismic stations were left on the Moon by the Apollo astronauts between 1969 and 1974. "One has failed, but the remaining four have operated since April 1974, and can detect miniscule lunar surface vibrations.

An impacting meteorite creates a distinctive signal, quite different from that of a moonquake, and boulders of a few kilograms can be detected landing anywhere on the Moon. From these impacts, variations in collision rates can be found (seen in figure____). The boulder impact rate is clearly non-random, and the seasonal peaks can be associated with known meteor streams."[160]

The outstanding peaks occurs in June and November, when the Earth-Moon system runs through the Beta Taurids and Leonid meteor showers. The Beta Taurid peak, a meteor stream associated with comet

159 Science, 15 November, 1991, p. 943.

160 Clube, The Cosmic Serpent, pg. 148

Encke, occurs June 25-30. The Leonid peak, a meteor stream associated with comet Tempel-Tuttle, occurs November 16-18.

Numerous impacts registered by the Moon's seismic monitors indicate both meteor streams contain large boulders along the track of each comet.

Great swarms of ice and rocky chunks of Genesis material still roam our solar system, debris left over from swirling clouds of gas and dust that formed our Sun and planets nearly 4 ½ billion years ago. And when one of those icy comets or space rocks hits a planet, bad things happen.

66 Impacts a Day

An estimated 24,000 meteorites weighing an average 3.5 ounces survive the fiery passage through the atmosphere and strike the Earth annually--nearly 66 per day! Fortunately, two-thirds of that number fall into the oceans. In addition, an estimated 100,000 tons of cosmic dust called Brownlee particles is deposited every year to the stratosphere, on polar ice caps, and in ocean sediments.

According to Gregory H. Canavan of the Los Alamos National Laboratory, "there's a one chance in 1,000 you'll be killed by a cosmic bomb. "That's equal to, or greater than, your chance of being killed in an airplane crash, averaged over your lifetime," he said. "So, this is not a negligible threat."[161]

Apparently not. In October 1953, a tiny meteorite about the size of a grapefruit streaked down over Alabama and killed a woman who was sleeping on her sofa. That meteor took almost the whole house with it. A tired farmer was taking an afternoon nap in a hayfield beneath his wagon in Kansas in 1890 when he was awakened by a brilliant flash, followed by a violent explosion and crash like thunder.

The ground erupted like a tidal wave, and he was buried in flying dirt. When the cause of the explosion was sought, investigators discovered that a meteorite weighing less than two hundred pounds had buried itself deeply in the earth about sixty feet from where he was sleeping under the wagon. At a place called Cabin Creek, Arkansas, another farmer's family was preparing to sit down for dinner when something roared out of the skies with such fury that it shook the entire house. The family ran outdoors and saw the tops of several pine trees flying through the air where a large meteorite had sheared them off. The meteor was later found in a nearby field.

Several times a year, meteorites and fireballs make the news. Late in 1994, a meteorite flashed through the daytime skies of France,

---

161 Virginia Pilot and the Ledger Star, It could Happen Here, July 19, 1994

smashed through a driver's windshield, bounced off his steering wheel, and broke his shoulder. The dazed Frenchman counted himself lucky to survive the impact. Closer to home, during the editing of this manuscript on March 29, 1995, one of my associates in upstate Wisconsin saw a brilliant fireball with a flaming tail roar over their house and splash down in a nearby river. The family was unable to recover the meteorite, but the close encounter served all of us a warning and an uneasy reminder of the unseen cosmic threat in the skies.

Nearly every week somewhere in the world small meteorites have destroyed cars, houses, impacted airplanes, and from time to time, punched holes in spacecraft, including the Space Shuttle. Daily meteor strikes are unpredictable and can occur at any time.

——— 15 ———

Super Bolide Impacts & Comet Explosions

David Asher and colleagues have looked at objects that are known to have come close to the earth in recent times. They conclude, based on various strands of evidence (for example, the number of meteorites discovered on earth that originated on the moon) that the average time between impacts on earth is no more than 300 years, probably less.162

Comet Hazard school scientists propose that the Tunguska event was due to a fragment of Comet Encke. These scientists also now have the fact of the fragments of Comet Shoemaker-Levy hitting Jupiter in July of 1994 to illustrate the problem we face. Comet Hazard, scientists also think, as mentioned above, that impacts are a lot more frequent than many people suppose. 163

Super Bolides

Super Bolides impact the Earth frequently. (get impact rates from Cosmic Winter. I just need to find them) The Barringer Crater in northern Arizona is familiar to many readers. Most scientists today think it was caused by an iron Super-Bolide. Published results of its age have varied drastically between 800-200,000 years ago for the crater's

---

162 (Baillie 2006, 199) Knight-Jadczyk, Laura (2012-07-26). The Apocalypse: Comets, Asteroids and Cyclical Catastrophes. Red Pill Press. Kindle Edition. p. 16.

163 Knight-Jadczyk, Laura (2012-07-26). The Apocalypse: Comets, Asteroids and Cyclical Catastrophes. Red Pill Press. Kindle Edition. p. 19.

time of formation. This large depression averages 1200 meters (about 4,000 feet) in diameter; the crater floor lies 175 meters (570 feet) below the rim.

The Age of the Barringer crater ranges from 800 years to 200, 000 years ago. Even though many scientists have settled on an impact date of 50,000 years ago for the Barringer Crater, why the great disparity in dating its age?

A closer look by Don B. DeYoung, in his 1994 abstract, *The Age of the Arizona Meteor Crater*, reveals why: *"This confusion is typified by a 1992 college text which gives two different ages for the crater on the very same page: 22,000 and 50,000 years (Payne, et al., 1992, p. 398). Thomas Arny also states that the crater is both 10,000 and 50,000 years old (1994, p. 261, 279).* 164

The Barringer meteor crater is a perfect example of a super bolide impact where scientists hotly debate its age, and even disregard American Indian myths describing its impact. Either the American Indians have been around much longer than we think, or the Barringer crater is much younger than we are willing to believe.

For this reason, I want to spend a page or two explaining this dilemma, because as we shall find, mythology and religion continue to clash with science in dating past impacts that have caused frequent periods of fire, floods, darkness and general destruction upon the Earth below.

Dendrochronology Dating

So why the great disparity of age ranging from 800–200,000 years ago? One method of dating uses tree rings, called Dendrochronology. Daniel Barringer, the owner of the Barringer crater property, noticed hundreds of large Juniper or Cedar trees growing around the crater rim. Some revealed as many as 700 growth rings. Since these trees must have begun growing after the explosion, their lifetime of 800 years becomes a minimum age for the crater (Blackwelder, 1932, p. 559).165

Today, a century later, the tree ring figure can be rounded to 800 years, surely the youngest age estimate for the Barringer Crater. "Erosion Benjamin Tilghman reported further evidence for a recent

---

164 Don B. Deyoung 1994. AGE OF THE ARIZONA METEOR CRATER DON B. DEYOUNG* Received 15 November 1993; Revised 18 March 1994. p. 3.

165 Don B. Deyoung 1994. AGE OF THE ARIZONA METEOR CRATER DON B. DEYOUNG* Received 15 November 1993; Revised 18 March 1994. p. 4.

crater formation in 1905. First, he was impressed with the still-sharp edges of ejected boulders (Tilghman, 1905)."166

Barringer's Sister Crater found in Odessa, Texas

After Barringer, further study in 1928, discovered a second terrestrial crater at Odessa, Texas. "It is a miniature twin of Barringer, 163 meters across (530 feet) and 5.5 meters deep (18 feet). Iron meteorites found at Odessa are very similar in structure and chemical composition to the Barringer site." (Heide 1964). *The Odessa crater is located 540 miles southwest of Barringer. It is generally thought that both meteorites arrived from the north. The Odessa crater was considered very recent in age until the fossilized remains of a "primitive" horse and an elephant were found buried within it (Mark, 1987, p. 44). The typical age given today is 10,000 years. Investigations of Barringer might well include Odessa as another possible source for origin dating.* 167

Witness Dating: Could the American Indians have observed the impact of the Meteor that Caused the Barringer Crater?

The current popular impact date of 49,000 B.C. advertised at Barringer may be far too old.

Why? While anthropologist agree the Clovis Indians arrived in North America some 15,000 years ago, (cite), but even more interesting are more recent Legends of the Hopi Indians that have been published which support the "observation" idea:

"They use of the pure white rock flour in Hopi religious ceremonies [finely ground silica from the crater edge] links the crater with a legend current in the tribe. Three of their gods, the Hopi believe, came down from the clouds on to the desert. One made his abode in Meteor Crater...That Meteor Crater should have a place in the legends of the Hopi indicates a fairly recent origin. (Boutwell, 1928, pp. 729-730). It is possible that the fall of the giant meteorite was observed by the local Indians. Three of their legends concern the crater. According to them, one of their gods came down from the sky, accompanied by thunder and lightning, and buried himself at this spot...

---

166 Don B. Deyoung 1994. AGE OF THE ARIZONA METEOR CRATER DON B. DEYOUNG* Received 15 November 1993; Revised 18 March 1994. p. 4.

167 Don B. Deyoung 1994. AGE OF THE ARIZONA METEOR CRATER DON B. DEYOUNG* Received 15 November 1993; Revised 18 March 1994. p. 4.

"Even today, Indians still following tribal customs are not permitted to visit the crater; it is considered taboo. It is also significant that (early) Indians did not participate in the search for meteor iron in the crater vicinity (Heide, 1964, p. 32). - VOLUME 31, DECEMBER 1994 157. It is interesting that the Hopi people speak of a fiery descent from heaven, i.e., a meteorite impact, long before modern geologists recognized this extraterrestrial origin. The Indian observation idea has been discounted by several writers. Perhaps this negative conclusion is based on the presupposition that the crater must greatly predate Indian presence in the region. (Blackwelder, 1932, p. 559; Heide, 1964, p. 32; Ley, 1966, p. 244). Noted here below.

Don B. Deyoung 1994. *AGE OF THE ARIZONA METEOR CRATER DON B. DEYOUNG** Received 15 November 1993; Revised 18 March 1994. pp. 6-7. See also References: Amy, Thomas. 1994. Explorations: an introduction to Astronomy. Mosby-Yearbook, Inc., St. Louis, MO. Barringer, D. M. 1905. Coon Mountain and its crater. Academy of Natural Science of Philadelphia 57:861-886. Bjork, R. L. 1961. Analysis of the formation of Meteor Crater, Arizona: a preliminary report. Journal of Geophysical Research 66(10):3379-3387. Beals, C. S. 1958. Fossil meteorite craters. Scientific American 199:32-39. Blackwelder, Eliot. 1932.

Heide, Fritz. 1964. Meteorites. University of Chicago Press. Chicago. Hoyt, William Graves. 1987. Coon Mountain controversies. See also, Blackwelder, Eliot. 1932. The age of Meteor Crater. Science 76:557- 60. Boutwell, William D. 1928.

The age of Meteor Crater. Science 76:557- 60. Boutwell, William D. 1928. The mysterious tomb of a giant meteorite. The National Geographic Magazine 53(6):721-730. Brown, Martin. 1993. The age of Meteor Crater. Science 772:239-40. Buddhue, J. D. 1961. The Age of Barringer Meteorite Crater. Paper presented to the 1961 Meteoritical Society meeting of Nantucket, MA (Unpublished). Burnham, R. 1988. The Rosetta Stone of Meteor Craters. Astronomy 16(6):38-42. Chao, E. C. T., E. M. Shoemaker, and B. M. Madsen. 1960. First Natural Occurrence of Coesite. Science 132:220-222. Colvocoresses, G. M. 1925. Private correspondence with D. M. Barringer. In Hoyt, W. G. 1987. Coon Mountain Controversies. University of Arizona Press. Tucson, AZ. DeYoung, Don B. 1974. Geochemistry of the stable isotopes. CRSQ 11(1):32-36. DeYoung, Don B. 1976. The precision of nuclear decay rates. CRSQ 12(1):38-41. Engelbrektson, S. 1994. Astronomy through space and time. Wm. C. Brown Publishers. Dubuque, IA. Fairchild, Herman L. 1907. Origin of Meteor Crater (Coon Butte), Arizona. Bulletin of the Geological Society of America 18:493- 504. Foote, A. E. 1891. A new locality for meteoric iron with a preliminary notice of the discovery of diamonds in the iron.

American Journal of Science. 42 (3rd Series):413-417. Gilbert, G. K. 1906. The origin of hypotheses, illustrated by the discussion of a topographic problem. Science 3:1-13. Hager, D. 1953. Crater Mound, Arizona, a geologic feature. Bulletin of the American Association of Petroleum Geologists 37:821-57. Hawkins, Gerald S. 1964. Meteors, comets, and meteorites. McGrawHill Book Co. New York. Heide, Fritz. 1964. Meteorites. University of Chicago Press. Chicago. Hoyt, William Graves. 1987. Coon Mountain controversies. University of Arizona Press. Tucson, AZ. Jakosky, J. J. 1930. Lowell Geophysical survey of Meteor Crater. Observatory Archives. October 13. Kaufmann, William J., III. 1993. Discovering the Universe. W.H. Freeman and Co. New York. King, Elbert A. 1976. Space geology. John Wiley and Sons, Inc. New York. Kuhn, Karl. 1994. In quest of the Universe. West Publishing Co. New York. LeMaire, T. R. 1980. Stones from the sky. Prentice-Hall, Inc. Englewood Cliffs. NT. Lewis, Richard S. 1983. Illustrated encyclopedia of the Universe. Harmony Books. New York. Ley, Willy. 1966. Watchers of the Skies. Viking Press. New York. McCall, J. H. Ed. 1977. Meteorite craters. Dawdon, Hutchingson and Ross, Inc. Stroudsburg, PA. McCall, G. J. H. 1973. Meteorites and their origins. John Wiley and Sons. New York. Mark, Kathleen. 1987. Meteorite craters. University of Arizona Press, Tucson, AZ. Merrill, G. P. 1908. The Meteor Crater of Canyon Diablo, Arizona: its history, origin and associated meteoric irons. Smithsonian Inst. Misc. Coll. 1. pp. 461-498. Moulton, F. R. 1929. Several informal letters and reports. On file in the Lowell Observatory Library. Flagstaff. Nininger, H. H. 1952. Out of the sky. Dover Publishing, Inc. New York. Payne, C. A., W. R. Falls, and C. J. Whidden. 1992. Physical science. Wm. C. Brown. Dubuque, IA. Pasachoff, Jay M. 1993. Astronomy. Saunders College Pub. New York. Reger, Richard D. and George L. Batchelder. 1971. Late Pleistocene mollusks and a minimum age of Meteor Crater, Arizona. Journal of the Arizona Academy of Science 6:190-195. Ronan, Colin A. 1991. The natural history of the Universe. Macmillan. New York. Sagan, Carl, and Ann Druyon. 1985. Comet. Random House. New York. Shoemaker, E. 1983. Asteroidal and comet bombardment of the Earth. Annual Review of Earth and Planetary Science 11:461- 494. Sutton, S. R. 1984. Thermoluminescence measurements on shockmetamorphosed sandstone and dolomite from Meteor Crater, thermoluminescence age of Meteor Crater. Unpublished paper referred to by Hoyt, W. G. p. 418. Tilghman, B. C. 1905. Coon Butte, AZ. Proceedings of the Academy of Natural Sciences of Philadelphia 57:887-914. Williams, E. L., J. R. Meyer, and G. W. Wolfrom. 1992. Erosion of the Grand Canyon of the Colorado River. Part III—Review of the possible formation of basins and lakes on Colorado Plateau and different climatic conditions of the past.

CRSQ 29(1):18-24. Wood, John A. 1979. The solar system. Prentice-Hall, Inc. Englewood Cliffs, NJ.

The Hopi legend becomes increasingly tantalizing because their traditions are handed down from the Anasizi, an ancient Pueblo People (Hopi: Hisatsinom or Navajo: Anasazi) who constructed large apartment-house complexes in northeastern Arizona, northwestern New Mexico, and southwestern Colorado along the Mogollon Rim, from the 12th–14th century—800 years ago—the earliest date assigned to the crater from dendrochronology dating of tree rings around the Barringer Crater.

But is also a time when anthropologists noted that towards the end of the 1300's, they abandoned their villages postulating drought? No researchers have been able to determine the reason, although it is likely that a drying of water sources would have forced the people away.

Events relegated to "myth status" around 1348 A.D. in Europe indicate the US impacts in New Mexico and Texas may have been a part of a world-wide impact event.

Laura Knight-Jadczyk, quotes **Baillie in** *The Apocalypse: Comets, Asteroids and Cyclical Catastrophes* contemporary mythical evidence from around the globe that indicate that the earth was, indeed, subjected to bombardment from space during the 14th century and that this may very well have been not only the cause of the 25 January, 1348 earthquake, but also the cause of the Black Death. *Baillie quotes a great selection of material from contemporary accounts including the work of Ziegler cited above: "droughts, floods, earthquakes, locusts, subterranean thunder, unheard of tempests, lightning, sheets of fire, hail stones of marvelous size, fire from heaven, stinking smoke, corrupted atmosphere, a vast rain of fire, masses of smoke" (Ziegler, paraphrased in Baillie 2006, 87). Ziegler discounts entirely reports of a black comet seen before the arrival of the epidemic but records: "heavy mists and clouds, falling stars, blasts of hot wind, a column of fire, a ball of fire, a violent earth tremor, in Italy a crescendo of calamity involving earthquakes, following which, the plague arrived" (Baillie 2006, 87).*[168]

As it happens, in the 1340s there was a veritable rash of earthquakes. In Rosemary Horrox's book, The Black Death, we find that a contemporary writer in Padua reported that not only was there a great earthquake on 25 January 1348, but it was at the twenty-third hour: In the thirty-first year of Emperoro Lewis, around the feast of the Conversion of St. Paul [25 January] there was an earthquake throughout Carinthia and Carniola which was so severe that everyone feared for their lives. There were repeated shocks, and on one night the earth shook 20 times.

168 Knight-Jadczyk, Laura (2012-07-26). The Apocalypse: Comets, Asteroids and Cyclical Catastrophes (pp. 34-35). Red Pill Press. Kindle Edition. p. 34-35.

Sixteen cities were destroyed and their inhabitants killed …. Thirty-six mountain fortresses and their inhabitants were destroyed and it was calculated that more than 40,000 men were swallowed up or overwhelmed. (Horrox, quoted in Baillie 2006, 87)

The author goes on to say that he received information from "a letter of the house of Friesach to the provincial prior of Germany": It says in the same letter that in this year [… 1348 …] fire falling from heaven consumed the land of the Turks for 16 days; that for a few days it rained toads and snakes, by which many men were killed; that a pestilence has gathered strength in many parts of the world (Horrox, quoted in Baillie 2006, 88) From Samuel Cohn's book: … a dragon at Jerusalem like that of Saint George that devoured all that crossed its path … a city of 40,000 … totally demolished by the fall from heaven of a great quantity of worms, big as a fist with eight legs, which killed all by their stench and poisonous vapours. (Cohn, quoted in Baillie 2006, 88) A story by the Dominican friar Bartolomeo: … massive rains of worms and serpents in parts of China, which devoured large numbers of people. Also in those parts fire rained from Heaven in the form of snow (ash), which burnt mountains, the land, and men. And from this fire arose a pestilential smoke that killed all who smelt it within twelve hours, as well as those who only saw the poison of that pestilential smoke. (Cohn, quoted in Baillie 2006, 88)[169]

***1348 Pestilence caused by Light from the stars**…Jon Arrizabalaga compiled a selection of writings in an attempt to understand what educated people were saying about the Black Death while it was happening. Regarding the terms used by doctors and other medical professionals in 1348 to describe the plague, he writes: One … Jacme d'Agramaont, discussed it in terms of an 'epidemic or pestilence and mortalities of people' … which threatened Lerida from 'some parts and regions neighbouring to us' …. Agramont said nothing concerning the term epidímia, but he extensively developed what he meant by pestilència. He gave this latter term a very peculiar etymology, in accordance with a form of knowledge established by Isidore of Seville (570 – 636) in his Etymologiae, which came to be widely accepted throughout Europe during the Middle Ages. He split the term pestilència up into three syllables, each having a particular meaning: pes (= tempesta: 'storm', 'tempest'), te (= temps: 'time'), and lència (= clardat: 'brightness', 'light'); hence, he concluded, the pestilència was 'the time of tempest caused by light from the stars.' (Arrizabalaga, quoted in Baillie 2006, 102)*[170]

---

169 Knight-Jadczyk, Laura (2012-07-26). The Apocalypse: Comets, Asteroids and Cyclical Catastrophes (pp. 34-35). Red Pill Press. Kindle Edition. p.34-35.

170 Knight-Jadczyk, Laura (2012-07-26). The Apocalypse: Comets, Asteroids and Cyclical Catastrophes (p. 37). Red Pill Press. Kindle Edition. p. 37.

I will talk more about the Anasazi and their legends later in the book.

Modern Super Bolide Impacts

Great swarms of ice and rocky chunks of Genesis material still roam our solar system, debris left over from swirling clouds of gas and dust that formed our Sun and planets nearly 4 ½ billion years ago. The biggest and most numerous of the bodies that pose a danger to our planet are remnants from the asteroid belt that crosses Earth orbit known as Apollos. In 1962, the asteroid *Icarus* sailed so close to Earth that, for the first time, scientists began to think seriously about sending up a rocket armed with thermonuclear weapons to nudge such an object into a new orbit. In 1989, another asteroid 300 meters across, missed the Earth by a mere 1,100,000 kilometers (5 earth to moon distances). The asteroid was later appropriately named 1989C. In January 1991, asteroid 1991 BA passed only 170,000 kilometers from Earth, half the distance from Earth to the Moon! The fact that the asteroid nearly collided with the Earth should concern us, but more importantly, it was not observed until hours after it had flashed by and would have impacted the Earth with little or no warning.

Such was the case of the meteor explosion over Chelyabinsk, Russia, on February 15, 2013. A picture taken of the smoke trail with the double plumes clearly visible either side of the bulbous "mushroom cloud" cap. The meteor exploded at high altitude over Chelyabinsk, Russia, leaving more than 1,000 people injured by flying glass and debris, apparently collided with another asteroid before hitting Earth. A mineral called jadeite that was embedded in fragments recovered after the explosion shows the asteroid's parent body struck a larger asteroid at a relative speed of 4,800 kilometers per hour. But many small fragments survived. The asteroid was traveling almost 60 times the speed of sound and exploded about 30 kilometers above ground with a force nearly 30 times as powerful as the atomic bomb dropped by the United States on Hiroshima, Japan, in 1945 in World War II.

Lead research Shin Ozawa, with the University of Tohoku in Japan, wrote in a paper published in the journal Scientific reported, *"This impact might have separated the Chelyabinsk asteroid from its parent body and delivered it to the Earth."* He further stated, *"The object was undetected before its atmospheric entry, in part because its radiant was close to the Sun."*

Its explosion created panic among local residents and about 1,500 people were injured seriously enough to seek medical treatment. All of the injuries were due to indirect effects rather than the meteor itself, mainly from broken glass from windows that were blown in when the shock wave arrived, minutes after the super bolide's flash.

Some 7,200 buildings in six cities across the region were damaged by the explosion's shock wave, and authorities scrambled to help repair the structures in sub-zero (°C) temperatures. With an estimated initial mass of about 12,000–13,000 metric tonnes (13,000–14,000 short tons, heavier than the Eiffel Tower), and measuring about 20 metres in diameter, *it is the largest known natural object to have entered Earth's atmosphere since the 1908 Tunguska event that destroyed a wide, remote, forested area of Siberia. The Chelyabinsk meteor is also the only MODERN meteor confirmed to have resulted in a large number of injuries."* (citation needed)

Fragmenting Comets & Comet Trains: A New Theory

Is it fear mongering that swarms of comets roam the solar system wreaking havoc bringing destruction and death down upon the earth? I posited my original of theory of catastrophic comet train impact in the form of meteor showers in 1995. Most meteor showers are tiny particles of cosmic material whose impact is insignificant. But history records many catastrophic exceptions, because occasionally,

*"in these trains of debris, there are chunks measuring between one and several hundred meters in diameter. When these either strike the earth or explode in the atmosphere, there can be catastrophic effects on our ecological system. Multi-megaton explosions of fireballs can destroy natural and cultural features on the surface of the earth by means of tidal-wave floods (if the debris lands in the sea), fire blasts and seismic damage leaving no crater as a trace, just scorched and blasted earth. In the case of a significant bombardment, an entire small country could be wiped out, completely vaporized.*171

There have been four regional atmospheric comet explosions that have been uncovered since Christ's Crucifixion in 30 A.D.

#1: 441 A.D.

After the collapse of Rome in 410 A.D., the British Isles suffered a major catastrophe around 441 A.D., followed by a darkened sky which essentially precipitated the pre-mediaeval Dark Age.

---

171 Knight-Jadczyk, Laura (2012-07-26). The Apocalypse: Comets, Asteroids and Cyclical Catastrophes (p. 51). Red Pill Press. Kindle Edition.

Such very limited accounts as we have of this catastrophe (nothing contemporary) speak of it as the "ruin of Britain" as well as explicit references to "fire [that] fell from heaven" and that "[the fire] did not die down until it had burned the whole surface of the island."[172] Even a century later the level of destruction and depopulation was such that many cities were still in a state of ruin. That same year Chinese and Spanish commentators observed a strange comet lighting up the skies.

#2: 526 A.D.

Baillie mentions that one obvious prospect is the great Antioch earthquake of A.D. 526, which was described by John Malalas (E. Jeffreys, M. Jeffreys and R. Scott, The Chronicle of John Malalas [Byzantina Australiensia 4], Melbourne: Australian Assoc. Byzantine Studies 4, 1986): ... those caught in the earth beneath the buildings were incinerated and sparks of fire appeared out of the air and burned everyone they struck like lightning. The surface of the earth boiled and foundations of buildings were struck by thunderbolts thrown up by the earthquakes and were burned to ashes by fire ... it was a tremendous and incredible marvel with fire belching out rain, rain falling from tremendous furnaces, flames dissolving into showers ... as a result Antioch became desolate ... in this terror up to 250,000 people perished. (Malalas, quoted in Baillie 2006, 130)[173]

#3: 530 A.D. Halley's Apparition

 # 4: 540 A.D. Comet of Gaul - As it happens, Isidore of Seville lived not long after another period of cometary bombardment over Europe that is also evident in the tree-ring and ice-core studies. On 17 August 1999, the Knight Ridder Washington Bureau published an article by Robert S. Boyd entitled 'Comets may have caused Earth's great empires to fall', which included the following: Analysis of tree rings shows that in 540 A.D. in different parts of the world the climate changed. Temperatures dropped enough to hinder the growth of trees as widely dispersed as northern Europe, Siberia, western North America, and southern South America. A search of historical records and mythical stories pointed to a disastrous visitation from the sky during the same period, it is claimed. There was one reference to a "comet in Gaul so vast that the whole sky seemed on fire" in 540-41. According to legend, King Arthur died around

172 V.Clube & Ralph Napier, The Cosmic Serpent

173 Knight-Jadczyk, Laura (2012-07-26). The Apocalypse: Comets, Asteroids and Cyclical Catastrophes (p. 33). Red Pill Press. Kindle Edition.

this time, and Celtic myths associated with Arthur hinted at bright sky Gods and bolts of fire. In the 530s, an unusual meteor shower was recorded by both Mediterranean and Chinese observers. Meteors are caused by the fine dust from comets burning up in the atmosphere. Furthermore, a team of astronomers from Armagh Observatory in Northern Ireland published research in 1990 which said the Earth would have been at risk from cometary bombardment between the years 400 and 600 A.D..[174]

5: 1014 A.D
#6: 1095 A.D.

A brilliant red comet associated with a swarm of falling meteors observed in medieval Europe in 1095 A.D., triggered a similar mass hysteria. A member of the council of Claremont who witnessed the spectacular event wrote:

Stars in the sky were seen throughout the whole world to fall towards the Earth, crowded together and dense, like hail or snowflakes. A short while later, a fiery way appeared in the heavens; and after another short period half the sky turned the color of crimson and earthquakes occurred in divers places. After the sensational event, the council of Claremont convened to discuss the widespread expectations that the Final Judgment was about to occur."[175]

7: 1348 A.D.
#8: 1872 A.D. The Chicago Fire (pull up research)

Dating the next train of comet fragments?

Inherent danger of multiple comet impacts for the next 400 years: Physicist Richard Firestone and geologists Allen West and Simon Warwick-Smith write in their book, The Cycle of Cosmic Catastrophes: In 1990, Victor Clube, an astrophysicist, and Bill Napier, an astronomer, published The Cosmic Winter, a book in which they describe performing orbital analyses of several of the meteor showers that hit Earth every year. Using sophisticated computer software, they carefully looked backward for thousands of years, tracing the orbits of comets, asteroids, and meteor showers until they uncovered something astounding. Many meteor

---

174 Knight-Jadczyk, Laura (2012-07-26). The Apocalypse: Comets, Asteroids and Cyclical Catastrophes (p. 38). Red Pill Press. Kindle Edition.

175 Censorinus, Liber de die natali xviii

*showers are related to one another, such as the **Taurids, Perseids, and Orionids.***

In addition, some very large cosmic objects are related: the comets Encke and Rudnicki, the asteroids Oljato, Hephaistos, and about 100 others. Every one of those 100-plus cosmic bodies is at least a half-mile in diameter and some are miles wide. And what do they have in common? According to those scientists, every one is the offspring of the same massive comet that first entered our system less than 20,000 years ago! Clube and Napier calculated that, to account for all the debris they found strewn throughout our solar system, the original comet had to have been enormous. ... Clube and Napier also calculated that, because of subtle changes in the orbits of Earth and the remaining cosmic debris, Earth crosses through the densest part of the giant comet clouds about every 2,000 to 4,000 years. When we look at climate and ice-core records, we can see that pattern.

For example, the iridium, helium-3, nitrate, ammonium, and other key measurements seem to rise and fall in tandem, producing noticeable peaks around 18,000, 16,000, 13,000, 9,000, 5,000, and 2,000 years ago. In that pattern of peaks every 2,000 to 4,000 years, we may be seeing the "calling cards" of the returning mega-comet. Fortunately, the oldest peaks were the heaviest bombardments, and things have been getting quieter since then, as the remains of the comet break up into even smaller pieces. The danger is not past, however. Some of the remaining miles-wide pieces are big enough to do serious damage to our cities, climate, and global economy. Clube and Napier (1984) predicted that, in the year 2000 and continuing for 400 years, Earth would enter another dangerous time in which the planet's changing orbit would bring us into a potential collision course with the densest parts of the clouds containing some very large debris. Twenty years after their prediction, we have just now moved into the danger zone. It is a widely accepted fact that some of those large objects are in Earth-crossing orbits at this very moment, and the only uncertainty is whether they will miss us, as is most likely, or whether they will crash into some part of our planet. [emphasis added] (Firestone et al. 2006, 354-355)[176]

176 Knight-Jadczyk, Laura (2012-07-26). The Apocalypse: Comets, Asteroids and Cyclical Catastrophes (p. 50-51). Red Pill Press. Kindle Edition.

——— 16 ———

What Happens During an Impact?

Impact Effects are CLASSIFIED BY ORDERS OF MAGNITUDE

What would happen if a comet or cometary debris struck the earth? If the comet were of sufficient size and mass, the main body would not have to impact the Earth to wreak havoc. Since comets are known to have a weak cohesive structure, any nearby approach to a strong gravitational field will often break the comet up into large masses of loose debris -- stony and iron asteroids, chunks of water ice, methane, and other frozen poisonous gases, and attract enough of the loose cosmic debris to rain down catastrophe below.

If a cosmic bomb one half mile in diameter struck the Earth, traveling at 24,500 miles per hour relative to Earth, it would pass through the lower atmosphere in less than a second. Shock waves would then drive both into the Earth and into the asteroid, causing the asteroid to vaporize, changing from solid to liquid to gas in a fraction of a second.

The explosion would generate the equivalent of a 1,000-megaton bomb explosion and temperatures of 20,000° C. Hot gas from the vaporized object would shoot into the sky and drag the air with it. Another shock wave would then spread away from the impact site and everything within a hundred kilometers would burst into flames from the intense heat of the blast. As far as 500 kilometers away the temperature would still reach a scalding 100° C. The blast would travel outward at 35,000 kilometers per hour and level everything for 250 kilometers. Material from the impact would rain down in the form of molten droplets of rock. A crater about ten times the diameter of the asteroid would be left behind. A city the size of New York would have vaporized nearly instantaneously.

The impact will kill tens of millions, trigger world-wide tsunami's flooding coastlines, volcanic eruptions, mountains rising, lands sinking (Japan), and an initial increase of atmospheric temperature to over 200-600 degrees that will melt polar ice caps, singe crops world-wide with destructive acid rain, followed by a three-year winter killing plants, humans and animals. The poles may shift several degrees.

The dust and soot raised by a cosmic impact would cut out sunlight for months. The result, according to a small team of atmospheric scientists led by Richard Turco of R&D Associates in Marina del Ray, California, would be a long "impact winter" lasting for as much as three years. During such a period of drastic climate change, large numbers of species could be wiped out. Impact winter is but one consequence of a collision with a large comet or asteroid. Following such an event, the plume that would rise into the atmosphere would carry a wide variety of chemicals that extended periods of acid rain would result, destroying crops, forests, and marine life in the oceans.

Tsunami

Quoted. Cite. "The oceans cover about 75% of the Earth's surface, so it is likely the asteroid will hit an ocean. The amount of water in the ocean is nowhere near large enough to "cushion" the asteroid. The asteroid will push the water aside and hit the ocean floor to create a large crater. The water pushed aside will form a huge tidal wave, a tsunami. The tidal wave height in meters $=10.9 \times$ (distance from impact in kilometers)$^{-0.717} \times$ (energy of impact in megatons TNT)$^{0.495}$. What this means is that a 10-km asteroid hitting any deep point in the Pacific (the largest ocean) produces a mega tsunami along the entire Pacific Rim.

Some values for the height of the tsunami at different distances from the impact site is given in the following table. The heights are given for the two typical asteroids, a 10-kilometer and a 1-kilometer asteroid.

Distance (in km)	10 km asteroid	1 km asteroid
300	1.3 km	42 m
1000	540 m	18 m
3000	250 m	8 m
10000	100 m	3 m

The steam blasts from the water at the crater site rushing back over the hot crater floor will also produce tsunamis following the initial impact tsunami and crustal shifting as a result of the initial impact would produce other tsunamis---a complex train of tsunamis would be created from the initial impact (something not usually shown in disaster movies).

Global Firestorm

The material ejected from the impact through the hole in the atmosphere will re-enter all over the globe and heat up from the friction with the atmosphere. The chunks of material will be hot enough to produce a lot of infrared light. The heat from the glowing material will start fires *around the globe*. Global fires will put about 7×10^{10} tons of soot into the air. This would "aggravate environmental stresses associated with the ... impact."

Are Earthquakes caused by atmospheric comet explosions?

Knight-Jadczyk: "As it happens, the ammonium signal in the ice-cores is directly connected to an earthquake that occurred on 25 January 1348 – and Baillie discovers that there was a fourteenth-century writer who wrote that the plague was a "corruption of the atmosphere" that came from this earthquake. How could a plague come from an earthquake, one may ask? Baillie points out that we don't always know if earthquakes are caused by tectonic movements; they could be caused by cometary explosions in the atmosphere or even impacts on the surface of the earth. In Rain of Iron and Ice, John Lewis, Professor of Planetary Sciences at the Lunar and Planetary Laboratory, co-director of the NASA/University of Arizona Space Engineering Research Center, and Commissioner of the Arizona State Space Commission, tells us that the earth is regularly hit by extraterrestrial objects and many of the impacting bodies explode

Baillie points out that we don't always know if earthquakes are caused by tectonic movements; they could be caused by cometary explosions in the atmosphere or even impacts on the surface of the earth. In Rain of Iron and Ice, John Lewis, Professor of Planetary Sciences at the Lunar and Planetary Laboratory, co-director of the NASA/University of Arizona Space Engineering Research Center, and Commissioner of the

*Arizona State Space Commission, tells us that the earth is regularly hit by extraterrestrial objects and many of the impacting bodies explode.*177

*...The point of this is that there is almost no way to monitor whether or not any given disaster or catastrophe is definitively an impact as opposed to a violent earthquake. As Baillie points out: there are many earthquakes recorded in history, but no impacts. And yet, there is the evidence that the impacts have happened – on the ground, and in the ice cores. And there is Tunguska.*178

Earthquake Reports of the Tunguska event in 1908 tell us that the ground shook around the impact/explosion zone for a radius of about 900 km. At the time of any larger impact event, the earthquake would be proportionally more severe. Any individuals far enough away to survive such an event, would only have seen a flash, felt a tremor, and heard a loud rumbling noise. If they were too far away to see the flash, or were indoors, they would only report an earthquake. In short, what the work of Lewis brings to the table is the idea that some well-known historical earthquakes could very well have been impact events. 179

Mountain Building
Massive Volcanic Eruptions
Acid Rain

The heat from the shock wave of the entering asteroid and reprocessing of the air close to the impact produces nitric and nitrous acids over the next few months to one year. The chemical reaction chain is:

$N_2 + O_2 ,> NO$ (molecular nitrogen combined with molecular oxygen produces nitrogen monoxide)

$2NO + O_2 ,> 2NO_2$ (two nitrogen monoxide molecules combined with one oxygen molecule produces two nitrogen dioxide molecules)

NO_2 is converted to nitric and nitrous acids when it is mixed with water.

These are really nasty acids. They will wash out of the air when it rains---a worldwide deluge of acid rain with damaging effects:

destruction or damage of foliage;

great amounts of weathering of continental rocks;

the upper ocean organisms are killed. These organisms are responsible for locking up carbon dioxide in their shells (calcium carbonate) that would eventually become limestone. However, the shells will dissolve in the acid water. That along with the "impact winter" (described below) kills off about 90% of all marine nanoplankton species.

177 Knight-Jadczyk, Laura (2012-07-26). The Apocalypse: Comets, Asteroids and Cyclical Catastrophes. Red Pill Press. Kindle Edition. p. 31.

178 Knight-Jadczyk, Laura (2012-07-26). The Apocalypse: Comets, Asteroids and Cyclical Catastrophes (pp. 32-33). Red Pill Press. Kindle Edition

179 Knight-Jadczyk, Laura (2012-07-26). The Apocalypse: Comets, Asteroids and Cyclical Catastrophes (p. 33). Red Pill Press. Kindle Edition.

A majority of the free oxygen from photosynthesis on the Earth is made by nanoplankton.

The ozone layer is destroyed by O_3 reacting with NO. The amount of ultraviolet light hitting the surface increases, killing small organisms and plants (key parts of the food chain). The NO_2 causes respiratory damage in larger animals. Harmful elements like Beryllium, Mercury, Thallium, etc. are let loose.

Temperature Effects

All of the dust in the air from the impact and soot from the fires will block the Sun. For several months you cannot see your hand in front of your face! The dramatic decrease of sunlight reaching the surface produces a drastic short-term global reduction in temperature, called *impact winter*. Plant photosynthesis stops and the food chain collapses. The cooling is followed by a *much more prolonged period* of increased temperature due to a large increase in the greenhouse effect. The greenhouse effect is increased because of the increase of the carbon dioxide and water vapor in the air. The carbon dioxide level rises because the plants are burned and most of the plankton is wiped out. Also, water vapor in the air from the impact stays aloft for a while. The temperatures are too warm for comfort for a while. The researchers studied lipids produced by the microbe Thaumarchaeota. The composition of these lipids changes as ocean temperatures change. An examination of lipids preserved in sediment at the K-Pg boundary revealed that after the impact, ocean temperatures fell an average of two degrees Celsius, with drops of up to seven degrees Celsius in some places. This decrease in temperature lasted up to several decades, a timescale supported by models and by evidence of species migration.

Global Heat followed by Global Cooling

The sudden cooling would have caused a great amount of stress on living things and therefore been a key contributor to mass extinction. When dust injected into the atmosphere rained out, the ocean surfaces would have become acidic, resulting in yet more stress for surface-dwelling organisms. Vellekoop's team found that a stable warm period followed this short period of global cooling. Large-scale mortality, forest fires and the vaporization of rock would have released greenhouse gases into the atmosphere, causing global warming. The team thinks the research should help increase understanding of the effects of rapid climate change. Read more at: http://phys.org/news/2014-05-proof-global-cooling-chicxulub-asteroid.html#jCp

Impact Timeline – Averages – A new impact science

Clube and Napier disagree with the current estimate of asteroid impact averages. While throughout time, like evolution, these averages may be accurate. However, the insertion of a comet into near-earth orbit will skew these averages creating a new "catastrophism" model. *(see Club & Napier model)*

Large Impacts causing Impact Winter

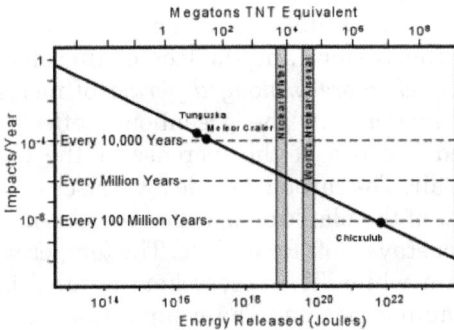

An impact like the one that struck the Yucatan Peninsula, in Mexico about 65 million years ago, thought responsible for the extinction of the dinosaurs and numerous other species, created the Chicxulub Crater, 180 km in diameter and released energy equivalent to about 100 million megatons of TNT.

For comparison, the amount of energy needed to create a nuclear winter on the Earth as a result of nuclear war is about 8,000 megatons, and the energy equivalent of the world's nuclear arsenal is about 60,000 megatons.

——— *17* ———

Search for the Armageddon Stones

"I would more easily believe that two Yankee professors would lie than that stones would fall from heaven."
~President Thomas Jefferson

The Meteor Showers

The business-like arrival of meteor showers heralds the potential for the larger, more dangerous variety of cosmic debris such as the one which occurred on October 10, 1992, when a huge green fireball was seen speeding through the early evening skies by several thousand awed spectators along the East coast of the United States. Reports of the fireball poured in from North Carolina, West Virginia, Ohio, Pennsylvania and New York. A historian in Virginia Beach, Virginia, wrote about the spectacular visitor from space; "It's frightening, but at the same time it's so outstandingly beautiful," he said. "You literally freeze when you see it." The brilliant fireball was "as large as the moon" and appeared to "burst into flames" leaving a brilliant white tail behind it. Reports the following day indicated a chunk from the meteor destroyed a parked car in upstate New York.

The Federal Aviation Administration identified the fireball as the Giacobini-Zinner Comet which leaves behind cometary debris known as the Draconid meteor shower which passes through the Western Hemisphere every October 9. Jon U. Bell, an astronomer at the Virginia Living Museum, said he thought the fireball was not the comet itself, but a chunk that had broken off. "I think if the comet itself had come in, I dare say we wouldn't be having this conversation," Bell said. "We would be in a very large, scarred surface of the earth with devastation all around."[180]

Some meteors shower! The Draconian fireball is still talked about today. What if the parent comet had descended on that self-same hour in a blaze of consuming fire? How would the survivors have written about it? As it stands, the description of the North American comet fall as both "frightening and beautiful," a fire which triggers a paralyzing fear that "literally freezes" one in their tracks, perfectly describes the terrible serpent haired "Medusa" of Greek mythology discussed in Serpents, Dragons, and sky-monsters; the frightening, yet beautiful goddess, who if seen, turned her victims into stone. And is this not the same phenomena described by Abraham when the angel of the Lord told his family not to look at the descending rain of fire which consumed the ancient city of Sodom and Gomorrah, lest they be turned into a pillar of salt?

Dangerous fireballs associated with meteor showers and the appearance of comets have been documented by Charles Fort, in the Book of the Damned. Researching anomalous celestial phenomena from newspapers and scientific journals from around the world between 1860 and 1910, Fort uncovered dozens of reports related to comets, meteoric explosions, and catastrophes on the earth within the last one hundred years alone. An entire gamut of strange, albeit dangerous events have been related to comet sightings including falls of ash, mud, blood red skies, earthquakes, volcanic eruptions, and even fires and floods. It's often difficult to see the connection between these disparate phenomena, but if we take an objective look at historical reports, we may begin to get a glimpse of the unseen danger lurking in the heavens.

In the journal Cosmos,[181] on September 4, 1868, it was written that something "comet-like" was seen in the heavens over South America. At that time, a dense, towering cloud appeared over Callao, Peru. Simultaneously, meteors and fireballs were seen. Then an earthquake shook the land. Thirty- seven years later on September 8, 1905, a comet [182] with a long, luminous tail was seen in the vicinity of the moon, over Calabria, Italy. This sighting was followed by a tremendous fall of meteors, and then three quarters of an hour later, to the same place on the earth came a massive meteor which exploded deep in the ground. A shock wave spread outward and utterly demolished 4600 buildings and

---

180 The Daily Press, Space Fireball sparks crash..., October 10, 1992

181 Cosmos, (n.s., 69-422)

182 Cosmos, (n.s., 69-422)

taking nearly four thousand lives. At the same time, more meteors streaked in, followed by a tremendous fall of red dust from the sky at Tiriolo, Italy."

Was this a return of cometary debris from the comet of 1868?

The Bulletin of Astronomical Societies of France said that the Calabrian catastrophe was preceded by the brief appearance of a comet. But comets don't appear and then capriciously disappear, unless of course, it "crashed" onto the Earth. Scientists explained that the Stromboli volcano had erupted. But Monsieur Lacrois, who was living quite near Stromboli, denied the volcano had been active.

On April 12, 1901, a new comet was discovered. Soon after the comet was sighted, clouds of dust appeared in Africa, and mud fell from the sky in torrents in Pennsylvania, New York, New Jersey, and Connecticut. On the same day, a violent earthquake struck Siberia. Meteors, bolides and fireballs fell in Alabama, Georgia and Florida, as though a comet had disintegrated high in the earth's atmosphere.

From Hong Kong to the Philippines and from the Philippines to Australia in early October 1901, huge columns of smoke obscured the entire Pacific Ocean.

Nothing of terrestrial origin had ever had such a widespread effect. The smoke from the eruption of mighty Krakatoa in August 1883, was only a haze compared to this. Even Vesuvius has never been known to smoke up the entire Mediterranean.

As it stands, the description of Firefall as both "frightening and beautiful," a fire which triggers a paralyzing fear that "literally freezes" one in their tracks, perfectly describes the terrible serpent haired "Medusa" of Greek mythology, the frightening, yet beautiful goddess, who if seen, turned her victims into stone. And is this not the same phenomena described by Abraham when the angel of the Lord told his family not to look at the descending rain of fire which consumed the ancient city of Sodom and Gomorrah, lest they be turned into a pillar of salt?

There is mounting evidence that comets and their progeny, the great meteor showers, have caused disasters far greater than Tunguska in historical and pre-historic periods. Nearly a dozen major meteor showers bombard the earth every year. Meteor "shower" is a misnomer, because they require an average person to strain to see, if one is lucky, a meteor or two every minute, and only if you are quick enough to catch the streak of light as it races across the heavens. From time to time a meteor storm, like the Draconids in 1940 and the Leonids in 1966, treats us with a celestial fireworks display, filling the sky with hundreds of thousands of falling stars.

There is mounting evidence that comets and their progeny, the great meteor showers, have caused disasters far greater than Tunguska in historical and prehistorical periods. According to astronomer Victor Clube, the Earth is in much greater danger than previously acknowledged. Two professors of Astrophysics at Oxford University and the Royal Edinburgh Observatory have staked their reputations on it. In their recently published book, The Cosmic Serpent, Victor Clube and William Napier propose that unseen masses of cosmic debris orbit the sun within the inner planets of the Solar system and collide with the Earth more frequently than ever suspected.

If they are correct in their calculations, fifty cometary impacts ranging from 1 to 100 megatons have occurred in the last 5,000 years; about five impacts ranging from 100 to 1,000 megatons have occurred; and there is an even chance that there has been an impact in the range of 1,000-10,000 megatons during this period.

Their unsettling conclusion is that a 10,000-megaton strike can be expected to hit the earth an average of once every 3,500 years! A 1,000-megaton strike will occur every 500 years; a 500 megaton strike every 100 years; and a 100-megaton strike, a large explosion by any standards, can be expected every 30 years! Their main focus, however, is collisions producing explosions in the 10,000-megaton range that could have caused a catastrophe of global proportions. This remarkable information tends to be greeted with incredulity, but we evidently cannot exclude the possibility that cultural dark ages are astronomically induced. "It became fashionable to assume that the world is safe, when in fact, multiple Tunguska bombardments globally releasing over 5000 megatons of explosive energy, the equivalent of a full-scale nuclear war, may happen at internals of about 1,500 years producing a dark age."[183]

Nearly a dozen major meteor showers bombard the earth every year. Meteor "shower" is a misnomer, because they require an average person to strain to see, if one is lucky, a meteor or two every minute, and only if you are quick enough to catch the streak of light as it races across the heavens. From time to time a meteor storm, like the Leonid's over many centuries, have treat us with a celestial fireworks display, filling the sky with hundreds of thousands of falling stars.

There are several dozen meteor showers a year. I have narrowed down the following four meteor showers as Armageddon Stone Candidates.

183 Catastrophism 2000 by Victor Clube, 1991

The Beta Taurids
Parent Comet
Firefall Dates
1908 Tunguska Explosion

On the early morning of June 30, 1908, the skies over Central Siberia played a grim host to such a cosmic blast. On the banks of the Stony Tunguska River, Russian villagers were hurled to the ground as a tremendous fireball more brilliant than the sun split the heavens sending violent shock waves in every direction.

Witnesses four hundred miles from the scene testified that as the flaming object plunged to the earth a dazzling white light illumined the entire northern horizon, followed by three explosions of unimaginable magnitude.

The ensuing conflagration from the blast spread a wall of fire which ignited the clothing and skin of Russian peasants, forcing many into the nearby icy river. Their brief spell of remission fatally ended, however, as a wall of water, generated by a massive blast of hurricane-force wind, swept upstream drowning those who sought refuge from the inferno.

One eyewitness, having observed the event from about 240 kilometers from the site of impact, reported:

"At the time I was ploughing my land at Narodima When I sat down to have my breakfast beside my plow, I heard sudden bangs, as if from gunfire. My horse fell on its knees. From the north side above the forest a flame shot up. I thought the enemy was firing, since at this time there was talk of war. Then I saw that the fir forest had been bent over by the wind and I thought of a hurricane. I seized hold of my plow with both hands, so that it would not be carried away. The wind was so strong that it carried off some of the topsoil from the surface of the ground, and then the hurricane drove a wall of water up the [river] Angara. I saw it all quite clearly because my land was on a hillside."[184]

Sweeping outward from the initial explosion, the blast waves rounded the earth twice, distributing enormous quantities of meteoric dust into the atmosphere. Silver clouds which formed from the debris cast a crimson pale over the light of day as they spread swiftly from western Siberia across the entire region of northern Europe. It is recorded that so intense was the reflected light from the atmospheric anomaly that on the following night astronomers were unable to photograph the stars. As far away as Scotland it was reported that newsprint could be read at midnight with no artificial lighting. The

184 E. L. Krinov, Giant Meteorites (Elmsford, N. Y.)

phenomena of "brilliant light at night" was also reported during the Exodus tragedy.

Had the fireball descended four hours earlier it would have obliterated the Russian capital of St. Petersburg. At the point of impact, where the entire kinetic energy was instantly converted into heat, a column of fire immediately rose 12 miles (20 km). Strangely there is no impact crater or meteoric debris at the site. Subsequent studies have not revealed any evidence of radiation, which might indicate a radioactive source. A 1927 scientific expedition to the remote region noted, however, that vast tracts of forestry had been completely leveled within a radius of 30 miles. The crowns of the huge Siberian larch and pine trees pointed outward, away from the epicenter of the explosion, and within a radius of 6 miles even the soil was scorched, and in placed melted and glazed.

The explosion at Tunguska was not unlike the celestial flares that mark the fiery death of many ordinary meteors. When a meteor body plunges into the earth's atmosphere, the pressure that builds on its frontal cortex generally crushes the body into fragments that disintegrate into trails of concandescent vapor. The process happens so rapidly -- typically on the order of 0.02 to 0.04 milliseconds—that it resembles an explosion. Microscopic grains of gold, silver, and other precious metals have been found embedded in tree rings dating to 1908 serve to indicative that the Tunguska fireball might have been a metallic meteorite of appreciable size.

Writing in the June 30, 1986, issue of Meteoritics, Russian astronomers B. Yu. Levin and V.A. Bronshten challenged the meteoric explanation of the Tunguska explosion. By comparing how the breakup altitudes of large meteors are affected by factors such as approach angle, mass, and terminal velocity, the two astronomers found that the 1908 fireball had a density around 1 gram per cubic centimeter, eliminating any possibility that the Tunguska fireball was a stony or iron meteor. Their conclusion: in 1908 a piece of a comet hit the Earth.

The Russian findings mirror the current community-wide consensus among astronomers. The Tunguska fireball was, in fact, an icy cometary fragment about 100-300 meters across --weighing approximately one million tons. Traveling at 130,000 kilometers per hour, the frozen mass was instantaneously heated to a gaseous state by the friction of Earth's atmosphere, exploding nearly 10 kilometers above the Earth's surface with the force of a **40-100 megaton hydrogen bomb.**

June 30, 1908, was the day of the Beta Taurid meteor shower,[185] a band of meteors associated with the orbit of comet Enke. Astronomers are now virtually certain the Tunguska collision was caused by a chunk of comet Enke, an ice fragment substantially larger than those flashing grains of sand we normally see during a meteor shower.[186] Another suspicious remnant of the Beta Taurid meteor shower appears to have hit the Moon just over 815 years ago during the month of June when Earth passes through the remains of comet Enke. Astronomer Carl Sagan notes this amazing event, "On the evening of June 25, 1178, five British monks reported something extraordinary, which was later recorded in the chronicle of Gervase of Canterbury, generally considered a reliable reporter on the political and cultural events of his time, after he had interviewed the eyewitnesses who asserted, under oath, the truth of their story. The Chronicle reads:

'There was a bright new Moon, and as usual in that phase its horns were tilted towards the east. Suddenly, the upper horn split in two. From the midpoint of the division, a flaming torch sprang up, spewing out fire, hot coals, and sparks.'

Astronomers Derral Mulholland and Odile Calame have calculated that a lunar impact would produce a dust cloud rising off the surface of the Moon with an appearance corresponding rather closely to the report of the Canterbury monks."[187] In fact, a prominently rayed crater lies exactly in the region of the Moon referred to by the Canterbury monks. According to astronomers, such a rayed crater is only visible if it is recently formed within the last thousand years.

The Draconids
Parent Comet: Comet Giacobini-Zinner
Firefall Dates: Oct 9-10

The Federal Aviation Administration identified a brilliant fireball that exploded over the East Coast in 1992 as a fragment Giacobini-Zinner Comet which leaves behind cometary debris known as the Draconid meteor shower which passes through the Western Hemisphere every October 9-10. Jon U. Bell, an astronomer at the Virginia Living Museum, said he thought the fireball was not the comet itself, but a chunk that had broken off. "I think if the comet itself had come in, I dare

---

185 Where the meteor stream intersects the Earth, meteors and icy fragments will appear to diverge from one central point in the sky called a radiant. The Beta Taurid meteor shower emanates from the constellation Taurus.

186 Note: where a meteor stream intersects the Earth, meteors and icy fragments will appear to diverge from one   central point in the sky called a radiant. The Beta Taurid meteor shower emanates from the constellation Taurus.

187 Carl Sagan, Cosmos, Random House. p. 85

say we wouldn't be having this conversation," Bell said.[188] Some meteor shower! The Draconian fireball is still talked about today. What if the parent comet had descended on that self-same hour in a blaze of consuming fire? How would the survivors have written about it?

The Leonids
Parent: Comet Tempel-Tuttle
Firefall Dates: Nov 17-18

In the Preface, Abraham Lincoln in 1833 stepped outside his home to witness one of the greatest meteor storms of all time, "It would seem as if worlds upon worlds from the infinity of space were rushing like a whirlwind to our globe.... and the stars descended like a snow fall to the earth..."[189]

The spectacular cosmic display observed by a Georgia Courier journalist on the night of November 12, 1833, described the impact thousands upon thousands of comet fragments associated with the Leonid meteor shower. A few times a century, hundreds of thousands of grain-sized meteors smash into the upper atmosphere in the space of a few minutes, filling the sky for several hours with a dazzling display of celestial fireworks. Astronomers call this rare event the Leonid Meteor Storm.

The Leonid's, named after its radiant point in the Constellation, Leo the Lion, appear to have fallen with noise, or great noise, on several occasions, indicating the meteor stream may have much larger fragmented debris colliding without atmosphere. This is characteristic of fireballs seen in 1799, 1666, 1566, 1533, and 1002. The Leonids we will discover, have a usually fiery history and have a legion of violent phenomena associated with them.

Persistent legends, stretching back to the beginning of recorded history and even into prehistory, identify the Leonid Meteor Storm with worldwide fires and floods— an indication that masses of a large, fragmented comet remain in orbit, and where that orbit intersects with the Earth, disaster awaits.

The Leonid Meteor Storm is a cosmic blast not unlike the 1994 collision of comet Shoemaker-Levy 9 with Jupiter. In modern times, the Leonid meteor shower occurs every November 17, but regresses through the calendar about thirty days every 1000 years, or one day every 33.33 years. For instance, the Leonid shower recorded in 902 A.D., occurred October 13. Usually, the Leonids put on a mediocre display, averaging about 60 meteor trails per hour, but every 33 years the shower can turn

---

188 The Daily Press, Space Fireball Sparks Crash, October 10, 1992
189 Ibid, p. 25.

into a full-blown meteor storm. Most meteor showers contain small granular fragments no larger than a pin head. However, the Leonid meteor stream may be larger, unseen elements may be lurking out in the cosmos waiting for their periodic return.

On August 31, 1886, three years after the great volcanic eruption of Krakatoa in the South Pacific, brilliant bolides were seen streaking over the skies of Charleston, South Carolina: "Just before the sun dipped behind the horizon it was eclipsed by a mass of inky black clouds." There followed a display of luminous clouds such as those seen during a volcanic eruption. Meteors appeared in the sky, moving slowly at first, then streaking in at fantastic speeds. Suddenly, directly beneath the streaking meteors, Charleston, was devastated by a massive earthquake that covered millions of square miles. On September 4th, the ground still shook, and the New York World carried stories about "volcanic dust' sifting down "from somewhere' out of a blackened sky at Wilmington, North Carolina. Every newspaper of the time reported another severe after-shock on September 5th at Charleston. The same day two brilliant "fireballs" with luminous tails were seen. At Columbia, South Carolina, two more brilliant meteors were seen at the same time.

"A strange, dark and heavy cloud hung off the South Carolina coast," according to the Charleston News and Courier of September 8th. "Meteors continued falling throughout September and October as the Charleston area continued quaking." On the 22nd of October, about fifty meteors fell at Charleston during a particularly severe aftershock. The next night a bright comet was seen over the southeastern United States. At midnight something exploded with titanic force over Atlanta, Georgia. "A meteor of huge dimension," was one guess. Whatever it was, it was brilliant enough to cast shadows and to read newspapers by. It happened again, almost identically, on the 28th of October. One physics professor suggested the meteors "were shooting out from an apparent radiant near the constellation Leo." But why would the Leonids appear in late October when they were due to appear November 13? Is the meteor stream much larger than previously observed?

Earthquakes were associated with the Leonid meteor storm of 1799. Alexander von Humboldt, a scientist-author-explorer, happened to be in Cumana, Venezuela in the fall of 1799, and witnessed the entire event. Humboldt was ardently interested in comets and meteors and wasn't so steeped in tradition that he failed to notice some far-fetched but rather obvious correlations. On November 4th, he observed some unusual things happening with the weather beginning with the sudden appearance of a slight red haze in the atmosphere. There followed a violent thunderstorm which rose suddenly out of the roseate sky. The downpour itself was unprecedented, according to Humboldt's diary, and in addition to a series of powerful electrical explosions that shook earth and sky, Cumana and the surrounding area for many miles were rocked by two earthquakes. Afterward, the sky turned blood red and stayed that way for two days.

Then on the morning of November 12th, there was a great shower of meteors, the number of which, estimated by Humboldt to have been about 300 shooting stars per second, or close to a million an hour. "At about 2:30 AM. The most extraordinary, luminous meteors began rising out of the sky from the east and northeast." Within a short period of time -- perhaps ten or fifteen minutes -- there wasn't a place in the sky larger than two full moons that wasn't completely filled with falling stars and exploding bolides. The falling debris left traces across the sky, and many of them had an apparent brilliance and diameter as large as Jupiter, "from which darted vivid sparks of light." For almost a half hour after the full sunrise the falling stars continued to be seen.

"Almost all the inhabitants of Cumana witnessed these phenomena," Humboldt wrote, "Because they had left their houses before four o'clock to attend early morning mass. They did not behold these bolides with indifference; the oldest among them remembered that the great earthquakes of 1766 were preceded by similar phenomena."

On November 12, 1833 the shower was repeated with 'meteors as thick as snowflakes' streaming from a point in the constellation Leo; it must have been an amazing sight. The connection of the Leonids with a comet was established in 1866, with the discovery of a faint comet by William Temple and the realization that it had the same orbit as the Leonid swarm.

The Leonid meteor shower on November 12th, 1901, turned into more than a fireworks display. Everywhere in Australia, except Queensland, showers of dust and mud fell copiously from the sky. The debris was illuminated. Glaring light penetrated the densest darkness as gout of fire fell from the sky with the dust and mud.

Balls of fire erupted in every district in Victoria. They fell into cities and set fire to houses. At Whycheproof the whole air seemed on Fire." Red dust sifted down all over Australia through the 12th as fires fell from the sky. The continued until the 13th in Queensland. On the 14th, heavy clouds of smoke rolled in from the sea onto northern Australia. In one place a "sticky, bituminous substance" fell from the skies, in another, "a light fluffy, gray material," according to the Sydney Daily Telegraph of November 18th, 1901.

The Sydney Herald also reported fire balls that fell from the sky. The British scientific magazine, Nature, (vol. 67) also reported 50 darkened, stifled towns. "Business was suspended. There has been nothing like it before in the history of the colony... people stumbling around with lanterns [in midday]."

On November 13th, there was a terrific meteoric explosion at Parrmatta, according to the Sydney Herald and the Melbourne Leader. On the 18th at a place called Murrumburrah (New South Wales) a huge fireball came down in a rain of red dust from the skies. Then on the 22nd, another "fireball" with a long luminous tail flew slowly over the town of Nyngan, "intensely illuminating the sky and ground."

Two nights before the Adelaide Observatory astronomer, Sir Charles Todd, watched a large "fireball" slowly crossing the sky. The next night at 11:00 pm. another fireball the apparent size of the sun itself was observed at Towitta. An hour later, a whole string of towns and cities were brilliantly illuminated by the bolide. Twenty-four hours later there was a terrific explosion and as something blew itself apart over Ipswich, Queensland.

The Otago Witness of November 19, 1901, reported heavy falls of sulfurous- smelling ashes in New Zealand on the 13th. Most of the correspondence was carried by ships in those days, so news traveled slowly. Balls of fire and rains of ashes burned down whole neighborhoods in Allendale, Deniliquin, Boort, Chiltern, and Langdale. During the same time period heavy smoke poured across the island of Java while the earth quaked violently there. At Mysore (Kamsagar) India, a huge meteorite fell during a disastrous deluge and seven bridges were washed out along one river in Malay States.

There were no volcanic eruptions anywhere on earth at this time.[190] Scientists went about their business as usual, each attending to his own specialty, perceiving no connection among the diverse cosmic terrestrial phenomena. No one saw any relationship between the fireballs and expected arrival of the Leonid meteor stream. Were these exploding bolides fragments of the Leonid Meteor stream?

If so, they appear to be quite deadly. Myths and legendary motifs recorded in the sacred texts of ancient cultures throughout the world chronicle just such extraordinary events as these.

The next expected return of the Leonid Meteor storm is November 19, 2034.

Predictions are calling for 20 to 1000 meteors an hour. Will the comet of the apocalypse roar out of the Lion's mouth, and pulverize the Earth with a hoard of comet fragments during the expected meteor storm in 2034? Or is the world destroying firefall caused by another comet swarm storm?

---

190 Journal of the Royal-Meteorological Society, 30-285

The Orionids

Meteor Shower: Orionids: Oct 7 – peak Oct 20
Eta Aquarids: Early May
Parent Comet: Comet Halley
Annual Return: 75 years
Next Return: 2061.

Despite the vast size of its coma, Halley's nucleus is relatively small: barely 15 kilometers long, 8 kilometers wide and perhaps 8 kilometers thick. Its shape vaguely resembles that of a peanut. When Halley's sweeps by Earth in July 2061, the comet will be on the same side of the sun as Earth and will be much brighter than in 1986. One astronomer predicted it could be as bright as apparent magnitude -0.3. This is relatively bright, but well below that of the brightest star in Earth's sky: Sirius, at magnitude -1.4 as seen from Earth.[191]

Inescapable Conclusion: The Core of the Meteor Storms are Coming

Using sophisticated computer software, Victor Clube, an astrophysicist, and Bill Napier, an astronomer, looked carefully backward for thousands of years, tracing the orbits of comets, asteroids, and meteor showers until they uncovered something astounding.

Many meteor showers are related to one another, such as the **Taurids, Perseids, and Orionids. They determined that there is an i**nherent danger of multiple comet impacts for the next 400 years. (Firestone et al. 2006, 354-355)[192]

An Uncomfortably CLOSE Fifth Possibility: The Asteroid Apophis:

On Christmas Eve 2004, Paul Chodas, Steve Chesley and Don Yeomans at NASA's Near-Earth Object Program office calculated a 1-in-60 chance that 2004 MN4 would collide with Earth. Impact date: April 13, 2029. Asteroid 2004 MN4 had been discovered in June 2004, lost, then discovered again six months later. With such sparse tracking data, it was difficult to say, precisely, where the asteroid would go. A collision with Earth was theoretically possible. "We weren't too worried," Chodas says, "but the odds were disturbing."

191More: http://www.space.com/19878-halleys-comet.html#sthash.5yNXMcMK.dpuf

192 Knight-Jadczyk, Laura (2012-07-26). The Apocalypse: Comets, Asteroids and Cyclical Catastrophes (p. 50-51). Red Pill Press. Kindle Edition.

On 13 April 2029 an asteroid named Apophis, will pass by the earth at a distance of less than 19,000km. If you're alive at the time, and it is not cloudy, you'll be able to see it pass with the naked eye. Apophis is more than 300m (1000 feet) in diameter.

David J. Tholen and Tucker—two of the co-discoverers of the asteroid—are reportedly fans of the TV series Stargate SG-1. One of the show's persistent villains is an alien named Apophis.

In the fictional world of the show, the alien's backstory was that he had lived on Earth during ancient times and had posed as a god, thereby giving rise to the myth of the Egyptian god of the same name.

Apophis is the Greek name of an enemy of the Ancient Egyptian sun-god Ra: Apep, the Uncreator, an evil serpent that dwells in the eternal darkness of the Duat and tries to swallow Ra during his nightly passage. Apep is held at bay by Set, the Ancient Egyptian god of storms and the desert.

Check out Wikipedia Apophis ref 18 copyrighted material below in red

According to scientists, during its closest approach to Earth, it Apophis, designated 99942 Apophis, will pass through a certain narrow window called a "keyhole" in space, deflecting the asteroid just enough to cause an orbital change that could on the second day of Passover, April 13, 2036, cause it to return and hit the earth. If Apophis hits the Earth, the impact will be in the 3000-megaton class. Such an impact, taking place anywhere on the planet, would collapse our current civilization and return the survivors, metaphorically speaking, to the Dark Ages. An impact this large, would cause globalized institutions, such as the financial and insurance markets, to collapse, bringing down the entire interconnected monetary, trade and transport systems. Perhaps it is why most people have chosen to avoid, or ignore, the issue. They don't want to think about it.

Apophis, a 325-meter diameter near-Earth asteroid has been the focus of considerable attention after it was found in December 2004 to have a significant probability of Earth impact in April 2029. While the 2029 potential impact was ruled out within days through the measurement of archival telescope images, the possibility of a potential impact in the years after 2029 continues to prove difficult to rule out. A new report, which does not make use of the 2013 radar measurements, identifies over a dozen keyholes that fall within the range of possible 2029 encounter distances. Notably, the potential impact in 2036 that had previously held the highest probability has been effectively ruled out since its probability has fallen to well below one chance in one million.

Indeed, only one of the potential impacts has a probability of impact greater than 1-in-a-million; there is a 2-meter wide keyhole that leads to an impact in 2068, with impact odds of about 2.3 in a million.[193] Apophis has a seven-year orbital period around the sun. It's last orbital flyby of the Earth on January 09, 2013, missed the Earth by more than 9 million miles (14.5 million kilometers).

Next time we won't be so lucky. On April 13, 2029, Apophis will come so close that it may destroy satellites in orbit. On its 2013 flyby, The European Space Agency's Herschel space observatory has acquired new images of the asteroid and their new data is conclusive. First, Apophis is much bigger than NASA's previous estimation. According to the new images, this rocky beast has a diameter of 1,066 foot (325 meters), with a margin of error of ±49 feet (±15 meters). According to team leader Thomas Müller of the Max Planck Institute for Extraterrestrial Physics in Garching, Germany, "the 20% increase in diameter, from 270 to 325 meters, translates into a 75% increase in our estimates of the asteroid's volume or mass."

What that means is if it hits Earth, its destructive power will be much higher than what scientists originally expected. Based on previous data, NASA estimated an impact of 510-800 megatons for Apophis. That's more than two times the energy released by the Krakatoa eruption of 1883, an event that changed Earth's global climate for five years.

There is always a possibility we don't have these measurements exactly right. Something could happen at any point in Apophis' orbit to modify its course, just a smidgen. A tiny collision with another object, way out beyond Mars? What could change between now and 2029, or during any orbit thereafter? Apophis masses at more than 20 million tons. If it hit Earth, the impact would unleash a blast the equivalent of over a billion tons of TNT. That's not an extinction event, but it could easily cause billions of deaths and months, if not years, of climate disruption. The potential risk is huge.

At the moment, Apophis is not predicted to hit the Earth, but the close approach of such large asteroid is rare, occurring only on 1000-year intervals, on average. "The future for Apophis on Friday, April 13 of 2029 includes an approach to Earth no closer than 29,470 km (18,300 miles, or 5.6 Earth radii from the center, or 4.6 Earth-radii from the surface) over the mid-Atlantic, appearing to the naked eye as a moderately bright point of light moving rapidly across the sky. This is within the distance of Earth's geosynchronous satellites.

---

193 See report @ http://arxiv.org/abs/1301.1607

However, because Apophis will pass interior to the positions of these satellites at closest approach, in a plane inclined at 40 degrees to the Earth's equator and passing outside the equatorial geosynchronous zone when crossing the equatorial plane, it does not threaten the satellites in that heavily populated region.

On April 13, 2029, asteroid 2004 MN4 will fly past Earth only 18,600 miles (30,000 km) above the ground. For comparison, geosynchronous satellites orbit at 22,300 miles (36,000 km). "At closest approach, the asteroid will shine like a 3rd magnitude star, visible to the unaided eye from Africa, Europe and Asia--even through city lights," says Jon Giorgini of JPL. This is rare. "Close approaches by objects as large as 2004 MN4 are currently thought to occur at 1000-year intervals, on average."

Above: The trajectory (blue) of asteroid 2004 MN4 past Earth on April 13, 2029. Uncertainty in the asteroid's close-approach distance is represented by the short white bar. The asteroid's trajectory will bend approximately 28 degrees during the encounter, "a result of Earth's gravitational pull," explains Giorgini. What happens next is uncertain. Some newspapers have stated that the asteroid might swing around and hit Earth after all in 2035 or so, but Giorgini discounts that: "Our ability to 'see' where 2004 MN4 will go (by extrapolating its orbit) is so blurred out by the 2029 Earth encounter, it can't even be said for certain what side of the sun 2004 MN4 will be on in 2035. Talk of Earth encounters in 2035 is premature." The closest encounter of all, Friday the 13th, 2029, will be spectacular to the naked eye--wow! No one in recorded history has ever seen an asteroid in space so bright.

Velocity and Energy Release of Incoming Objects

The velocities at which small meteorites have impacted the Earth range from 4 to 40 km/sec. larger objects would not be slowed down much by the friction associated with passage through the atmosphere, and thus would impact the Earth with high velocity. Calculations show that a meteorite with a diameter of 30 m, weighing about 300,000 tons, traveling at a velocity of 15 km/sec (33,500 miles/hour) would release energy equivalent to about 20 million tons of TNT.

Such a meteorite struck at Meteor Crater, Arizona (the Barringer Crater) about 49,000 years ago leaving a crater 1200 m in diameter and 200 m deep. The amount of energy released by an impact depends on the size of the impacting body and its velocity. $E = \frac{1}{2} MV^2$ where E = Energy, M = Mass (depends on size and density of the object), and V = Velocity.

http://rack.0.mshcdn.com/media/ZgkyMDEzLzAxLzEwL2Q4L0Fzd
GVyb2lkSWRhLjNkMmVmLmpwZwpwCXRodW1iCTk1MHg1MzQjCm
UJanBn/ccf63ae3/f82/Asteroid-Ida.jpg

The Slooh Space Camera caught a glimpse of the much-talked-about
Apophis on Jan. 9 as it passed 9 million miles away from Earth. While
the asteroid's size isn't that large — about three-and-a-half football fields
— it's notable because it will come in very close contact with Earth when
it circles back around in 14 years. In Egyptian myth, **Apophis** was the
ancient spirit of evil and destruction, ... from **Apophis**, which has an
outside chance of hitting the Earth in **2036**, ... said: "When it does pass
close to us on **April** 13, 2029, the Earth will deflect it ...

The Keyhole:

During the 2029 flyby, it was predicted that Apophis might pass
through a gravitational keyhole, a narrow region of space (just 600 wide
meter wide in this instance) where Earth's gravity alters the asteroid's
path in such a way as to set it on a future collision course. While remote,
the possibility was enough to cause concern. –

See more at: http://astrobob.areavoices.com/2013/01/11/take-a-
breath-apophis-wont-hit-earth-in-2036/#sthash.gPmsMOZC.dpuf

On that date, Apophis will pass only 19,400 miles from the surface
of Earth and shine brightly enough (mag. 3.5) to be visible with the
naked eye from suburban skies. Viewing conditions will be ideal in
Europe, Africa and the Middle East observers with dark skies during
closest approach on the 13th; U.S. observers won't be as fortunate. When
Apophis is nearest and brightest, the sun will still be up in the sky. The best views for the States will happen on the evening of the 12th, when you'll see it in binoculars as a 7th magnitude point of light low in the southern sky. When darkness falls on the 13th in the U.S., you'll need an 8-inch or larger scope to spot it. Speed and proximity are behind Apophis' dramatic change in brightness over such a short span of time. It tears by Earth at such a high rate of speed that it brightens and fades in hours instead of days or weeks.

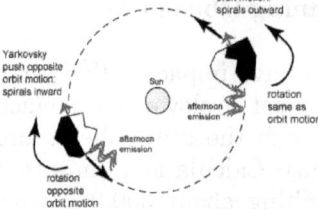

Because any material takes some time to heat up, it emits more infrared energy in the afternoon than in the morning. This extra afternoon emission provides a *small* push like a tiny rocket (from Newton's 3rd law) that's always 'on' called the 'Yarkovsky effect'. The Yarkovsky effect can make the object spiral outward away from the Sun if the object is spinning in the same direction as its orbit motion or spiral inward toward the Sun if the object is spinning in the opposite direction as its orbit motion.

Changes in an asteroid orbit -The Yarkovsky Effect

Adding to the uncertainty is the extent to which a subtle force,
known as the Yarkovsky effect, might be altering the asteroid's orbit.

This effect is caused by the uneven way that a spinning body absorbs sunlight and then reradiates it back to space.

Ground-based observers determined that Apophis rotates in 30½ hours, but it likely has more than one period involving multiple spin axes. The object's shape and spin orientation are unknown — and might remain open questions until 2029. "We might get coarse-resolution images that barely resolve the object and indicate its orientation," explains Benner, "but even that could be optimistic." Conceivably, gentle but persistent nudging from the Yarkovsky effect might have pushed Apophis straight through the 2029 keyhole. But again, says Giorgini, there's no longer any chance of that. The Goldstone observations have "shrunk the orbital uncertainties so much that, regardless of what the still-unknown physical parameters of Apophis might be, radiation pressure can't be enough to move the measurement uncertainty region enough to encounter the Earth in 2036." - See more at: http://www.skyandtelescope.com/astronomy-news/asteroid-apophis-takes-a-pass-in-2036/#sthash.Eqrmwga1.dpuf

One process that affects the orbits of asteroids and, therefore, introduces uncertainty in whether a particular NEA will hit the Earth is the Yarkovsky effect. In the Yarkovsky effect, there is a slight misalignment of the energy emitted by the asteroid and the energy it receives from the Sun. Because any material takes some time to heat up, the asteroid's afternoon side emits more infrared energy than the morning side. The afternoon emission of infrared energy from solar heating is not pointed right at the Sun, so the thermal radiation from the asteroid is not exactly balanced by the solar photons.

This results in a push that can move the asteroid inward toward the Sun or away from the Sun. If the asteroid is rotating in the same direction that it moves in its orbit around the Sun ("prograde rotation"), the asteroid will be pushed away from the Sun; if the asteroid is rotating in the opposite direction from its orbital motion ("retrograde rotation"), the asteroid will be pushed toward the Sun. The effect is very small, but it is continually acting on the asteroid so over many years it can have a measurable influence on the asteroid's motion.

Unfortunately for us, the Yarkovsky effect depends on all sorts of features about the asteroid itself that we do not know things like the asteroid's size, mass, how the material in the asteroid responds to heat, and most importantly on the orientation of the asteroid's spin axis (remember the misalignment of the noon Sun energy input direction vs. the afternoon asteroid energy output direction). If the asteroid is tumbling about two or more axes, then the afternoon infrared radiation it emits from solar heating is randomly directed instead of being in one consistent direction. Without a consistent direction of the infrared emission, the Yarkovsky effect is eliminated. A related effect called the "YORP effect" arises from the non-uniform shape and reflectivity of various parts of the asteroid that can cause one side of the asteroid to be a better emitter of its infrared energy than another part. The YORP effect can speed up or slow down the asteroid's spin rate and also change the spin axis direction leading to a possible increase in the Yarkovsky effect.

Future Apophis Impact possibilities

But the worry about Apophis has only been postponed, not eliminated. Its orbit is not all that different from Earth's, and some day in the distant future the two bodies will either have a catastrophic collision — or an encounter so close that Earth's gravity will yank Apophis onto a new and significantly different interplanetary path.

In fact, "The 2068 impact probability for Apophis is now one in 189,000," notes Rusty Schweickart (cofounder of the asteroid monitoring B612 Foundation), "which is higher than the 2036 impact probability was." - See more at: **http://www.skyandtelescope.com/astronomy-news/asteroid-apophis-takes-a-pass-in-2036/#sthash.Eqrmwga1.dpuf**

Part VI

The Future - Armageddon

18

Ancient Prophecies of Modern Apocalypse

Mayan Baktun 4: 1993 - 2012

Mayan Baktun 2: 2012-2032

Mayan Baktun 13: 2032 -2052

The Apocalypse of Ezra

The seer speaks of the signs which precede the End Days come when the inhabitants of the earth are seized with great panic and the way of truth hidden and the land will be barren of faith.

Then the sun will suddenly shine by night and the moon by day. Blood will trickle forth from wood and stone speak its voice. People will be confounded, and stars change course. An unknown force will wield sovereignty and birds take general flight and the sea hurl up its fish. An unknown voice will be heard by night, and all will hear, and the earth will break open over vast regions and fire explode interminably.

Wild beasts will desert their haunts, and women bear monsters. One-year-old children will speak, pregnant women will bring forth at three or four months, and these will live and dance. Sown fields will dry up, full storehouses be empty, saltwaters turn sweet, friends attack each other fiercely, then intelligence will hide and wisdom withdraw to its chamber where none can find it. Unrighteousness and lust will cloud the earth and lands will ask each other: "Has righteousness come your way?

"And the answer will be NO! In that time all hope will fail, all labor fail. These signs I tell you, but if you pray, weep, and fast for seven days, you will hear wondrous things. *Chapters 5:1,5:4-13. All selections in this chapter are revisions of selections from the translation in R. H. Charles, The Apocrypha and Pseudepigrapha of the Old Testament, vol.2 (Oxford: Clarendon Press)

Some people ask, "What about free will and our ability to change the course of events, and therefore, change the world ending prophecy?" The entire argument between free will and predestination occupied many thinkers and writers for centuries and was sometimes a raging argument indeed. Had God predestined the "end of the world" or could humanity, through free will or perhaps the redemption of a Savior, escape -- or, better still, transform that terrible fate?

Hopi Prophecy:

The Signs that reveal the Coming Age:

`Someday there will be a road in the sky and a machine will ride this road and drop a gourd of ashes and destroy the people and boil the land.'

`Men on earth will make a small ball that shall contain the elements of their earth... and it shall cause a great destruction and a great cloud over all people.'

And yet another says:

`In a future day, there will be seen houses in the sky, and they will have no support from the earth... and some will be joined together.'

The Hopis claim there will be two forerunners to the `true white brother' who will witness for Him. One messenger will carry a Swastika (Moon?) and the other one will carry a Sun-Disc (Sun?). In this connection, it is interesting to read Revelation 11:3:

`And I will give power unto my two witnesses, and they shall prophesy...'

`And there shall be signs in the sun, and in the moon, and in the stars; and upon the earth distress of nations with perplexity; the sea and waves roaring; Men's hearts failing them for fear...' (St. Luke 21:25,26)

*MAT 24:29 "Immediately after the distress of those days "'the sun will be darkened, and the moon will not give its light; the stars will fall from the sky, and the heavenly bodies will be shaken.' (Isaiah 13:10; 34:4)

One of the Hopi prophecies says:

`In the last days, strange lights will be seen in the sky and they will be watching the Hopi people to see if they are following the Life Plan and these strange lights will report to the "true white brother" in the east and they will tell Him when it is time for Him to come again.'

Another prophecy says:

`When the "true white brother" returns, all forms of transportation will be stopped and man will not be able to move about the land...weapons will be useless.' (note: Angela's dream of no cellphones or technology.)

The sign of the "Blue Star":

Hopi people will migrate to the safety of Oraibi.

According to Hopi Prophecy, their leaders have watched for three world shaking events, accompanied by the appearance of certain symbols that describe the primordia forces that govern all life. The Gourd rattle is a key symbol they expected to see. Drawn on it is a "Swastika" (meaning "seed" forces) surrounded by a ring of red fire. This possibly shows the encircling penetration of the sun's warmth which causes the seed to sprout and grow.

The first two world-shaking events would involve the forces portrayed by the swastika and the sun. Out of the violence and destruction of the first, the strongest elements would emerge with still greater force to produce the second event. When the actual symbols appeared (In world war I & World War II) this stage of the prophecy was being fulfilled.

Eventually a "gourd full of ashes" would be invented, which if dropped from the sky would boil the oceans and burn the land, causing nothing to grow there for many years. Apparently, the gourd of ashes is the atomic bomb, or a comet, or large meteorite shower.

A final stage called the "great day of purification," has also been described as a "mystery egg," in which the forces of the swastika and the sun, plus a third force symbolized by the color red, culminate either in total rebirth, or total annihilation. War and natural catastrophe may be involved. In this crisis, rich and poor will struggle as equals to survive.

The Hindu Mahabarata

Hindu prophecy written 4000 years ago, describes the end of the Age of Kali, which we are now approaching, "And always oppressed by bad rulers with burdens of taxes, the foremost of the best classes will, in those terrible times, take leave of all patience and do improper acts... And the low will become high, and the course of things will look contrary. And renouncing the gods, men will worship bones and other relics deposited in walls.... These all will take place at the end of the Yuga. And when men become fierce and destitute of virtue and carnivorous and addicted to intoxicating drinks, then does the Yuga come to an end. And when flowers will be begot within flowers, and fruits within fruits, then will the Yuga come to an end...

And the course of the winds will be confused and agitated, and innumerable meteors will flash through the sky foreboding evil. And then the Sun will appear with six others of the same kind. And all around will be din and uproar, and everywhere there will be conflagrations... And fires will blaze up on all sides. ... And even the foremost of the best classes, afflicted by robbers, will, like crows, fly in terror and will speed and seek refuge in rivers and mountains and inaccessible regions. And, when the end of the Yuga comes, crows and snakes and vultures and kites and other animals and birds will utter frightful and dissonant cries.... And people will wander over the Earth, uttering, "Oh father! Oh son!" and such other frightful and heart-rending cries."

Tibetan Buddhist

The Tibetans say we are living at the end of a 26,000-year period of darkness. A series of global catastrophes, accompanied by political strife, will initiate Purification and a new era of spirituality and light. The Shambhala tradition of Tibet is preserved in numerous sacred texts and oral teachings that tell of a mystical kingdom hidden behind the snow peaks somewhere to the north. There, a line of enlightened kings is guarding the innermost teachings of Buddhism for a time when all truth in the world outside is lost in war and greed. Then, according to prophecy, the King of Shambhala will emerge with a great army to destroy the forces of evil and bring in a Golden Age.

The final battle is to be expected shortly after the barbarians of the outer world fly over the protective snow mountains in "vehicles made of iron" in an attempt to invade Shambhala.194

---

194 See Edwin Bernbaum, The Way to Shambhala (Los Angeles: Tarcher, 1989)

The Andaman Islanders

The world will come to an end in a great earthquake, which will destroy the barrier between Heaven and Earth. The spirits of the dead will then be reunited with their souls, and human beings will lead happy lives without sickness, death, or marriage. Even now, they say, the impatient spirits of the underworld are beginning to shake the roots of the palm tree that supports the Earth, so as to bring the end more quickly.

The Aborigines of Australia

The end of the world will come when the Dreamtime Law the code of rituals established by the Creator Ancestors is no longer kept. Among the many Aboriginal tribes, the last members initiated into these codes of ritual are growing old, with no young initiates to take their place. The Dreamtime Law is being forgotten, and the elders anticipate dire consequences for the entire world.

The Mortlock Islanders of the South Pacific

They foretell when the day comes in which people no longer worship the Creator God Luk, when they wage wars and commit sins, the Lord of the World will put an end to them. Everything will go to ruin; only the gods will live on in their heavenly places.

Pawnee Myth

There will come a termination to all earthly life, which will be preceded by horrifying portents:

1. The Moon will turn red, and the Sun will die.
2. The North Star will preside over the great destruction. "When the time comes for all things to end," say the Pawnee prophets, "Our people will turn into small stars and will fly to the South Star, where they belong."195

The Sibylline Oracles

Five-hundred years before the Revelation of St. John Divine was written, the Sibylline Oracles prophesied from distant Iceland, a

---

195 Harley Burr 1916. North American Mythology, in the Mythology of All Races, vol. 10 (Boston: Marshall Jones, 1916), p. 116.

similar "wormwood-like" star that falls into the ocean causing an immediate ensuing winter:..."Be afraid ye Indians and high hearted Ethiopians: for when the fiery wheel of the ecliptic....and Capricorn....and Taurus...among the Twins encircles the mid-heaven, when the Virgin ascending and the Sun fastening the girdle round his forehead dominates the whole firmament; there shall be a great conflagration from the sky, falling on the earth; And then in his anger the immortal God who dwells on high shall hurl from the sky a fiery bolt on the head of the unholy; and summer shall change into winter in that day."

The Icelandic prophetess, Sibyl, says the beginning of the end would find all men embroiled in the Age of the Wolf, a "terrible time" Celtic Legends, that will be proceeded by five signs:

1. War & butchery.
2. Sordid greed.
3. Followed by the Monstrous Winter when the world would have no summertime for three years in a row.
4. Theme of sinful betrayal. Warriors would betray one another, and men would be thieves.
5. The coming Waste Land, which was also a kind of apocalypse brought on by human faults. According to the Irish Fate Goddess Babd, the "Boiling One" of the cauldrons, the Waste Land would come upon the world when "old men would give false judgments and legislators make unjust laws; and there would be no more virtue left in the world.196

The final stanzas of the Voluspa, the "Sibyl's Prophecy," paint an idyllic picture of restored Paradise on Earth:

Now do I see the earth anew.
Rise all green from the waves again;
the cataracts fall, and the eagle flies, and the fish he catches beneath the cliffs.
The gods in Idavoll meet together,
Of the terrible girdler of Earth they talk.
And the mighty past they call to mind,
And the ancient runes of the Ruler of Gods.
In wondrous beauty once again shall the golden tables stand mid the grass,
Which the gods had owned in the days of old.
Then fielded unsowed bear ripened fruit,

---

196 Teutonic Norse: Ragnarök "The doom of the gods."

All ills grow better, and Baldr comes back:
More fair than the sun, a hall I see,
Roofed with gold, on Gimle it stands;
There shall the righteous rulers dwell.
And happiness ever there shall they have."197

In the **Ragnarök** version of the Volupsa legend, six signs that precede the End of the World, are caused by Surt

1. Preceded by a period of anarchy in which human beings will perform every kind
of foul crime.
2. Then, the sky will be rent, the stars will fall.
3. The mountains will be shattered with earthquakes.
4. All the gods will die, with the exception of Surtr (comet) who will cause the
Earth to be engulfed by flames, destroying all humankind.
5. As the flames rise to Heaven, the Earth will sink into the sea. But then it will rise from the waters renewed, fresh and green, to be re-peopled again.
6. A New Earth.

The **Ragnarök** prophecy describes the great battle in detail, "He will go to battle against the Æsir, he will do battle with the major god **Freyr**, an afterward the flames that he brings forth will engulf the Earth. In chapter 51 of Gylfaginning, High describes the events of Ragnarök. High says that "amid this turmoil the sky will open and from it will ride the sons of Muspell. Surt will ride in front, and both before and behind him there will be burning fire. His sword will be very fine. Light will shine from it more brightly than from the sun."

High continues that when the sons of Múspell ride over the bridge **Bifröst** it will break, and that they will continue to the field of **Vígríðr**. The wolf **Fenrir** and the **Midgard Serpent** will also arrive there. By then, **Loki (Satan)** will have arrived with "all of Hel's people", **Hrym**, and all of the frost jötnar; "but Muspell's sons will have their own battle array; it will be very bright".

Further into the chapter, High describes that a fierce battle will erupt between these forces and the Æsir, and that during this, Surt and Freyr will engage in battle "and there will be a harsh conflict before Freyr falls." High adds that the cause of Freyr's death will be that Freyr is lacking "the good sword" that he once gave his servant **Skírnir**.

197 E.O.G. Turville Petre 1964, translated, Myth and Religion of the North. London: Weidenfeld and Nicolson, pp. 42-66.

As foretold by High further into chapter 51 Gylfaginning, once **Heimdallr** and **Loki** fight (and mutually kill one another), Surt "will fling fire over the earth burn the whole world." As the flames rise to Heaven; the Earth will sink into the sea. But then it will rise from the waters renewed, fresh and green, to be repeopled. High quotes ten stanzas from Völuspá in support, and then proceeds to describe the rebirth and new fertility of the reborn world, and the survivors of Ragnarök, including various gods and the two humans named **Líf and Lífthrasir** that have hidden from "Surt's fire" in the wood **Hoddmímis holt**."

The Paraphrase of Shem

I interpreted the following non-Christian Gnostic work discovered in the Nag Hammadi Library as a typical example of comet body accompanied with Firefall, its progeny of serpents—super bolides exploding in the atmosphere and impacting Earth below causing tsunami, oceanic floods, earthquakes, hurricanes, fires, high atmospheric heat, acid rain, destruction of humans, plants and animals, destruction of crops, darkness from impact and volcanic ash, famine, cannibalism, and darkness. The only effect missing is the inevitable impact winter:

And when the era of Nature is approaching destruction, darkness will come upon the earth. [Darkness will envelop the Earth soon after meteor/comet impact] The number [left remaining] will be small. [majority of life on Earth will be wiped out.] And a demon will come up from the power who has the likeness of fire. [Impact fire heats atmosphere to several hundred degrees igniting anything flammable. ire and heat spring up] He will divide the heaven, and he will rest in the depth of the east. [comet will split the heaven as it rises from the East] For the whole world will quake. [massive world-wide earthquakes]. And the deceived world will be thrown into confusion. [Fear] Many places will be flooded because of envy of the winds and the demon. [Hurricane Winds] who have a name which is senseless: Phorbea and Chloerga. [govern floods, hurricanes, teaching the world fear] They are the ones who govern the world with their teaching. And they lead astray many hearts because of their disorder and their unchastity. [Sin] Many places will be sprinkled with blood. [common first sign of Firefall] And five races by themselves will eat their sons. [Famine drives people to cannibalism] But the regions of the south will receive the Word of the Light. [South regions less affected by Firefall.] But they who are from the error of the world and from the east will not. [Humans living in the East covered by darkness.] A demon will come forth from the belly of the serpent.

[Serpents are meteors from the primary comet body.] He was in hiding in a desolate place. [Parent Comet and Serpents of stone impacting Earth appeared suddenly, without notice.]

He will perform many wonders. Many will loathe him. A wind will come forth from his mouth with a female likeness. Her name will be called Abalphe. [See Revelation:]

He will reign over the world from east to west. [Effects of impacts will cover entire planet] Nature will have a final opportunity. And the stars will cease from the sky. [Impact ash, volcanic smoke will darken the Earth.] The mouth of error will be opened in order that the evil Darkness may become idle and silent. And in the last days, the forms of Nature [humans, plants, and animals] will be destroyed with the winds and all their demons; they will become a dark lump, [will be covered up and lost to the ages] just as they were from the beginning...[198]

---

198 The Paraphrase of Shem: A non-Christian Gnostic work The Conflict of Light and Darkness, Codex VII 1, 1-49, 9. Translated by Frederick Wisse. From James M Robinson, ed,,The Nag Hammadi Library (San Francisco: Harper and Row, 1977) pp. 309_329.(From the Other Bible)...

—— **19** ——

Wheels of the Apocalypse

Origins of the Destroyer

Concerning the Great Comet described then by Clube and Napier, that they determined first entered our solar system some 20,000 years ago, one is bound to consider the gravitational capture of the object by the sun. Indeed, were this to have occurred with some level of stability achieved, with an orbital period established within a range of some several thousand years or so, there is the distinct possibility that one or more of the destructive events noted in this presentation, deep in antiquity, were caused by this very comet. Clube and Napier calculated that, to account for all the debris they found strewn throughout our solar system, the original comet had to have been enormous. ... Clube and Napier also calculated that, because of subtle changes in the orbits of Earth and the remaining cosmic debris, Earth crosses through the densest part of the giant comet clouds about every 2,000 to 4,000 years. When we look at climate and ice-core records, we can see that pattern.

For example, iridium, helium-3, nitrate, ammonium, and other key measurements seem to rise and fall in tandem, producing noticeable peaks around 18,000, 16,000, 13,000, 9,000, 5,000, and 2,000 years ago. In that pattern of peaks every 2,000 to 4,000 years, we may be seeing the "calling cards" of the returning mega-comet. Fortunately, the oldest peaks were the heaviest bombardments, and things have been getting quieter since then, as the remains of the comet break up into even smaller pieces. The danger is not past, however. Some of the remaining miles-wide pieces are big enough to do serious damage to our cities, climate, and global economy...

Figure 1. Temperature fluctuations over the past 17,000 years showing the abrupt cooling during the Younger Dryas. The late Pleistocene cold glacial climate that built immense ice sheets terminated suddenly about 14,500 years ago (1), causing glaciers to melt dramatically. About 12,800 years ago, after about 2000 years of fluctuating climate (2-4), temperatures plunged suddenly (5) and remained cool for 1300 years (6). About 11,500 years ago, the climate again warmed suddenly, and the Younger Dryas ended (7).

And in this regard, it is quite uncanny that the description given by Clube and Napier, as to what the comet would have looked like viewed from the earth, is very close to that contained within the ancient Egyptian books of the Kolbrin concerning the passage of The Destroyer; the name assigned to a great red comet that passed close to the earth causing death and destruction in some remote age; a comet with a destroying rain of meteors and red dust described in mythology and legend and prophecy...

Mythical Beliefs: Sun/Epoch Ages by Culture

Why did ancient peoples of both hemispheres substitute the word "sun" for "epoch?" Did the reason lie in the changed appearance of the luminary and its changed path across the sky in each world age?

The traditions of the world ages are the advent of a new sun in the sky at the beginning of every age. The word "sun" is substituted for the word "age" in the cosmological traditions of many peoples all over the world. The Mayas counted their ages by the names of their consecutive suns. These were called Water Sun, Earthquake Sun, Hurricane Sun, and Fire Sun. "These suns mark the epochs to which are attributed the various catastrophes the world has suffered."

Important Note: The Brahmans called the epochs between two destructions "the great days."

South American/ Ixtlilxochitl (c. 1568-1648)

The native Indian scholar, in his annals of the kings of Tezcuco, describe the world ages by the names of "suns. "The Water Sun (or Sun of Waters) was the first age, terminated by a deluge in which almost all creatures perished; the Earthquake Sun or age perished in a terrific earthquake when the earth broke in many places and mountains fell. The world age of Hurricane Sun came to its destruction in a cosmic hurricane. The Fire Sun was the world age that went down in a rain of fire.

South American / Culhua / Mexico

"The nations of Culhua or Mexico," Humboldt quoted Gomara (accent over the a), the Spanish writer of the sixteenth century, "believe according to their hieroglyphic paintings, that, previous to the sun which now enlightens them, four had already been successively extinguished. These four suns are as many ages, in which our species has been annihilated by inundations, by earthquakes, by a general conflagration, and by the effect of destroying tempests. Every one of the four elements participated in each of the catastrophes; deluge, hurricane, earthquake, and fire gave their names to the catastrophes because of the predominance of one of them in the upheavals. Symbols of the successive suns are painted on the pre-Columbian literary documents of Mexico.

Tibet

"Analogous traditions of four expired ages persist on the shores of the Bengal Sea and in the highland of Tibet the present age is the fifth."199

India / Bhagavata Purana

"The sacred Hindu book Bhagavata Purana tells of four ages and of pralayas or cataclysms in which, in various epochs, mankind was nearly destroyed; the fifth age is that of the present. The world ages are called Kalpas or Yugas. Each world age met its destruction in catastrophes of conflagration, flood, and hurricane. Ezour Vedam and Bhaga Vedam, sacred Hindu books, keeping to the scheme of four expired ages, differ only in the number of years ascribed to each." 200

India / Visuddhi-Magga

In the chapter, "World Cycles," in Visuddhi-Magga, it is said that "there are three destructions; the destruction by water, the destruction by fire, and the destruction by wind," but that there are seven ages, each of which is separated from the previous one by a world catastrophe. After the catastrophe of the deluge, 'When now a long period has elapsed from the cessation of the rains, a second sun appeared." In the interim the

---

199 Immanuel Velikovsky, Worlds in Collision, p. 30.

200 Immanuel Velikovsky, Worlds in Collision, p. 31.

world was enveloped in gloom. [author's note: An excellent description of the beginning of the Younger Dryas, a 1500-year period of cold that followed collision / oceanic flood].

When this second sun appears, there is no distinction of day and night, but "an incessant heat beats upon the world." When the fifth sun appeared, the ocean gradually dried up; when the sixth sun appeared, 'the whole world became filled with smoke.'

'After the lapse of another long period, a seventh sun appears, and the whole world breaks into flames.' This Buddhist book refers also to a more ancient "Discourse on the Seven Suns." 201

North America / Hopi Indians

The following account of the Second World is from The Book of the Hopi. It clearly describes the 10,890 B.C. Pole Shift/Flood event:

This is when the trouble started. Everything they needed was on this Second World, but they wanted more... The people began to quarrel and fight, and then wars between villages began. Still there were a few people in every village who sang the song of their Creation. But the wicked people laughed at them until they could sing it only in their hearts. Even so, Sotukknang [Lord of the Universe] appeared before them.

'Spider Woman tells me your thread is running out on this world,' he said, 'I have decided we must do something about it. We are going to destroy this Second World just as soon as we put you people who still have a song in your hearts in a safe place.'

So again, as on the First World, Sótukknang, called on the Ant People to open up their underground world for the chosen people. When they were safely underground, Sótukknang commanded the twins, Pöqánghoya and Palöngawhoya, to leave their posts at the north and south ends of the world's axis, where they were stationed to keep the Earth properly rotating.

The twins had hardly abandoned their stations when the world, with no one to control it, teetered off balance, spun around crazily, then rolled over twice. Mountains plunged into seas with a great splash, seas and lakes sloshed over the land; and as the world spun through cold and lifeless space it froze into solid ice. This was the end of Tokpa, the Second World." 202

201 Immanuel Velikovsky, Worlds in Collision, p. 34

202 Oswalk White Bear Fredericks recounted by Frank Waters, The Book of the Hopi,

Kuskurza, the Third World

The Hopi third world, Ksukurza, describes evil men who built tall buildings: pyramids, and Ziggurats. Before destroying men, he sent the Hopi on ships to the four corners of the world:
Man built cities and tall buildings, and grew so evil that this time, before destroying the world, Taoiwa sent the Hopi on a long journey to the four corners of the earth. Afterwards [after the destruction], they were to settle in the dry barren land where they live today, there to await the coming of their white brother Pahana, who upon united with the, would usher in a new age of brotherly love.
The white brother, Christ of the Americas, is a common story told throughout the America's. The white man with a T or Cross on his palms.

Tuwaquchi, The Fourth World, Present Age.

Persian / Zend-Avesta

Sacred scriptures of Mazdaism, the ancient religion of the Persians, describe, "Bahman Yast," one of the books of Avesta, counts seven world ages or millennia. Zaratthustra (Zoroaster), the prophet of Mazdaism, speaks of "the signs, wonders, and perplexity which are manifested in the world at the end of each millennium."203

Chinese Tradition

The Chinese call the perished ages kis and number ten kis from the beginning of the world until Confucius. In the ancient Chinese encyclopedia, Sing-li-ta-tsiuen-chou, the general convulsions of nature are discussed. Because of the periodicity of these convulsions, the span of time between two catastrophes is regarded as a "great year." As during a year, so during a world age, the cosmic mechanism winds itself up and "in a general convulsion of nature, the sea is carried out of its bed, mountains spring out of the ground, rivers change their course, human beings and everything are ruined, and the ancient traces effaced." (citation needed)

South American / Aztec / Mayan

An old tradition, and a very persistent one, of world ages that went down in cosmic catastrophes was found in the Americas among the Incas,

203 Immanuel Velikovsky, Worlds in Collision, p. 31.

the Aztecs, and the Mayas. A major part of stone inscriptions found in Yucatan refer to world catastrophes. "The most ancient of these fragments [katuns or calendar stones of Yucatan] refer, in general, to great catastrophes which, at intervals and repeatedly, convulsed the American continent, and of which all inundations of this continent have preserved a more or less distinct memory."

Codices of Mexico and Indian authors who composed the annals of their past give a prominent place to the tradition of world catastrophes that decimated humankind and changed the face of the earth. In the chronicles of the Mexican kingdom, it is said; "The ancients knew that before the present sky and earth were formed, man was already created and life had manifested itself four times."

A tradition of successive creations and catastrophes is found in the Pacific -- on Hawaii   and on the islands of Polynesia; there were nine ages and, in each age, a different sky was above the earth.

Icelanders, too, believed that nine worlds went down in a succession of ages, a tradition that is contained in the Edda.

Jewish Ages

The rabbinical conception of ages crystallized in the post-Exilic period. Already before the birth of our earth, worlds had been shaped and brought into existence, only to be destroyed in time. "He made several worlds before ours, but He destroyed them all." This earth, too, was not created at the beginning to satisfy the Divine Plan. It underwent reshaping, six consecutive remoldings. New conditions were created after each of the catastrophes. On the fourth earth lived the generation of the Tower of Babel; we belong to the seventh age. Each of the ages or "earths" has a name.

Seven heavens were created and seven earths were created; the most removed, the seventh, Eretz; the sixth, Adamah; the fifth, Arka: the fourth, Harabah; the third, Yabbashah; the second, Tevel; and "our own land called Heled, and like the others, it is separated from the foregoing by abyss, chaos, and water."

Wheels & Gears

#1:   Halley's   Comet   75-76   year/Michael-Gabriel/Sphinx/Orion/Pleiades Cycle

#2: Short term Phoenix cycle: 350-500 years. Tree Rings & Greenland Ice Core Connections

"Baillie, a dendochronologist, "compared tree rings to dated ice-core samples that had been analyzed and he discovered a very strange thing: ammonium. There are, as it happens, four occasions in the last 1,500 years where scientists can confidently link dated layers of ammonium in Greenland ice to high-energy atmospheric interactions with objects coming from space: 539, 626, 1014, and 1908 (the Tunguska event). In short, there is a connection between ammonium in the ice cores and extra-terrestrial bombardment of the surface of the earth.204

Insert short term impact evidence...

#3: 1460-1500 impact cycle *The Sothis Wheel: 1,460 to 1,500 years / Islamic Wheel*

The life of my community will not exceed 1500 years.

(Suyuti (1445-1505 A.D.), al-Kasfu an Mujawazati Hazihil Ummah al-Alfu, al-hawi lil Fetawi, Suyuti. 2/248, tafsir Ruh-ul Beyan, Bursewi 4/262, also Ahmad ibn Hanbal, Kitab al-Ilal, p.89)205

"The lord of time is close at hand, insha'Allah. The coming of the lord of time is nigh, and the world will change. We are approaching the Day of Judgment. Our master (missing quote)206

There appears to be a common consensus in Islam that the coming of the Madhi will happen sometime between 1400-1500 of the Islamic era— that is, sometime between 21st November 1979 and 28th November 2079.207

They ask you, [O Muhammad], about the Hour: when is its arrival? **Say, "Its knowledge is only with my Lord. None will reveal its time except Him.** *It lays heavily upon the heavens and the earth. It will not come upon you unexpectedly." They ask you as if you are familiar with it. Say, "Its knowledge is only with Allah, but most people do not know." (Surah: 7:187)208*

---

204 Knight-Jadczyk, Laura (2012-07-26). The Apocalypse: Comets, Asteroids and Cyclical Catastrophes. Red Pill Press. Kindle Edition. p. 31.
205 Ibid. p.29.
206 Ibid. p.29
207 Ibid. p.29
208 Ibid. p.29.

1460 Year Sothis Egyptian Cycle

Enke Beta Taurid 1500-year cycle

Astrophysicists - Astrophysicist, Victor Clube, in The Cosmic Winter, believes comet swarms have impacted the Earth in 1500-year cycles. *According to astronomer Victor Clube, the Earth is in much greater danger than previously acknowledged from meteoric impact: "It became fashionable to assume that the world is safe, when in fact, multiple Tunguska bombardments globally releasing over 5000 megatons of explosive energy, the equivalent of a full-scale nuclear war, may happen at internals of about 1,500 years producing a dark age."*

#4: 5126-year sun cycle: Mayan Hebrew Apocalypse Age-According to Enoch, the world is supposed to last 4,900 years.
(490 x 10) 1Enoch 93:3-10, 91:12-16 (Close to 5,126 Mayan age)
See chart of 6000-7000 Life Span of the world. (Note: Jewish, Christian, and Muslim Holy books speak of the world lasting only 6000-7000 years. The Old Testament says the world began in 4006 B.C. If we look at Jewish Calendar, they have the world beginning (or ending) in 3766 B.C.
Islam says the world was created 5,600 years before Mohammed. Counting from his Birth to Creation, in 570 A.D. it would be 5030 B.C., or from Creation to his death in 622 B.C., it would be 4,468 B.C.
Islam: The lifespan of the world is seven days in the days of the hereafter. Almighty Allah says;
One day in the sight of your Lord is like one-thousand of your years...Whoever meets the needs of his brother in the religion on the path of Allah, Allah will give him merit of the seven thousand-year life span of this world..." (Al-Muttaqu a-Hindi, Al-Burhan fi Alamat al-Mahdi Akhir al-Zaman, 88)209. *The age of the world up until the time of the Prophet Mohammed is given 5600 years in the work of ahmad Ibn Hanbal (780-855A.D.), the founder of the Hanbili school, in a hadith transmitted from the Prophet relates:*
Five thousand six hundred years have passed from this world." (..." (Al-Muttaqu a-Hindi, Al-Burhan fi Alamat al-Mahdi Akhir al-Zaman, 89)210

South American Urghamahas: 4,166 B.C.
Mayan Estimates: 3300 B.C.
Mayan Calendar: 3,114 B.C

209 Adam to Apophis, quote taken from p.29
210 Ibid. p.29.

Adjusted Catastrophe coinciding with Noah's Flood/Flood of Deucalion & comet impact – 3,700 B.C. – 2,750 B.C.

#5: 12,960-year return of **Quetzlcoatl / Sphinx of Leo Lion / Aquarius Man**

The ancient Sumerians, who thrived in Mesopotamia in 3,500 B.C., devised this unique system of measurement which we still use even today. We have inherited the Sumerian 12 in our counting of daily hours, 60 in our counting of time (60 seconds in a minute, 60 minutes in an hour), and 360 in geometry (360 degrees in a circle). This system, called "sexagesimal" is still the only perfect one in the celestial sciences, in time reckoning, and in geometry (Where a triangle has angles adding up to 180 degrees and a square's angles add up to 360 degrees).

When record keeping began, a stylus with a round trip was used to impress on wet clay the various symbols that stood for the numbers, 1, 10, 60, 600, and 3,600. The ultimate numeral was 3,600, signified by a large circle; it was called SAR --the "royal" number, and according to the Sumerians, the number of years it takes "Niberu," the "planet of the crossing," to complete one orbit around the sun.

Although dubbed "sexagesimal," the Sumerian system of numeration and mathematics was in reality not simply based on the number 60, but on a combination of 6 and 10. While in the decimal system each step up is accomplished by multiplying the previous sum by 10, in the Sumerian system the numbers increased by alternate multiplications; once by 10, then by 6, then by 10, then by 6. The decimal system is obviously geared to the ten digits of the human hands, so the 10 in the Sumerian system can be understood; but where did the 6 come from, and why?

Among the thousands of mathematical tablets from Mesopotamia, many held tables of ready-made calculations. Surprisingly, however, they did not run from smaller numbers up (like 1, 10, 60, etc) but ran down, starting from a number than can only be described as astronomical; 12,960,000:

1. 12960000 its 2/3 part = 8640000
2. its 1/2 part = 6480000
3. its 1/3 part = 4320000
4. its 1/4 part = 3240000
5. its 2160th part = 6000
6. its 216,000th part = 60

The number 12,960,000 is literally astronomical. It likely stems from the phenomenon of Precession, which retards the zodiac constellation against which the Sun rises by a full House once in 2,160 years. The complete circle of twelve Houses, by which the Sun returns to its original background spot, thus takes 25,920 years. The number 12,960,000 represents exactly 500 such complete processional circles. If the number 12,960,000 does indeed represent 500 processional cycles, why did they choose this starting point?

If 5 was an important number among the ancients, then if we divide a circle by 5, the number of degrees per section is 72. This proportional formula was often symbolized by the ancients as a Pentagram within the circle. It is also the inspiration for Leonardo Di Vinci's division of a circle by the five appendages of man within a circle.

It is interesting to note that every 72 years a gradual shift of one degree of precession was observed by ancient astronomers. Every 3,600 years a gradual shift of 50 degrees. Could it be more than coincidence that the Hebrew Jubilee year is celebrated on the 50th year? And that there are 72 Jubilee "Years" every 3600 years?

If we compare 3600 (One Great Year):2160 (One Zodiac Age) as a ratio, and reduce this "formula" to its lowest common denominator, it reduces down to 10:6. 10:6 is the Golden number ratio, and its reciprocal is the Divine number.

Here's where it gets interesting:

If 12,960,000 is divided by 2,160 (1 Zodiac Age) the result is 6,000. 1000 X 6.

Six as a number of "days" is not unfamiliar--the number of days of creation.

This may explain the puzzling statement of the Psalmist,

"A thousand years in thine eyes are but a day, a yesterday past."

Could the psalmist have seen the mathematical tablets which he would have found the line listing "12, 960,000 the 2160th part of which is 6000?

The Chaldeans, Egyptians, Chinese, and others, always used this method of division for purposes of measurement and number. For this reason, it must have been known and well understood by Moses, as he "was learned in all the wisdom of the Egyptians," and also by Daniel who lived among the Chaldeans in Babylon. Both of these servants of God knew that a "time" stood for 360 degrees, parts, days or years.

Like the ancients we divide the circle into 360 parts. There is no astronomical measurement without this mathematical division, and in this respect, we are but following the Chaldeans who long before us divided the Zodiac into 360 degrees or parts.

For the same purpose of ensuring accuracy, the navigator divides the equator into 360 degrees, and investigation shows that the Chaldean and ancient Hebrew solar cycle was reduced from 21 years to 15 (6) in order to make it subservient to the spherical computation of 360 years. This solar cycle was used by both Moses and Daniel, as well as John in the Book of Revelation.

For these reasons I feel assured that as a "time" is 360 years, so "time, times, and half a time" are 1260 years; even "seven times" are 2520 years, or twice 1260, the number used in Revelation. Multiplying by 10, a common practice by the Apocryphal writers, especially Daniel, "seventy times" equals 25,200 years.

Previously, we have mentioned the formula 10:6, A "Great Year of 3600 years" to one Age of the Zodiac of 2160 years. It incorporates the numbers 10 and 6. However, another more common formula used is 12:7. Again, a Precessional Age, consists of 12 ages of 2,160 years each totaling 25,920 years. 25,920 is not evenly divide by 7, however, if we take the opposite approach and multiply 3,600 X 7, it equals 25,200, our "seventy times" number. Thus:

One Precessional Rotation 25,920 = 12 (Divide by 2160 years)
7 Periods of "Great Years" 25,200 = 7 (Divide by 3600 years)

25,200 is 720 years short of a precessional age! Why is this interesting? 720 years is exactly (2X) 360, and is numerically related to 72, the number of years it takes for the sun to transit one degree of the Equinox. (out of 360 degrees). Therefore, 720 years represents 10 degrees of transit "short" of an even division into 25,920. This missing "10" degrees of transit is referred to in Isaiah 38:8.

"Behold, I will bring again the shadow of the degrees, which is gone down in the sundial of Ahazten ten degrees backwards. So, the sun returned ten degrees by which degrees it was gone down."

Author's Note: Isaiah spoke this prophecy around 700 B.C., predicting a return of Ahazten's sundial's ten-degree backward movement in a future apocalypse involving the Sun standing still over Gibeon and Moon standing still over Mt. Perazim causing the Lord God of Hosts (Lord God of Firefall comet swarms) a consumption (destruction) upon the entire planet:

Therefore, thus saith the Lord God, Behold, I lay in Zion for a foundation a stone, a tried stone, a precious corner stone, a sure foundation: he that believeth shall not make haste. Judgment also will I lay to the line, and righteousness to the plummet: and the hail shall sweep away the refuge of lies, and the waters shall overflow the hiding place. And your covenant with death shall be disannulled, and your agreement with hell shall not stand; when the overflowing scourge shall pass through, then ye shall be trodden down by it.

From the time that it goeth forth it shall take you: for morning by morning shall it pass over, by day and by night: and it shall be a vexation only to understand the report. For the bed is shorter than that a man can stretch himself on it: and the covering narrower than that he can wrap himself in it. For the Lord shall rise up as in mount Perazim, he shall be wroth as in the valley of Gibeon, that he may do his work, his strange work; and bring to pass his act, his strange act. Now therefore be ye not mockers, lest your bands be made strong: for I have heard from the Lord God of hosts a consumption, even determined upon the whole Earth. (Isaiah 28:9-22).

10 Degrees of retrograde movement by the sun in a day equals 40 minutes. It is my belief that "10 degrees"/ 40 minutes/ is another numerical relationship particularly relating to the 40 days and nights of Noah's Flood, the 40 years of the wandering in the wilderness during the Exodus, and a future event of retrograde movement of the sun is expected, a 40 (day, year) event which is the time of the "Great and terrible day of our Lord."

Moses & Joshua Story/Enoch Story a retelling of Noah's Flood/Deucalion Story of 2,750 B.C. impact and temporary cessation of Earth's Rotation / change of calendar from 360 to 365 days as recorded in Enoch's Age. It would have been noted by perplexed astronomers and by those who worked the land, for within a short period of only forty years - a period a person could readily observe -- the seasons would have become displaced by more than two hundred days. Around 700 B.C. the calendars of the Hindu, Persian, Chaldean, Assyrian, Chinese, Greek, Roman, and Mayan all had this 360-day count. Then something mysterious began to happen. Cultures widely separated, indeed, in some cases unaware of one another's existence, began to reform their calendars, change the day count to 365. 360 days from the old calendar added 5 "unlucky" days to meet the new calendar. When did this change from 360 days to 365 days occur? Enoch was 365 years old when God took him to Heaven. He did not die.

Revelation Math: 26,920 years = 1 Precession = 7 years

7 Heads of the Dragon. One gravely wounded. The Eight goes to Perdition.
10 Horns (comets)
1 horn = 1,290 years
12,960 years = 3.5 years

42 months = Judgment
1260 days. 2160 years. Notaricon.
570 A.D. - Mohammad's birthday: 570 A.D.

12 A.D. – 2029 =   Revelation Jesus Christ name repeated 37x
Gematria in Book of Revelation:

37 x 37=1369 years.
570 A.D. – 2039 A.D = 1369 years.

570 A.D. – 2029=   1459 years.
    19 periods of 75.5 Halley's cycles

The Question: When do the four wheels of the Apocalypse converge
with Prophecies of the Apocalypse?

20

Have we entered the final countdown to the Apocalypse?

Overview: Have we entered the convergence of Prophecy and the convergence of the arrival of the Armageddon Stones? Input verbal overview here: **#1:** We talked about "signs" of the End Times in Chapter 3.

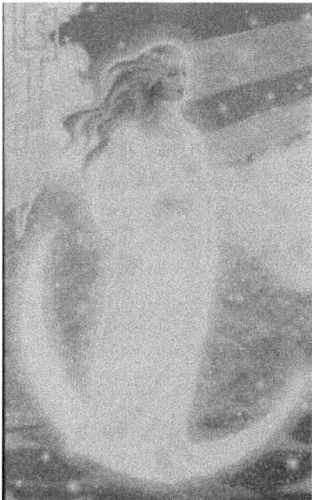

1987- End of Aztec

Signs the Magnetic Pole is Wandering

Ancient astronomers twice a year observed the perfect alignment of summer and winter solstices, assuring the steadfastness of the stations of the poles. The serious nature regarding even the slight movement of the polar axis was hammered out in stone over two thousand years ago.

According to Barbara Walker, author of the Crone: "The Greek sect of Cynics founded by the famous sage Diogenes in (c.

412- 323 B.C.), considered themselves the world's 'watchdogs.' Their name came from kynikos, 'dog-like ones.' They carefully watched the north star, Polaris, which they called Kynos oura, 'The Dog's Tail."211

They saw the constellation Ursa Minor as a dog or wolf since it was (and is) the axis point of the turning the heavens, the Dog's Tail became the 'cynosure' or focus of attention. The Cynics said, 'the end of the world would be at hand when they saw this star begin to move from its fixed place in the middle of the heavens.' The Cosmology of the Lapps also indicated that 'the end of the age will be marked by the movement of the Pole Star, and as a result "heaven will fall," and the earth will be destroyed in a great conflagration.' "212 The Slavs' Triple Goddess of Fate, the three Zorya, kept this mighty canine "fastened by an iron chain to the hub of the heavens, the north celestial pole, marked by the star Polaris in the constellation Ursa Minor. It was said that the end of the world would arrive when the wolfdog was released from his chain.' "213

Early stellar observation by Sumerian and Egyptian astronomers is not satisfactorily explained by scholars today. It seems we need to look deeper into the stellar mythos to uncover their original purpose.

Anasazi.

The Magnetic Poles Have Begun to Wander Erratically.

Physicist Arnold Zachow, electromagnetics consultant at Villanova and Drexel Universities, states that it is the locking of the Earth's core to the mantle that is the source of the geomagnetic field. As the Earth revolves, it generates a magnetic field along the axis of rotation that both radiates and attracts energy and serves to deflect cosmic rays away from the planet.

The strength and direction of the magnetic field are influenced by the speed of the Earth's rotation, and also by solar cycles. The greatest perbutations of Earth's magnetism always accompany the appearance of large sunspots as they pass through the central meridian of the sun.

Increased sunspot activity is directly proportional to the strength of magnetic influences that stream to the Earth on the solar wind. Of note, is the theory of Astronomer Michael Papagiannis of Boston University, "who postulates that the constant stream of energized solar particles that produced the solar wind thrusts on earthward, causing a torque that can influence the magnetic poles and their meanderings."214

---

211 Barbara Walker 1988. The Crone: Woman of Age, Wisdom, and Power. Harper & Row; 1st HarperCollins. p. 149.

212 "The Earth's Magnetic Hiccup," Science news (October 1985), p. 218.

213 De Santillana and von Dechend, Hamlet's Mill, p.383

214 Science News (January 5, 1974), p.5

This is one of several reasons for recent rotational slowing of the Earth, a phenomenon also related to the extreme wandering of the magnetic poles over the past three decades.

Are we Headed for a Magnetic Pole Shift?

According to Timm's research, "Earthquake energy and mean daily shift of the magnetic polar wobble has a definite correlation. Magnetic Wobble and Earthquake energy coincide every 7 years. According to physicist Heirtzler, even minor changes in the Earth's axis of rotation can affect "to a surprising extent both the climate at the surface of the Earth and the forces and stresses within it."[215]

"Since 1900, this rotational irregularity has been measured by astronomers, who say that it reaches a maximum every seven years—a periodicity correlated with peaks of earthquake activity."[216]

During the last 1,800 years, the strength of the Earth's magnetic field has decreased and is estimated to have fallen to less than two-thirds of its original strength. [217] One fact Timms' discovered in her research, is that magnetic field reversals were accompanied by declines in intensity. In the last one hundred years alone, the field strength has decreased by more than ten percent![218]

Decrease in magnetic field strength increases the amount of radiation now reaching the surface from the sun and the cosmos.

Is the Earth's rotation Slowing Down?

The Paris Observatory, which uses atomic clocks to keep official time for the whole world began delaying the New Year in 1971.

At the end of 1979 the United States Commerce Department issued a statement about an annual loss of time. A spokesperson said a discrepancy existed between solar and atomic time and that "the earth doesn't spin on its axis at a constant rate. It slows down and speeds up at unpredictable times and for the past few decades it has been slowing down. [219]

215 Scientific American (December 1968) p. 62.

216 Science News (14 August 1971), p 108

217 "The Mysterious Earth," Science Digest (December 1960)

218 Moira Timms...

219 "Last day to be a second longer," Press Democrat (31 December 1979)

Signs of the Prophets

1987 – End of (9 Hells) At the end of the nine hells would come a time of supreme cleansing and purification, when cities and mountains would collapse and most of the world would be reduced to rubble by fire. Quetzalcoatl promised to return at that time to initiate a Golden Age of spiritual renewal.

1988 A.D. – Ritchie & Christ 'You have 45 years.' In 1943, at the age of 20, Dr. George G. Ritchie, Jr., M.D., (September 25, 1923 – October 29, 2007) was a private in the Army stationed in Texas awaiting a transfer to Richmond to study medicine at the Medical College of Virginia to become a doctor for the military. However, he got sick with pneumonia and died. Jesus said to him, in October 1943..." You have 45 years."

1988 A.D. Harmonic Convergence

1991 January 15, Gulf War ends: (PS reading-Beginning of WWIII)

1991 Capricorn, 7 planet alignment: 4,300 years ago, The Astrologer Berössos is reported by Seneca to have learned that around 2300 B.C., the priesthood of King Sargon of Akkad compiled a seventy-two-volume astronomical treatise called the illumination of Bel. Bel was the dominant god of the fertile crescent. In their work, the scholars predicted that "when all the planets meet in Capricorn the world will be destroyed by fire. And that when all the planets meet in Cancer, the world will be destroyed by the deluge."220
It is likely the tropics of Capricorn and the Tropic of Cancer were named from this early prophecy. The northern latitudes tropic of Capricorn is the time of fire or heat. The Tropic of Cancer is a water sign, representing the southern latitudes.

When did the seven planets meet in Capricorn?

They already did. On January 15, 1991, a grand conjunction of six planets occurred in Capricorn, with Venus, Jupiter and Pluto heavily aspected and squared to each other, combined with a very potent annular solar eclipse. What happened on that day? Significantly, this was the date set by the United Nations, a line drawn in the sand, for Iraq's withdrawal from Kuwait."

220 See Moira Timms...

A Paul Solomon Reading in Nagoya, Japan, on September 1, 1991, stated, "Input ps reading from email...

1993-2012- Baktun 4 (Insert Gary Daniels notes)
September 11, 2001
2012-32 - Baktun 2
2014-2015 Blood Moons
2015- June – ISIS Caliphate & Muslim Brotherhood
2016 Cold War
2017 August 20 Total Solar Eclipse
2017 September 22 Virgin Crowned with Stars   seen   in   the heavens
2022 Apophis Returns for astronomical sighting

Economic Collapse (there will be a trigger for the collapse of the US Dollar) – Barter Economy...Dead Saint visions

War on Homeland

George Washington's Vision
Dead Saints Visions
Nuclear Catastrophe: NYC & DC **2028**
Extreme world-wide drought and famine -DSC quote
Angela's dream. 2028. Just before the Apocalypse.
Pre-millennial Rapture. Before Impact. What is it?
One Dead Saint's Dream- Mother Ships taking 1/3 of population of Earth during Pole Shift. Reads like Close Encounters of the 3rd Kind.
Paul Solomon Source Reading. UFO ships.

Visions from Children's NDEs 2013-2039

P.M.H. Atwater noticed a 2013-2029 timeframe for Earth Changes/Second Coming/return of the Christ in children's near-death experiences: In essence, the years between 2013 and 2029 were the time they felt when the United States would be faced with the greatest challenge to its continuance ever, a time of great achievement and indescribable tragedy. They warned that the energy within this timeframe could equal that of the major wars we have faced, all put together. It is a sobering passage - WHICH MATCHES THE SAME TIMEFRAME THE CHILDREN DESCRIBED The New Children and Near-Death Experiences").

The result was the span of years from 2013 through 2029 when they said the biggest earth changes would occur." "Of the 277 in my research base, a little less than a third of them were quite adamant about why they were here. To a child they would say, "I'm here for the changes." When I pressed them as to what that might mean, they would describe times of change so tremendous that the earth would need all the help it could get, and so would the people who lived on earth. You can't get timing with any reliability with kids, so I asked them to describe what they would look like when these big changes came. They would either say they were adults with older children of their own or grandparents with grandchildren. I pulled out my handy-dandy calculator and did some figuring, projecting ahead in concert with the age they were now and the age they would probably be when a scenario like this could be expected.

~Chronicle December 17, 2015 – "Woke up with the words, "The Kings of the People will become the people." I spoke them to Delynn, and then went back to sleep." ~ **Chronicle 920**

See Chart 1 – currently unable to fit this in chart properly – saved under

Years since Creation Creation Year Creation to Judgment Wheel/Gear Cycle Birth of Prophet/ Messiah Death of Prophet 2016 Projected End Gregorian Year Notes

Quetzcotl Age 12, 890 B.C. 12,960 years 2070 A.D.

Jewish 3761 B.C, October 6 6000 5776 5790 2029 AM 5776 began at sunset on 13 September 2015 and will end at sunset on 2 October 2016

2! Signs and Coincidences from God, by Michael Flipp, claims Jesus will return in the 69th week the Jubilee in 2029, (calculated from Jesus' crucifixion in 29 A.D. to 2029 A.D. (2000 years) as "the year of Jubilee." (50th) That this is the Acceptable Year of the Lord prophesied in Isaiah. 2029

Islam 5000 B.C 7000 1,450–1,500 years 570 A.D. 622 A.D.

1437 AH 1450 AH 2029-2070 A.D. Halley's Comet's next visit. 2061 A.D.

(See Islam prophecy chart)

Christian 4004 B.C 6000 12 B.C 30 A.D. 2028 2041 2029

Halley's 1986 2028 Michael 42 years

Apophis 2004, Dec/Jan 5, 2005 2029 April 3, 2029

Comet Machholtz/ Orion Pleiades 2004, Dec/ Jan 5, 2005 – 12,500year periodicity 2004 Comet Machholtz Periodicity

Angela's Dream   2013, July 16   15 years 2028   July   16,   2028, Ben, Patrick, Angela, Mel. Also, present in the dream, were Delynn, Mom, Dad, Melissa, David B, uncles, Ray, Dead. Hot. Famine. July 2016 Event causes Migration to desert place. Flagstaff? No Cell phones.

The Dead Saints Prophecies

Ancient legends, the Bible included, provide a general formula for recognizing the End Times. Moral decay, earthquakes, wars and rumors of wars, signs in the heavens, and the re-establishment of Israel. I won't rehash what is commonly available in bookstores. The purpose of this book is to look at the Dead Saints Prophecies to see if they provide a backdrop against traditional prophecies about the future.

<u>September 11, 2001</u>: Destruction of World Trade Centers (Dead Saint Report)

<u>June 15, 2002:</u> Mary Anne's Dead Saint experience. She asks to see God and then seem come down an elevator as Jesus:
I remember being thrown from the car and my head hit the divider. I was knocked unconscious. My first memories were when I was looking down from the ceiling and seeing my parents crying. I thought I was awake. The next thing I was in bed and I could see a light at the middle of the room. An Irish nurse came over to me and she said, 'look at this pretty darling. I am going to clean her up.' Then the other nurse called to her and told her there's nothing we could do for her so take care of the others first.
  I said to myself, 'Am I dead?' I was so scared. Then I found myself going fast up this dark tunnel. I was so scared, but I could see a very small light at the very end of the tunnel. Then I was in this beautiful light. My girlfriend's sister, Kathleen, greeted me. She was wearing white and all aglow. She told me not to be afraid. I asked her if I could speak to God. Then God came down, on an elevator, from a higher Heaven and sat down in front of me. God looked like Jesus in an all-white gown and tall, about 6 foot 5 inches. I know I could not see myself.
  God asked if I wanted to stay or go back while he let me look down and see my mother and sisters, who were small at the time, in the kitchen. My father was drunk. I was in such peace and surrounded by such beauty, I did not want to come back, but I told God, 'You see father, I have to go back.' God then showed me my whole life. He told me I would have 13 children, but three would not live. I would be married three times. The second marriage would be like going through hell with an abusive husband. The third man would change my whole life around and

would be my reward later in life. He showed me the hell I would be living in and before I come back to Heaven. He told me my daughter would come back to Heaven which has happened. Then I was taken into a beautiful garden where the flowers were indescribable. I have never seen flowers like that before. The aroma well let's put it like this, I don't know how to put it into words. Angels came and I felt like I was flying with them... then I was taken into this forest and shown a cure for cancer. I was so excited. I couldn't wait to get back to tell everybody.

I felt so smart because I was not very smart in school. I believe it had to do with flowers and leaves. God told me I would not be able to talk about it because that memory would be taken away from me until later in my life. God was telling me about my life. I was saying to myself, I would never do that, like being married three times and having 13 children. [authors note: Everything she was shown occurred later in life as Jesus showed her.]

Then Mary Ann witnesses a SECOND ATTACK on New York City sometime in our future over RELIGION:

I saw fire in the streets of New York and explosions. There were tornados in Queens, and I saw people running to get into shelters...I looked up and saw this big church and the next thing I know, I was sitting down and watching television. I was shown lots of disasters. I was so frightened. I saw fire and underground explosions in New York City. I did see peace after all of what I was shown. I was shown so much I did not like it. What I saw it was all over religion and at that time I couldn't believe that people would fight over God.[221]

[Author's Note: In October 1981, while living in New York City and working in the world Trade Center Collapse. Instead of collapsing in on themselves, they collapsed sideways...]

Author's Note: In 1999, while living in Olympia, Washington, I experienced an absolute KNOWING that a nuclear device will explode in both New York City and Washington D.C. It will happen as a judgment against the United States for dropping nuclear weapons on Hiroshima and Nagasaki in 1945.

April 4, 2003: Lynn's Dead Saint experience occurred in Venezuela. Major "Bubble" in the Middle East and around Africa:

Our earth planet is going through a shift in energy and so are we and all life on this planet. The earth has a crystalline structure around it like the one I was in, spinning so fast it appears to be a golden orb surrounding it, a light that keeps it intact. There was major undulation

221 Mary Ann F's NDE, #134, 6.15.02, NDERF.org

or bubbling around the area of the Middle East mainly, also around Africa."[222]

Sharon was struck by Lightning in 2005. saw six huge waves cover the land. Relate to other Dead Saints vision. Sharon focuses on the West Coast and Pacific Rim "waves." In another Dead Saint Vision of the East Coast and Europe suffer from devastating "waves." Both appear to be caused by multiple asteroid impacts. The impacts cause an oceanic Tsunami or Polar Shift Flood. Both Dead Saints visions describe volcanic eruptions and darkness, all effects of a Firefall/asteroid/comet swarm impact:

After God and I finished our conversation then we got up and started walking through the forest and were met by two beautiful ornately gowned women who led me to a calm, serene lake at the end of the wooded area. The two women I knew were angels and they began showing me what looked like moving pictures of future events that would take place on the earth. What was shown to me were the events surrounding the 911 attacks and other terrorist attacks against our country as well as our financial institution crumbling or better said our money not being worth the paper it is written on, I was shown silver and gold coins being used to purchase things, also they said that in time we would be going back to the barter system as we had done long ago in the past. They showed me many natural disasters, such as earthquakes, volcanoes, tornadoes, and storms, and six huge waves of water covering the land. I saw the one in Japan and Indonesia and also one in Chile. I also saw a woman in Canada who had a little boy in her car and her car went off the road because of flood waters and her car was immersed under the water and they were drowning. God sent angels in the form of people to pull them out, but the boy had already passed away. They told me he would survive however, and he did...

They showed me the government and how they are destroying the peace in our world and how corrupt they are, and they showed me the darkness that surrounds them, they showed me different governments being overthrown and huge riots in the streets. They showed me one particular riot where someone, a man, was throwing something through a store front window and there was a building nearby that was on fire, I also heard the sound of gunshots. They showed me the pockets of light that are still left in small sections called 'safe havens', mostly these areas are in the mountainous regions.

They showed me how to see the dark clouds around the lands to know where the safe havens are located and the last thing, they showed me was a silver ribbon splitting the United States apart, I was given

222 Lynn M's NDE, #264, 04.06.03, NDERF.org

knowledge that this ribbon was a river. I am assuming it was the Mississippi River, but they gave me no explanation as to the meaning of this ribbon other than the ribbon get larger. 223

Valerie had a heart attack after falling asleep, and was transferred to the local hospital ICU and saw the devastation from WWIII:

I don't know how long I had been asleep. When I got out of the body, [I] immediately appeared in the Third World War. The arms [weapons] I was seeing were completely strange, and rare conventional arms [weapons]. The devastation was horrible; dead everywhere, roads, buildings, children, trees and animals destroyed. The water contaminated. The world was deplorable, and I died. Then I came [back] to my body and told myself, 'I have to go back and see more.' I went out of my body again and I saw many wounded and mourning and much desolation.224

March 27th dream, the next night, I had another dream about Russian nuclear warheads being lifted through ice from icy artic waters:

On October 7th, 2015, I started for the first time in a year, my chemo treatments of Temozolide. As if he knew I needed encouragement, Dannion Brinkley called around 4pm. We had only communicated by email since July about his writing the Foreword to the Chronicles. Everyone was getting a little anxious up until his call. He had the first four chapters of the Dead Saints Chronicles and loved it. We spoke for nearly an hour about his life and mine. His sweet southern South Carolina voice made everything alright. I think he just wanted to hear my voice and get a sense of "me." We talked about prophetic dreams of the future and the 13 beings of Light he experienced during his NDE in 1975. I related a dream I had just a few nights before his call that reminded me of his vision of a "nuclear accident" in the Russian North Sea:

DREAM: I saw many dark green to black colored, cone shaped nuclear warheads pulled out of a 10-foot hole dug into thick arctic ice. They were heavy and from my view in the dream, they looked to be about four to five feet long. It's estimated because from my viewpoint in the dream, the warheads being pulled out by steel cable, were measured against the height of Russian men wearing wool black coats. The nuclear warheads were being retrieved through the hole as if they had been stored underwater, under the ice, and were now being retrieved. I don't know if I were witnessing the event real time, or that it was an event that would occur in the future...

223 Sharon M NDE, #4107, NDE 7925, NDERF.org

224 Valerie B NDE, #3432, 8.28.13, NDERF.org

Cities on Fire

"Before you go you may look into the future." He motions to his left. To my right, I notice a white table. It is stone, perhaps marble. On the table is a flat gold bowl filled with a liquid like oil or water. It's reflective and dark. Behind the table are three old men in white. Two are sitting. One is standing next to something like a pillar or a podium. There may be a book on it. They motion for me to come closer, and I do. "Look into the bowl and see the future of mankind."

I peer into the black liquid and see devastation. Cities are on fire. It's horrific. I turn away. I don't want to see this. Why are you showing me this? "We want you to take a message back with you. Man must change his ways." But I'm only one person. What can I do? "Spread the message." He continued, "You have a special ability." I know he was referring to my paranormal senses. "I am going to ask you a question. Whatever your first response is the one we will accept. You cannot change your mind afterwards. Do you understand?" Yes. I do. What's the question? "Do you want these powers you have and the ability to see into the future?" Immediately, I respond, no, I just want to be normal. "Very well then. It is done."[225]

Nuclear Attack against Iran.

Saint Lawrence had this Near-Death vision in 1973, reported to nderf.com in 2010. NDE was brought on by a medication overdose error. Notes: Explain the combination of War and Earth changes. Technology described appear to be in our future (medical discoveries and secret weapons possibly unknown today)

"As I looked on in amazement, a cinema screen appeared in front of me. I saw myself in a film of myself. It started when I was about 5 years old onwards, showing all the details from childhood right through to old age.

Then the film changed to people suffering in desert countries and I saw earthquakes and large water waves covering peoples' villages. **I saw the USA air fleet attacking an Arab country and devastating it like they did at Hiroshima and Nagasaki. I thought the country was Iran**. The USA had developed a secret weapon unlike anything ever seen on earth. Medical discoveries gave new treatments for ailments and an end to horrible diseases were the topic of the report I was hearing. After a while the screen vanished and I floated back down to the floor..... I went home and was ok the next day. Ever since then, I've had

---

225 By Anonymous on Thursday, March 4, 2004 · 08:58 pm, IANDS.org

premonitions that have all come true. I have been able to predict the future. I have also seen events weeks before they have happened - everything said here is the truth."[226]

Syria may obtain a Nuclear Weapon – Nuclear War
Paul Solomon Prophecy

When Saint Guenter was a child, he had an NDE during which Jesus had an extensive conversation with him about averting a nuclear war. "Jesus told Guenter that there were very powerful and cunning beings who were his enemies with whom he had been constantly fighting. He told Guenter to warn the world against them. Then Jesus told him to turn around. When he did, he heard people screaming and saw fire and smoke that gradually took on the shape of a mushroom [cloud]. Jesus told Guenter that he will do everything in his power to prevent this from happening.

Jesus told him that this war has been going on for a very long time and the evil beings had become more and more powerful. Jesus said that if this continued, He would not see any means of preventing them from taking over everything. He said that if the enemy had really succeeded in coming into his world, the war would have been lost. He would have been powerful enough, no doubt, to fight them off, but the war would have been lost in the long run. Jesus was afraid of losing this war. Then it was intimated to Guenter that Jesus needed help. Jesus told him that the only way he could help him was by going back and telling the world about his experience."[227]

Future Devastation

Recorded NDERF NDE 2013. Valerie later confirmed she had a heart attack after falling asleep and was transferred to the local hospital ICU.

"I don't know how long I had been asleep. When I got out of the body, [I] immediately appeared in the Third World War. The arms [weapons] I was seeing were completely strange, and rare conventional arms [weapons]. The devastation was horrible; dead everywhere, roads, buildings, children, trees and animals destroyed. The water contaminated. The world was deplorable, and I died. Then I came [back] to my body and told myself, "I have to go back and see more." I went out

---

226 Lawrence W's NDE, #2424, 10.11.10, NDERF.org

227 Near Death Experiences, near-death.com

of my body again and I saw many wounded and mourning and much desolation."[228]

Author's Dream: When I was a young man during the 1980's, I used to have a recurring dream of seeing a "mushroom cloud" from a nuclear detonation. It was one of my biggest fears about the future of mankind.

Spread the Word

"I was walking down Whitstable high street at night, when a mugger attacked me, I resisted his attacks, but he still pushed me into the road, where I was hit by a car. Doctors said I was dead on the scene, but one still attempted to revive me. He succeeded. Whilst I was dead, I saw the future. The world was covered in flames, and then Jesus spoke to me, he said "Spread the word and this will not happen, my son."[229]

Famine & Starvation / Ecological Collapse

Are we headed for the Sixth Extinction?
(See Book II: chapter xxx. The Light of Wisdom. Tamo-san vision)
Jean sees in a Dead Saint vision that the Earth is in trouble:
Suddenly I again felt the same sensation as at the beginning, I was shot at great speed and found myself floating in a sort of dark abyss, where I could hear heart breaking lamentations and suffering. I could feel thousands of souls who were suffering and lamenting, trying to cling to me, but not with the intention of taking me with them, rather they were trying to get out of the abyss. The feeling of fear and anxiety came back. Then suddenly I seemed to be floating in space with the image of planet Earth in front of me. Although it looked beautiful, the feeling was awful. I felt all the suffering which was in it. I felt how sick it was and that the sickness was growing. I felt egoism, violence. I felt how humanity was getting further away from the spiritual world and this was infecting it with a sort of cancer. **And that this would bring consequences, not as some kind of divine punishment, rather by its own process of cause-effect**.[230]

Jean makes an extremely important observation. It is not a judgment by God. Humanity will experience the cause/effect of moving away from a spiritual understanding OF THE GARDEN MA.D.E BY GOD. THE GARDEN OF THE EARTH. Misuse of Earth's resources, etc. (expand)

228 Valerie B NDE, #3432, 8.28.13, NDERF.org

229 England K's NDE, #506, 10.24.04, NDERF.org

230 Miguel G's NDE, #1938, 07.18.09, NDERF.org

Amy was told in her Dead Saint Vision that we are entering the 7th Day, a reference to the Seventh Day in the Book of Revelation. All world religions speak of Seven Cycles. Insert here.

I was also brought before what appeared to be a living picture of our planet. While I was looking at it, I saw a word above it. I believe it was "Novata." Then the whole planet seemed to open up like an eyelid that slowly awakens to dawn. It looked to be one eye opening up. There was a lovely, soft woman's voice who spoke the days of the week in a different language, and then said, 'Prepare for the Seventh Day.'

At this, I saw the curious image of a piano. There was something about music and octaves. The next thing I remember is traveling quickly over the Earth. It felt very surreal while doing so. It almost seemed that I was being shown a movie, and yet the movie seemed alive. Like flying over a panoramic film of a live scene on earth. I have lost much of what I saw, but I retain the main idea of what I was given while moving over the planet, or possibly having a movie of the planet being shown before me. There were fields of crops all over, in specific.

As I would zoom in and get close, for instance, to a field of wheat, I would be told, 'The food has been altered and poisoned. It is no longer pure. The people are consuming impure food. This is death.' I felt sad and concerned about this and wondered why, or how it was possible. How could a field of wheat or corn be "poisoned," and WHY?! I was told that man should return to the Earth or death would ensue everywhere. It was said again and again during this scene to 'Return to the Earth.'

I was told that upon my return that I should look for pure food, unadulterated, and to only consume that which is "clean," but I dismissed this somewhat because I had no intention of returning [to Earth].[231]

Financial Collapse of Paper Empires

Note: Accident/NDE occurred in 1981 and reported to nderf.org in 2012.

"The economic turmoil we are now going through is one of those "world events" that was pre-set. People have a choice as to how to react to these events. From what I was shown...the spiritual way is to help each other and help those in need. This is the ultimate act of love. But there is also the choice of becoming more protective and self-centered...less sharing and keeping one's own possessions as primary in one's reactions to what is there.

---

231 Amy C's NDE, #2382, 10.09.10, NDERF.org

This is a materialistic way of viewing it all...as if the material world is more than the connection between all of mankind. So, what choices will the majority make?

I was shown in 1981 that this time would come and that banks were paper empires, built on paper and nothing more. But, too, so are many other businesses...paper empires...built to collapse under pressure. How do people react to all of this? **This is the key event and will test many.** Will they reach out and take care of each other, or will they become more and more self-centered and protective of the material? There are always choices in this...just to determine which choices individuals will make."[232]

~Saw the movie "Knowing." Professor tries to teach the difference between Determinism & Randomness. Fate vs. chaos. Will God allow me to see the future? Can it be changed? Can a few men change the destiny of the planet? Even if we could we would never know. We cannot know our destiny, no matter how great, otherwise it would defeat our God given free will. We must follow the breadcrumbs of our Fate. ~Chronicle 252.

Can we change the Future?

Question #1: The Cassandra Effect-Can We Change the Future if we SEE it?

A theory of Quantum physics says if we look at a vision or particle from the future, we have already changed it.

Repeatedly, the Dead Saints and the prophets hint that decisions made by each of us during the near future are crucial to either a "planet-wide" destruction of our cities and humans, or a gentler transition toward a more "peaceful" planet, heralding a new era, a New Heaven and New Earth. But can we escape the birth of a New Era, a new age unscathed?

George Ritchie... In 1943, Jesus told him: "You have 45 years to change the future. That's 1988. Have we changed the future?

Question #2: Are some prophecies set in stone?

Question #3: Can prophets and the Dead Saints really see the future? Can our decisions change the Cassandra Effect of an assured catastrophic future? Tomorrow Land. A future event believed to be inevitable, cannot be stopped, even if the prophet tries to tell others.

---

232 Jean R NDE, #2932, 01.18.12, NDERF.org

The answer. It's a set of probabilities.

At other times, the Saints may see multiple probable or possible futures. Kenneth Ring reports one Saint describing, while his "consciousness split over his dying body, that she became aware of ... three lines of trajectories that would lead towards futures...each of them is an alternative arrangement of things I saw..."[233] She referred to these trajectories as A, B, and C, representatively.

Ring, in his 1984 book, Heading Towards Omega, goes on to describe her encounter with future trajectories.

"Future A was a future that would have developed if certain events had not taken place around the time of Pythagoras three thousand years ago. It was future of peace and harmony, marked by the absence of religious wars and of a Christ figure. Figure B was in effect, the classic PV (planetary vision—earthquakes, wars, catastrophe, financial collapse, followed by peace). Future C was an even more destructive version of Future B. Both futures B and C projected to her simultaneous images associated with these separate future tracks from about the end of the century backward to 1956; this was also true for the future-oriented events of the non-realized Future A except that these images were fewer and less detailed that those connected with futures B and C."[234]

According to Ring, the Saint who made the prophecy, feels we are headed to Future B.... earthquakes, wars, catastrophe, financial collapse, followed by peace. Dr. Kenning Ring, founder of IANDS.org, described those who have died and returned back life describe "Seeing the entirety of the earth's evolution and history, from the beginning to the end of time. **The future scenario, however, is usually of short duration, seldom extending much beyond the beginning of the twenty-first century.** The individual reports that ... there will be an **increasing incidence of earthquakes, volcanic activity and generally massive geophysical changes.** The **possibility of nuclear accident is very great** (respondents are not agreed on whether a nuclear catastrophe will occur). All of these events are transitional rather than ultimate, however, and they will be **followed by a new era in human history**, marked by **human brotherhood, universal love and world peace.** Though many will die, the earth will live."[235]

233 Heading Towards Omega, Kenneth Ring, pg. 218
234 Heading Towards Omega, Kenneth Ring, pg. 218
235 Ring. 1982,p55-56

It is important, however, to note the published 1984 date of Ring's book, it appeared most "prophecies of destruction" slated for 1984-2000 by the Dead Saints have not yet occurred.

This has been equally true of many other Seers and prophets, including Edgar Cayce, Paul Solomon, and many others, whose prophecies were made between 1920 and 1984, looking towards cataclysm and war around the year 2000...events which have not yet occurred.

Howard Storms Descent into Death, he writes:

"In my conversation with Jesus and the angels, they told me about God. I asked them what God is like and they told me this: God knows everything that will happen and, more important, God knows everything that could happen. From one moment to the next, God is aware of every possible variable of every event and each outcome. God doesn't control or dictate the outcome of every event, which would be a violation of God's creation."[236]

During the three decades of studying prophecy and discovering thousands of direct mind and heart encounters and conversations with God brought back by the Dead Saints,

What causes these confluence of events? Meeting a set of conditions.

Reason for Prophecy and Precognitions?

Make a change. Prophet Joel.

The Saints see the future a during their near-death experience as a possible future, if the current set of conditions during the time of their vision play out.

At best, even good readings are subject to change, especially since some aspects of the future are not pre-determined, but rather subject to change depending on the use of free will, thus can flux in accuracy a great deal over time. Such readings are only accurate a certain percentage of the time, thus, we think people should use their own intuition to determine such things (which is partly what the book is about - teaching you to become your own source of information from God, and not rely on others). In the old days, a false prophet didn't have a chance to make excuses, they were stoned to death. And a good prophet, an alive prophet, didn't make a prediction unless it was certain. Even Cayce would say that the time and places of many things could not be known, because of the free will factor.

---

236 Howard Storm, My descent into death, New York, Doubleday, 2005, pg 38.

Can we stop Nuclear Armageddon?

I believe it is our mission to change World War III prophecies by choosing love instead of fear in our own daily lives. We can make a difference.

We can choose life instead of death, love instead of hate, light instead of darkness.

According to the Dead Saints, the following prophecies of war WE CAN CHANGE:

I was walking down Whitstable high street at night, when a mugger attacked me, I resisted his attacks, but he still pushed me into the road, where I was hit by a car. Doctors said I was dead on the scene, but one still attempted to revive me. He succeeded. Whilst I was dead, I saw the future. The world was covered in flames, and then Jesus spoke to me, he said 'Spread the word, and this will not happen, my son.[237]

Many Dead Saints have been told the same message from Christ. They are shocked when they see American Cities on fire on a WWWIII scenario. We each have a vital, important role to play. They ask, what can I do? 'Man must change his ways. Spread the message.

Pre-impact Rapture

Rending of the Veil. Caught up to Heaven with Christ or Extra-terrestrial intervention? The disappearance is the mystery. Left Behind. But will the Rapture happen this way? Or the extra-terrestrial way. I'm sure that ET will not wait until the last minute to pick up 2 billion earthlings. Perhaps it's a "beam me up Scotty" situation. I can only speculate. The result is the same. Billions will mass exterminated by the Firefall impact from global Tsunami, erupting volcanoes, and earthquakes, etc. A few will survive in high plateaus such as Arizona and Africa. The movie 2012 paints a perfect picture of the Pole Shift, except the Human Arks save only a hundred thousand people, where the UFO arks save 2.1 billion, who will be returned to Earth when it has become safe to do so.

~Angela Solomon's dream of the future in 2028

The Scene. Angela is on a theater stage dressed as Velma, the scientist from the cartoon, Scooby Doo. She is being chased by a dark

---

237 England K's NDE, #506, 10.24.04, NDERF.org

man brandishing a sword. Angela defended herself with an old Arabian sword, balancing a "ball" (tumor shaped object) at the end of the flat edge of the sword.

On stage, she felt like "Spiderman" with Spiderman powers, and "flew around" to escape the dark man chasing her. Angela could also get away from him by going up and down in an elevator. [authors note: through dimension and time]

In the next scene, Angela finds herself in a desert. Three years have passed since Dad's brain tumor diagnosis. She is walking with some friends and a lot of people she didn't know. They appeared to be refugees migrating from many different places and had only the clothes on their backs.

No cellphones. No technology. They were migrating because of an "event" that had happened

Everyone was happy.

Besides the people she didn't know, the entire family walked with her including her Dad. In the dream, Dad looked healthy and alive. He had survived at least 3 years.

Angela looked behind her and saw Noel, and his younger brother Matthew. Angela ran to Noel and gave him a hug and a warm kiss on the mouth. She was so happy because she hadn't seen Noel in a few years.

The scene suddenly changes to her home at Akio Gardens in Olympia, Washington. It was still July 16, 2013. She climbs in bed and sleeps "for hours," but when she awakens she discovers 15 years had passed. Angela opens her bedroom door to find her Granny sitting on the floor in the hallway, her arms wrapped tightly around her knees. Granny's face had aged, as if she had grown 15 years older—90 years old. Granny told her that her mother, Anna, was dead. Her step-mother, Delynn, was dead. Many of her aunts and uncles, including Grandpa Ray were dead. (Ray died, June 18, 2015) Chynna, Angela's close friend in Washington State, her parents, Melissa and David, were dead.

Granny then told her to look for "Uncle Mel," then stood up, and walked towards the veil-like blue opening of the Peter Lik Desert photo hung on the wall over the stairway and walked on air (Granny appeared solid) towards "Lik" photograph, with the blue sky that opened like a portal that exposed a bright flashlight-like Light shining in Angela's eyes, and Granny disappearing through the light, and then she was gone.

Angela is then transported to the next scene in her dream. She finds herself standing outside in an old, dark wooden house in the desert-like place. There are no trees. The ground was dry and cracked. It was HOT.

In distance, Angela could see a hundred miles, Grand Canyon rock outcroppings (look up) and Arizona desert features, with red, orange, and browns hues.

Angela knocks on the door of what appears to be an old, dark wooden cabin hidden in the desert. It appeared one hundred years old. "Uncle Mel" opens the door and lets her in. He wore Star Trek Jordi "blind" techno gear—gear that allowed him to see computer images in 3-D, but because he was blind, the mask allowed him to see.

That's when Angela saw her brother Benjamin sitting at a long table to the left of the door, along was his life-long best friend, Patrick. Benjamin was 33 years old and brandished a darker clean-cut goatee, with a neatly trimmed beard grown along his cheek bones to his ears. Everyone appeared clean and well kept. The heat and Benjamin's age suggested Angela was seeing the future in the summer of the year 2028.

They were eating dinner. Patrick had a plate with one single bean and was cutting it like a juicy steak with a knife. Benjamin was eating a large bowl of mashed potatoes. Angela felt the entire scene described a severe famine.

Angela wakes up. It's one of the most vivid dreams she has ever had. July 16, 2013. **~Chronicle 34**

~January 25, 2016. Angela applies to attend Northern Arizona University in Flagstaff, AZ when she discovered her close friend, Noel in Tacoma, Washington, has been accepted there for a 4-year term beginning September 7, 2016. Angela didn't remember her dream about the Arizona "desert" place when she applied to NAU in January.

So, when the dream was pointed out to her again on February 25, 2016, she was a little shocked. Her response, "Oh...My...God!" She is truly a little Velma like scientist at heart. She needs proof. The Arizona Grand Canyon-like desert is the only desert that matches the description of her 2013 dream. It is not the high mountains of Flagstaff, Arizona...but close. The time frame in the future is 15 years later, July 16, 2028.

Dating: Angela's dream occurred July 16, 2013, thirty-four days after my Cancer discovery. On July 16, 2013, Benjamin was 19 years old. Fast forward 15 years. Benjamin is 33 years and 9 months old. Angela is 30 years old. Who is Uncle Mel? How will she find the cabin in the Arizona Desert? Is this part of the dream symbolic?

Dead Saint Visions from small asteroid impacting ocean on West Coast

I saw glimpses of the future, wars, famines tsunamis, world war, earthquakes etc. Japan, Eastern China and parts of the USA (Western USA) was destroyed by a tsunami over 150 meters high (450 feet).[238] I wasn't able to bring back exact years nor time as it could affect what should happen for our own evolution. I know, as I told my story to some people after 1990, the year of the accident, that there will be lots of conflicts involving the U.S.A. I remember that it will start right in New York City in a few years from now (1990). I had no information that it would be the World Trade Center, and what year exactly, but was sure for NYC, and that all the people of earth would be aware of what would happen.

It's now part of the past (9/11).

After that, the east coast mainly will be affected. There was a <u>major city in a north central state</u> (Illinois/Chicago) that would be affected. Almost nothing to the west coast would be affected. Because of that, as I am sure of it, I'll never move to a major city on the east coast even if some company were to entice me with a very good six-figure salary to begin with. I said to myself, now that I am dead, I can't change anything anyway...

So, I went to the end of the conflict... Even more frightening! Because of the conflict with different countries and less investment in space research and NASA, no one can see something happening and approaching earth... **Some small asteroid but big enough to cause substantial damage on earth.**

The good news in all that is **that because of this event,** all nations will stop all their conflicts and try to work together instead of fighting each other, but it will be too late to avoid what could have been avoided...Starting from that point, the different nations will realize the stupidity of war and will work together for many years to come, there will be finally peace on earth but millions of lives lost could have been avoided! **I know this is not too far away in time...**

Suddenly, I was back in the car, flames leaping all around. I turned and saw my friend's body and knew beyond doubt that she no longer inhabited it. The Light gave me just enough time to exit the vehicle before the flames reached me.[239]

238 Lawrence W NDE, #2424, 10.11.10, NDERF.org

239 Roger C's NDE, #253, 04.02.03, nderf.org

Pole Shift Dream: James Yax

~On February 28, 2016, James Yax visited Delynn and me for 3 hours. During his stay, he recalled a vivid dream he had in 1977 about a future Pole Shift:

I was in a large room of a home with about a dozen other people, some of whom I recognized as friends. On a small table next to my chair, I saw a star map. There were no familiar constellations on that map. The time was close to midnight. I then left the room and went outside, walking into an atmosphere of soft golden light. I knew it was 1 AM, and I saw the sun rising over the ocean to the East, but a sun many times larger than normal. My thought was: There must have been a pole shift.

In front of me stretched a meadow of emerald, green grass, and in the distance, I saw hundreds of UFOs rising above the horizon and descending through from on high. They were the "cigar" shaped craft-- huge "mother ships" from which smaller craft emerged. One of these flew toward me and landed on the grass about fifty feet away. As I walked toward it a small ET emerged. Another one remained at what appeared to be a control panel in the front of the ship. He (I could not swear to the gender) was small, in height a bit above five feet, and perfectly human in form. When I came up to this being, I asked if there had been a pole shift. He said yes. I asked then where the tidal wave was. His response was: "You have 15 minutes." I asked if they were here to save humans from the destruction. He said they would be able to take 1/3 of the people off planet, to be returned when things had settled down. I went back to the house and told my wife that we had to leave. I told everyone in the room, but none wished to leave.

When my wife and our baby and I were in the craft, it quickly ascended and flew over what looked like a harbour on the seashore, and in the distance, we could see an enormous wall of water approaching the land. At that point I woke from the dream. Sunday, February 28, 2016

Note: Beyond James Yax's remarkable dream, is the TIMING in which he revealed his dream to me— a dream he had not told anyone before. That same day, I had been feverishly putting the remaining pieces of the Armageddon Stones together and was working on the Pole Shift chapter. His dream helped me put dots into place and answer important questions about pre-impact phenomenon such as the rapture and extra-terrestrial assistance before Firefall Impact. For this reason alone, his dream should not be easily dismissed. Remember the "still small voice." A coincidence? I think by now, even the skeptic must be wondering. ~Chronicle 991.

Judgment Day: Impact Scenario

The Apocalypse of Thomas describes seven days of the End. Dating from the fifth century, or perhaps earlier, it follows the Revelation of John in certain imagery and is the only apocryphal apocalypse to assign seven days to the destruction of the End. These are the words of the "savior" to Thomas about the end of the world."

Day 1: The Rain of Blood

"... When the hour of the end draws near, there will be great signs in the sky for seven days and the powers of the heavens will be set in motion. Then at the beginning of the third hour of the first day there will be a mighty and strong voice in the firmament of heaven; a cloud of blood will go up from the north and there will follow it great rolls of thunder and powerful flashes of lightning and it will cover the whole heaven. Then it will rain blood on all the earth. These are the signs of the first day."

The Hope prophecies state, "The whole world will shake and turn red..."

And it is to be noted that Nostradamus, in his Prophecies of 1555 A.D., refers in several quatrains to falls of red rain associated with the appearance of a comet and fall of stones. He uses the words "bloody" and ferruginous" when speaking of the light of the comet which is to come, but it is never referred to as "fiery red." This is in perfect accord with ancient records on the Exodus comet: "Non igneo sed sanguineo rubore fuisse." (It was not the redness of fire, but the redness of blood.) When the tail of the Exodus comet crossed the path of the earth, a red dust, impalpable, like fine flour, began to fall. It was too fine to be seen, which is why it is not named in Exodus (7:21), but it colored everything red, and the water of the Egyptians was changed into "blood." The fish died and the water was poisoned by the decomposition of their flesh. It is for this reason that the Egyptians had to "scratch the earth," that is to say, to open new wells.

A similar occurrence was recorded in various parts of the world. After the fine rusty pigment fell over Egypt, there followed a coarser dust -- "like ash," this is recorded in Exodus, for then it was visible. This ash irritated the skin and eyes of both men and animals. They scratched themselves and sores formed; boils appeared and changed into pustules for want of being treated. Soon, the infection spread to the whole body and death followed. After that ash-like substance came a shower of fine sand, then coarse sand, grit, gravel, small stones, large stones, and finally, boulders. The narrative of the Book of Exodus confirms this and is in turn corroborated by various documents found in Mexico, Finland, Siberia, Egypt, and India. It is therefore certain that a comet crossed the path of the earth about (3,600) years ago, causing widespread destruction. This is the kind of phenomenon which is soon to strike the earth again.

Mountains will literally split open; the sea will overrun whole provinces and possibly even small nations such as Holland; some coastal plains will just collapse into the sea; the sky will be on fire in many places when the oxygen of the earth's atmosphere will ignite the hydrogen of the comet's tail. As a result, tremendous hurricanes will be induced, adding to the devastation. The oxygen supply being used up in those regions where the sky fire will rage, people who leave their houses will die from asphyxiation, which is probably why many prophecies say: "stay indoors; keep your windows shut." but the hurricanes will promptly bring back fresh supplies of oxygen with the effect that this ordeal will be of very short duration. Then torrential rains will fall, quenching the many fires on earth, but, at the same time, causing widespread flooding. After all these disasters, food will be so scarce that a general famine will follow. Millions of people will starve to death; their unburied bodies will cause pestilence and epidemics all over the land.

The Final Approach of the Comet:

    *   The Comet approaches the earth; Climatic changes take place. Droughts and floods occur. Summer is cold because the comet's gases interfere with normal solar radiation.

    *   The Comet is very close: Climatic changes worsen. Food shortage begins. Winter is hot because the comet is now close enough to radiate its own heat.

    *   Hail of stones; This occurs either before or after the sky fire, and is preceded by a rain of dust, caused by the comets tail.

    *   Sky fire: The hydrogen of the comet's tail mixes with the oxygen of the earth's atmosphere and is ignited.

    *   Violent lightning: Caused by electrical discharges between the earth and the comet's head.

    *   Poisoned atmosphere: Caused by a shortage of oxygen in the earth's atmosphere or by deadly gases such as methane.

    *   Fires on earth: The land has been heated by the sky fire. Forests and cities ignite.

    *   Droughts: No rain. Caused by the sky fire.

    *   Floods: The combustion of the comet's hydrogen in the sky causes not only a shortage of oxygen, but also the formation of huge clouds which condense into torrential rains.

    *   Hurricanes: Occur with the floods because the atmosphere has been superheated, lifting huge amounts of water vapor into the atmosphere, which condense into clouds, thunderstorms of ferocious intensity, and super hurricanes never before witnessed by mankind.

    *   Darkness: Caused by the thick dust clouds.

    *   Earthquakes, volcanic eruptions, and tidal waves: caused by the shifting of the earth's axis.

    *   Cold and famine: Especially in the Northern and Southern portion of each hemisphere.

Day 2: Great Fire from Heaven / Pole shift

"And on the second day a great voice will resound in the firmament of heaven and the earth will be moved from its place. The gates of Heaven will be opened in the firmament of heaven from the east. The smoke of a great fire will burst forth through the gates of Heaven and will cover the whole heaven as far as the west. In that day there will be fears and great terrors in the world. These are the signs of the second day."

In the Paraphrase of Shem, a non-Christian Gnostic work discovered in the Nag Hammadi Library, we find that "When the era of Nature is approaching destruction, darkness will come upon the earth..., and a demon will come up from the power who has the likeness of fire. He will divide the heaven, and he will rest in the depth of the East...[causing] the whole world to quake..."* (The Conflict of Light and Darkness Codex VII 1,49,9. Translated by Frederick Wisse. From James M. Robinson, ed, The Nag Hammadi Library (San Francisco: Harper and Row, 1977 pp 309-329)

Day 3: Volcanic Eruptions

"And on the third day at about the third hour there will be a great voice in heaven and the depths of the earth will roar out from the four corners of the world. The pinnacles of the firmament of heaven will be laid open and all the air will be filled with pillars of smoke. An exceedingly evil stench of Sulphur will last until the tenth hour. Men will say: We think the end is upon us so that we perish. These are the signs of the third day.

Day 4: The Great Earthquake

"And at the first hour of the fourth day the abyss will melt and rumble from the land of the east; then the whole earth will shake before the force of the earthquake. In that day the idols of the heathen will fall

as well as all the buildings of the earth before the force of the earthquake.
These are the signs of the fourth day.

>First, then: There is soon to come into the world a body.
>[See Par. 6 below]; one of our own numbers here that to many
>has been a representative of a sect, of a thought, of a
>philosophy, of a group, yet one beloved of all men in all
>places where the universality of God in the earth has been
>proclaimed, where the oneness of the Father as God is known
>and is consciously magnified in the activities of individuals
>that proclaim the acceptable day of the Lord. Hence that one
>John, the beloved in the earth - his name shall be John, and
>also at the place where he met face to face [Peniel].
>[GD's note: Could this mean that John, the beloved, had
>been Jacob? See 3976-15, Par. R2.] * *attached below**

>7. As to the material changes that are to be as an omen, as a
>sign to those that this is shortly to come to pass - as has
>been given of old, the sun will be darkened and the earth
>shall be broken up in divers places - and THEN shall be
>PROCLAIMED - through the spiritual interception in the hearts
>and minds and souls of those that have sought His way - that
>HIS star has appeared, and will point [pause] the way for
>those that enter into the holy of holies in themselves. For,
>God the Father, God the Teacher, God the director, in the
>minds and hearts of men, must ever be IN those that come
>to know Him as first and foremost in the seeking of those
>souls; for He is first the GOD to the individual and as He is
>exemplified, as He is manifested in the heart and in the acts
>of the body, of the individual, He becomes manifested before
>men. And those that seek in the latter portion of the year
>of our Lord (as ye have counted in and among men) '36, He
>[He, Christ Spirit?] will appear. [See 3976-1, Par R2
>(12/20/34 EC note in re '36), and 3976-10, Par. 4-A, 5-A on
>2/8/32 in re '36 changes.]

>8. As to the changes physical again: The earth will be broken
>up in the western portion of America. The greater portion of
>Japan must go into the sea. The upper portion of Europe will
>be changed as in the twinkling of an eye. Land will appear
>off the east coast of America. There will be the upheavals
>in the Arctic and in the Antarctic that will make for the
>eruption of volcanos in the Torrid areas, and there will be

shifting then of the poles - so that where there has been
those of a frigid or the semi-tropical will become the more
tropical, and moss and fern will grow. And these will begin
in those periods in '58 to '98, when these will be proclaimed
as the periods when His light will be seen again in the
clouds. As to times, as to seasons, as to places, ALONE is
it given to those who have named the name - and who bear the
mark of those of His calling and His election in their
bodies. To them it shall be given.

9. As to those things that deal with the mental of the earth,
these shall call upon the mountains to cover many. As ye
have seen those in lowly places raised to those of power in
the political, in the machinery of nations' activities, so
shall ye see those in high places reduced and calling on
the waters of darkness to cover them. And those that in the
in most recesses of their selves awaken to the spiritual truths
that are to be given, and those places that have acted in
the capacity of teachers among men, the rottenness of those
that have ministered in places will be brought to light,
and turmoils and strifes shall enter. And, as there is
the wavering of those that would enter as emissaries, as
teachers, from the throne of life, the throne of light, the
throne of immortality, and wage war in the air with those of
darkness, then know ye the Armageddon is at hand. For with
the great numbers of the gathering of the hosts of those
that have hindered and would make for man and his weaknesses
stumbling blocks, they shall wage war with the spirits of
light that come into the earth for this awakening; that have
been and are being called by those of the sons of men into
the service of the living God. For He, as ye have been told,
is not the God of the dead, not the God of those that have
forsaken Him, but those that love His coming, that love His
associations among men - the God of the LIVING, the God of
Life! For, He IS Life.

10. Who shall proclaim the acceptable year of the Lord in him
that has been born in the earth in America? Those from that
land where there has been the regeneration, not only of the
body but the mind and the spirit of men, THEY shall come and
declare that John Peniel is giving to the world the new ORDER
of things. Not that these that have been proclaimed have
been refused, but that they are made PLAIN in the minds of

men, that they may know the truth and the truth, the life, the light, will make them free.

11. I have declared this, that has been delivered unto me to

The weakling, the unsteady, must enter into the crucible and become as naught, even as He, that they may know the way. I, Halaliel, have spoken.240

Day 5: Darkness Covers the Sun, Moon & Stars

"But on the fifth day at the sixth hour suddenly there will be great thunderings in the heaven and the powers of the light will flash, and the sphere of the sun will be burst and great darkness will be in the whole world as far as the west. The air will be sorrowful without sun and moon. The stars will cease their work. In that day all nations will so see as if they were enclosed in a sack, and they will despise the life of this world. These are the signs of the fifth day."

Day 6: Heaven Split Open / World Consumed in Fire

"And at the fourth hour of the sixth day there will be a great voice in heaven. The firmament of heaven will be split from east to west.... Then all men will flee into the tombs (caves) and hide themselves from before the righteous angels, and say, "Oh that the earth would open and swallow us.... Then at my arrival the restraint on the fire of Paradise will be loosed, for Paradise is enclosed with fire. And this is the eternal fire which devours the earthly globe and all the elements of the world."

Day 7: The Destruction of the World / Elect are Saved

"And at the eighth hour of the seventh day, there will be voices in the four corners of Heaven. All the air will be set in motion and spaceship. Uncle Mel. Angela's dream 6 months before. Then all men will see that the hour of their destruction is come near. These are the signs of the seventh day.

World War stopped by small asteroid impact

---

240 : http://www.gatheringspot.net/topic/inspirationstransformational-stories/who-jon-peniel-amazing-story

—— **21** ——

The Messiah, the Tribulation & the Second Coming of Christ through the eyes of the Dead Saints.

His Return through the Visions of Near-Death Experiences.

The Tribulation Years: 2029 – 2061
    A New Heaven and New Earth. 2061 – 2100
    Halley's next visit in 2061.
    42 Years later 2100 A.D. Cayce's vision

Anna saw an Afterlife vision of the Earth with no borders, a future which could only happen if all governments world-wide volunteered to do so, or if the BIG EVENT set the stage for world peace:

Look again... 'What do you see NOW?'... Suddenly I saw what the voice saw. 'I see our planet and there are no borders dividing countries... The borders are gone!!'... He said 'This is why you're going back. You have a mission.' And that's how I came back. I became more conscious of the environment, of clean air, clean water, world hunger, wars and poverty.
241

---

241 Anna A's NDE, #3784, 10.27.14, NDERF.org

Messiah & Israel

1984 February Trip to Israel private tour with Nicole, Jennifer, and Paul. Legend of John Peniel. If he was King David, (see article 1961. David Morris) who spoke ancient Hebrew, he could be Messiah. Edgar Cayce made John Peniel's name famous forty years earlier in a few obscure readings that hardly made public. Many had searched for him, and a few had even stepped forward claiming to be "the one" whom it was said, "He will come." Now Paul says, "He's living in Israel," and will be the forerunner to the Messiah. Amazing. Somehow, being so close to the real thing is an experience I had always dreamed about, but never really thought would happen.

Will the long-awaited Messiah appear in our life time? When will the Second Coming of Christ occur?

Dead Saint Vision: 12-Year-old sees Second Coming of Christ

The experiences started when I was 12-13 years old and continued until I was 28 years old, ending with a real vision. I used to stop breathing at night shortly after settling in bed. Probably sleep apnea, plus I had what I thought was a mild case of asthma. My mother had died about a year before, I was in a sense, grieving. It all together has changed my life.

My first experience, I was lying in bed, trying to go to sleep. Suddenly I realized I could not breathe, my whole body felt like it went into instant paralysis. I tried to move and bang the wall and scream for help, to no avail. No voice, no movement or feeling. I then felt myself raising into the air, the ceiling getting closer, and I peered down for a moment, seeing my body just helplessly laying there in bed.

I arose fast through the dark sky to a bright light. When I got there, I saw others spirits. I was rising so fast, I passed them on the way up. I saw each and every one of their faces, people I did not know. I came to what seemed like a path, leading to some kind of structure in the distance.

I then slowed right down, to a very slow pace, and watched the spirits of the others I passed earlier, pass me by, going in through a gate in the structure. I came close to the humongous structure, which was very beautiful. The walls were very high and long. Everywhere was very well lit up, the light emanating from inside the structure, to the outside for some distance. But there was no sun, no moon or stars. It was so bright everywhere, it did not cast any shadows. The base of the wall had numerous shiny and beautiful gemstones.

I stopped at the gate, taking notice of it. It was huge, as high as was wide, and completely made of the most purest, whitist pearl. I'd never seen anything so bright here on earth. At the doorway, I saw inside by the doorway, the floor was clear like glass, yet could not see through it. Then as I looked up. I saw the Guardian Angel of the gate. I didn't want to die, and I tried to cry, but it would not come out of me. I was scared of dying. It seems as the angel understood. I had thought of my father and sister who would be left behind. She smiled, and pointed back in the direction from which I came. As soon as I looked that way, I felt relieved, and very peacefully happy, and returned back to my body faster than I'd come up.

Upon returning to my body, I gasped for air, choking. It wasn't until years later, I started reading the bible, that I figured out that where I'd been, was to Heaven. What I saw, is written in Revelations chapter 21. But I did not go directly into the structure.

The following 15 years, I had many more, I lost count. But I didn't see Heaven again, just tunnels with lights at the end. I just fought hard for my body to breathe again. I didn't want to die. They ended in 1993, with a vision. It was confirmed to me to be real. It was morning and I was lying in bed.

When it started, I could see myself walking on a street in my home town with my father. Then all of a sudden, there was such loud peals of thunder as I've never heard before in my life. Then the whole sky, clouds and all, parted wide open. A bright light appeared, such as I'd never seen in my life -- like looking into the sun, at close range, but white. I started to hear music, as like harps, trumpets, and singing. I was so excited. I was shouting happily that Jesus was here! It was Judgment Day! [242]

Judgment Day is a CELESTIAL EVENT.

The explosion in the heavens, the coming with clouds, loud explosions, will cause the whole world to cower in fear.

Then I saw something that bothered me, even to this day. People were running away scared, gathering their valuables and hiding! I turned my attention back upwards to the west in the sky (the direction from which He came). I saw the spirits of many of the dead rise into the sky. Then in a moment, I arose into the sky with some of the other living people. I first saw my mother, who had been deceased for 16 years. Everyone had a renewed perfect and healthy body. Then I saw Him. Jesus. He sat on a great throne, dressed in a gown that shone like white light. I did not see his face. But I saw His hands and feet. They were pierced, each with a hole, from the cross where He was crucified....

Being up there, I gained an overwhelming feeling of Peace and Happiness and Love, such as I'd never felt in my lifetime. The feeling is incomprehensible to describe. All who were in the air proceeded to follow

---

242 Linda's NDE, #40, NDE & a Vision, aleroy.com/board40.htm

Him towards Heaven. The vision ended there, but not my overwhelmed feeling of Peace and Happiness. It stayed for weeks, afterward. It was after this that I bought a Bible and started reading it. I really feel sadness for those who do not believe in Jesus. They will be left behind for Armageddon....

Linda describes a RENDING OF THE VEIL caused by RIPPING of the atmosphere during a comet fragment explosion. It is the CAUSE OF THE RAPTURE. WHAT IS THE RAPTURE? A TRANSFIGURATION OF THE PHYSICAL BODY TO THE BODY OF LIGHT SIMILAR TO THE RESURRECTION OF CHRIST 2000 YEARS AGO. WHEN THIS HAPPENS, WE RISE INTO THE AIR TO BE CAUGHT UP WITH HIM.

He really does exist. I believe what I have seen is what is to come. I look forward to the day when He returns."[243]

Prisoner says "Jesus is Coming Soon"

I was a withdrawing mess, suicidal, full of hate. Alone in the holding, or so I thought. Then it happened. In short, the cell took on a different light and color. There was a sound and intense feeling of rushing water, but it was not wet or physical, it was alive, moving and with voice, "living water". This gentle loving voice said it loved me and called me by name. By this time, I was pressed back on the bunk and could not move. Physical breathing became nil.

These waters intensified, at a seemingly high vibrating rate penetrating every fiber of my being. I found myself basking in this ocean of love, still held in total awe of what was happening to me. (there are no proper words to use here)

To the point. The message; 'to love one another.' That is what I am to tell people, especially Christians, and their leaders, also to government officials. But, again few to none will, or have the ability to listen. Who am I to them? Just another tithe, a vote, a number, etc. (It seems our leaders have their own agenda). He said to tell you to "make straight the path of the Lord," "there should be no divisions but to love one another". I am to tell you that" He is coming soon." That we have "all gone astray." I don't know anyone in Ireland, but He said to tell them "to seek peace, stop fighting." And that we are to be "specific in prayer."

There is much more but for the purpose of this I hope it will suffice. Now, several years later I am understanding more of why and what was said to me, or shown. I asked the Lord, 'Who do I follow? What church do I go to? He said, 'None teach the true gospel, but to love one another.'

243 Linda's NDE, #40, NDE & a Vision, aleroy.com/board40.htm

Starting my new Christian walk, unaware of all this denomination stuff, I found myself caught up in all the hatred, various "cults", materialism, power mongers, control freaks (like, we will help you "if" you believe this, or do that), the so called best or biggest, the celebrity manure, and all the rest of the typical arrogant, superficial stuff. Plus, being told by so called preachers that God would not use such a person as myself for such a message. Don't know much do they? He can and will use anyone he desires.

So I think you see I'm definitely not into the mainstream, or have a head full of the herding instinct, just to please, or to fit in. I don't "fit" in. I'm not a good game player "now," so don't expect me to be one to play that kind of game. Status or no.[244]

The War in Heaven (War between the Sons of Light and the Sons of Darkness)

Always wondered if this story was just a myth.... NOT!

Just before final edits were made on the Bibliography section of the *Chronicles*, I came across a strange comment by young six-year-old, in Todd Burpo's, *Heaven is for Real*, his son, Colton says to his mother Sonya after watching C.S Lewis', *Narnia, the Lion, the Witch, and the Wardrobe*, "Mom, angels carry swords in heaven."

His mother responded, 'Um...okay. Why do they need swords in heaven?"

"Mom, Satan's not in hell yet," he said, almost scolding. "Angels carry swords so they can keep Satan out of heaven."[245]

How did a six-year-old know this? Colton's Sunday School curriculum didn't teach this to young children.

Colton finished by saying, "Dad, did you know there is going to be a war?" ...There is going to be a war and it's going to destroy the world. Jesus and the angels and the good people are going to fight against Satan and the monsters and the bad people. I saw it."[246]

Jesus is coming to Earth again to fight against the Dragon.

Secret Service Agents stood nearby watching us.

I spoke to Paul about the creation of a Unified Mind caused by the descent of the Shekinah light resting over the head, creating a crown or halo of light depicted in renaissance paintings of Jesus, the apostles, holy

---

244 My Story, #92, aleroy.com/board92.htm

245 Todd Burpo with Lynn Vincent 2010. Heaven is for Real: A little boy's astounding story of his trip to heaven and back. Nashville, Dallas, Mexico City, Rio De Janiero: Thomas Nelson. pp. 132-133.

246 Ibid. p.136

family, and saints. I told Paul you can't teach a student how to do this, you can only prepare them to be chosen by God. **~Chronicle 505.**

An A.D.C (after-death communication) dream shed more light on "Secret Service agents/angels" on October 31, 2014, eighty-two days after Paul Solomon's visitation at the Fellowship on August 10th, 2014. In the dream, I saw myself in the "Presidents Mansion" – a Celestial White House, having a drink with him. Beautiful paintings covered every wall, just as you see in the White House in Washington D.C.

Another dream I had in December 2014 emphasized the idea of Secret Service Agent. I saw myself in the Washington DC/heavenly Capitol. I was chosen to be part of the Lord's Secret Service council called "Red," named after the blood of Christ. I was the 17th secret service agent. I had a feeling this Agent Council was part of heaven's "defense team." In the dream I called Mom and told her I wouldn't be seeing her much anymore. It's all part of the Battle between the Sons of Light and the Sons of Darkness.

Secret Service agents (a.k.a. angels) were always standing nearby holy places and near the celestial city. They were always nice guys, but they appeared on-guard and ready for action at all times. Were they there to protect from an incursion by evil? **~Chronicle 519**

I always wondered why dreams I had of Jesus and going to the Heavenly City had Secret Service Agents nearby...." Every time... All nice guys...angels, but angel who protect us...and keep them from infiltration. Dream being Agent #17 for the Red team....

Warrior Angels / Warring Angels/ *The Second Coming & War with the Dragon*

*I saw a group of about fourteen warrior angels who were coming from the direction of the Throne. The angels were a good 20 feet tall and 10 feet across the shoulders. Their eyes glistened with a fiery light from the altars of God, and their swords were flames of fire. The ground shook as they passed me. I stepped aside, and the angels with me bowed their heads in respect. I thought to myself, I would never want to be any demon who wants to fight with them. Just one of them could destroy an entire army. Then, I hear that strong, firm, yet so gentle voice. Jesus was behind me. He said, 'I wanted you to see them. They are being sent into your future. They will be there when you need them.*247

---

247 Richard Sigmund 2004 & 2010. My Time in Heaven: A True Story of Dying...And Coming Back. New Kensington, PA: Whitaker House. p. 47.

Note: I find this stuff really strange to me, but I have to admit it seems important to write into Book III about the War between the Sons of Light and the Sons of Darkness. This refers to the BIG event coming soon....

Then I read two hours later in Betty Eadie's, #1 National Bestseller, Embraced By the Light, who was told there are other types of angels called "Warring Angels:"

It was shown to me that their purpose is to do battle against Satan and his angels. Although each of us have protecting, or guardian, spirits to assist us, there are times when the Warring Angels are necessary to protect us, and I understood that they are available to us through prayer. I saw they are giant men, very muscularly built, with a wonderful countenance about them. They are magnificent spirits. I understood simply by looking at them that to struggle against them would be an act of futility. They are actually dressed like warriors, in head dress and armor, and I saw that they moved more swiftly than other angels. But perhaps what set them apart more than anything was their aura of confidence; they were absolutely sure of their abilities. Nothing evil could daunt them, and they knew it.[248]

4-year-old Kennedy who was dead more than 12 minutes underwater, relates his experience with an angel & Jesus:

"I went to heaven."
What did you see?
"I saw Jesus. I saw lots of people and angels. They were very happy."
Did you see Uncle Mark?
"Yes. He looked just like Jesus. All his boo-boos were gone. He was happy. I saw a door with jewels on it. There was snow on the other side when they opened it."
When he talked about what happened next, Kennedy got very quiet."
Craig said, "He spoke in a whisper."
I saw a volcano, and there was a Pokemon in it. No, it really wasn't a Pokemon. I've never seen that one. It was a dragon.
Was the dragon happy or sad?
He was happy. He looked at me and growled.
Were you scared?
No, I was with Jesus and Uncle Mark. I was standing on the grass and I was invisible.
Was the volcano part of heaven?

248 Betty J. Eadie 1992. Embraced By The Light. Placerville, CA: Gold Leaf Press. pp. 91-92.

No. (his tone seemed to imply, No, silly) It's not part of heaven. There were lots of people in the volcano. They were very sad.

How did you get back?

Uncle Mark pushed me down and the angel brought me back. You know, Mommy, Jesus is coming back here.

"I'll never forget how this little finger pointed down at the ground when he said, "Jesus is coming back here,"

Amy said, "And, no, he (Kennedy) didn't say when."

Appendix A

Interpretation of Revelations

Rev	Scripture Quote	INTERPRETATION
1:1	The Revelation of Jesus Christ, which God gave unto him, to shew unto his servant's things which must shortly come to pass; and he sent and signified it by his angel unto his servant John:	Shortly come to pass
1:2	Who bare record of the word of God, and of the testimony of Jesus Christ, and of all things that he saw.	Word of God
1:3	Blessed is he that readeth, and they that hear the words of this prophecy, and keep those things which are written therein: for the time is at hand.	This prophecy
	Commentary: Matthew 24:34: The apostles asked Jesus, "What will be the sign of your coming and of the end of the age?" "Verily I say unto you. *This generation shall not pass, till all these things be fulfilled.* This controversial verse is in all three of the Olivet Discourse accounts. (These accounts are to be found in Matthew 24:1-51, Mark 13:1-37, and Luke 21:5-33). For some time, critics of the Christian faith have argued that Jesus explicitly said here that all of the events prophesied in the Olivet Discourse, **including His return, would happen before the last person living at that time died. It could be why John of Patmos wrote, "for the time is at hand."** The generation Jesus was speaking about was the current one, but the generation living at the End	The Return or Arrival of the Christ—The Parousia –

	of the Age of Pisces. **The return of Christ is called Parousia, in Greek, meaning "arrival." The main use is the physical presence of a person, which where that person is not already present refers to the prospect of the physical arrival of that person, especially the visit of a royal personage, at the end of the PISCEAN AGE—THE AGE OF AQUARIUS.**	
1:4	John to the seven churches which are in Asia: Grace be unto you, and peace, from him which is, and which was, and which is to come; and from the seven Spirits which are before his throne;	Asia / Taurus
1:5	And from Jesus Christ, who is the faithful witness, and the first begotten of the dead, and the prince of the kings of the earth. Unto him that loved us, and washed us from our sins in his own blood,	Faithful witness- Jesus has been here from the beginning First Begotten of the Dead. Jesus is the first to resurrect himself from physical death Prince of Kings of the Earth. Jesus, Planetary Headmaster in charge of our Earth University Evolution. Jesus' Blood Sacrifice redeemed us from our Sins. We are already forgiven
1:6	And hath made us kings and priests unto God and his Father; to him be glory and dominion for ever and ever. Amen	Kings & Priests. A chosen generation, a royal priesthood, an holy nation, a peculiar people; that ye should shew forth the praises of him who hath called you out of darkness into his marvelous light. I Peter: 2:9
1.7	Behold, He cometh with clouds; and every eye shall see him, and they also which pierced him: and all kindreds of the earth shall wail because of him. Even so, Amen.	Clouds- World-wide explosions (See Russian comet explode. February 15, 2013 World-wide atmospheric explosions seen by all the Earth Reference to THE GENERATION WHO WITNESSED THE CRUCIFIXION 2000 years AGO. Comet also pierces and cleaves the heavens. It is a crucifixion of the world at the apocalypse. It will be sudden and expected. Everyone will be afraid. "Behold, I

		come like a thief." Everyone will cry out. Oh My God! .
1.8	I am Alpha and Omega, the beginning and the ending, saith the Lord, which is, and which was, and which is to come, the Almighty	
1:9	I John, who also am your brother, and companion in tribulation, and in the kingdom and patience of Jesus Christ, was in the isle that is called Patmos, for the word of God, and for the testimony of Jesus Christ.	
1:10	I was in the Spirit on the Lord's day, and heard behind me a great voice, as of a trumpet,	
1:11	Saying, I am Alpha and Omega, the first and the last: and, What thou seest, write in a book, and send it unto the seven churches which are in Asia; unto Ephesus, and unto Smyrna, and unto Pergamos, and unto Thyatira, and unto Sardis, and unto Philadelphia, and unto Laodicea.	Christ begins the age and ends the age. He is the christ comet heralding the apocalypse 12,960 years ago, [alpha], and he is the christ comet ending the age [omega] now!!! It is interesting that the seven candlesticks are seven stars (Pleiades) in Asia (Taurus). The Beta Taurid meteor shower issues forth from Taurus the Bull. Source of the Prometheus legend. He ridiculed the gods; surpassed all in cunning and fraud; suspected the trickery of Zeus in offering him Pandora, whom he refused to accept as a wife; stole fire from heaven, for which crime Zeus ordered Hermes to chain Prometheus to a rock on Mount Caucasus (Taurus mountain) where a vulture fed daily on his liver, which grew back daily for thirty years. He created mankind from clay; gave him fire, how to use plants for medicine. Note: Tartarus, is the gloomy region of the underworld, but also a Titan named Tartaros is the father by Gaea of Typhon (Typhoeus) who had a hundred heads like those of a dragon.

1:12	And I turned to see the voice that spake with me. And being turned, I saw seven golden candlesticks	
		Pleiades – 7 sister stars in Taurus (Asia)
		After Atlas was forced to carry the heavens on his shoulders, Orion began to pursue all of the Pleiades, and Zeus transformed them first into doves, and then into stars to comfort their father. The constellation of Orion is said to still pursue them across the night sky. The Greek myths say all seven sisters committed suicide because they were so saddened by either the fate of their father, Atlas, or the loss of their siblings, the Hyades. In turn Zeus, the ruler of the Greek gods, immortalized the sisters by placing them in the sky. There these seven stars formed the star cluster known thereafter as the Pleiades.[249]
1:13	And in the midst of the seven candlesticks one like unto the Son of man, clothed with a garment down to the foot, and girt about the paps with a golden girdle.	
1:14	His head and his hairs were white like wool, as white as snow; and his eyes were as a flame of fire;	
1:15	And his feet like unto fine brass, as if they burned in a furnace; and his voice as the sound of many waters.	
1:16	And he had in his right hand seven stars: and out of his mouth went a sharp twoedged sword: and his countenance was as the sun shineth in his strength	
1:17	And when I saw him, I fell at his feet as dead. And he laid his right hand upon me, saying	

249 Wikepidia: Pleiades.

Rev 1	Scripture	Interpretation / commentary
	unto me, Fear not; I am the first and the last:	
1:18	] I am he that liveth, and was dead; and, behold, I am alive for evermore, Amen; and have the keys of hell and of death.	
1:19	Write the things which thou hast seen, and the things which are, and the things which shall be hereafter;	
1:20	The mystery of the seven stars which thou sawest in my right hand, and the seven golden candlesticks. The seven stars are the angels of the seven churches: and the seven candlesticks which thou sawest are the seven churches.	

Rev 2	Scripture	Interpretation / commentary
2:1	Unto the angel of the church of Ephesus write; These things saith he that holdeth the seven stars in his right hand, who walketh in the midst of the seven golden candlesticks;	**Historical: Ephesus.** Ephesus was a city on the western coast of Asia Minor, near the mouth of the Cayster River. The city was famous for its temple of Diana (or Artemis, Acts 19:27), and pilgrims came to Ephesus from all over the Mediterranean world to worship the goddess. Angel of the Church 7 Stars – Pleiades 7 Golden Candlesticks – 7 planets / 7 Days - Sun - Moon - Mercury - Mars - Saturn - Jupiter
2:2	I know thy works, and thy labour, and thy patience, and how thou canst not bear them which are evil: and thou hast tried them which say they are apostles, and are not, and hast found them liars:	Try the Spirits to see if they are telling you the truth. Are you telling yourself the truth?
2:3	And hast borne, and hast patience, and for my name's sake hast laboured, and hast not fainted.	

2.4	Nevertheless, I have somewhat against thee, because thou hast left thy first love	You have fallen in love with things that distract you from God. Survival. Competition.
2:5	Remember therefore from whence thou art fallen, and repent, and do the first works; or else I will come unto thee quickly, and will remove thy candlestick out of his place, except thou repent.	You cannot change (repent) if you don't know what you are doing. Admitting the truth makes your light brighter (candlestick). Denying the truth, darkens your light.
2:6	But this thou hast, that thou hatest the deeds of the Nicolaitans, which I also hate.	**Historical:** the Nicolaitans are the followers of that Nicolas who was one of the seven first ordained to the diaconate by the apostles. They lead lives of unrestrained indulgence. The character of these men is very plainly pointed out in the Apocalypse of John, [when they are represented] as teaching that it is a matter of indifference to practice adultery, and to eat things sacrificed to idols.— Irenaeus, Adversus haereses, i. 26, §3. A Warning against sexual indulgences.
2:7	He that hath an ear, let him hear what the Spirit saith unto the churches; To him that overcometh will I give to eat of the tree of life, which is in the midst of the paradise of God.	If you learn to control (overcome excessive acts of sexuality-reference to Rev. 2:6 and the Nicolaitans) your physical vitality (chi) will not be wasted. You receive life and power from the Tree of Life Tree of Life is in the Midst of the Paradise of God. (Heaven) within and without you
2:8	And unto the angel of the church in Smyrna write; These things saith the first and the last, which was dead, and is alive;	**Historical Smyrna:** Smyrna was a large, important city on the western coast of Asia Minor, famed for its schools of medicine and science. This is one of just two of the seven churches where there is no condemnation, there is no discussion about sin, there is no word from the Lord about punishment. A humble city. Reference: rev 1:11 Alpha & omega: dead after first apocalypse 12,960 years ago (disappeared into the wilderness) but is now alive as the christ comet ending the age of pisces. Shing in the heavens.

2:9	I know thy works, and tribulation, and poverty, (but thou art rich) and I know the blasphemy of them which say they are Jews, and are not, but are the synagogue of Satan.	
2:10	Fear none of those things which thou shalt suffer: behold, the devil shall cast some of you into prison, that ye may be tried; and ye shall have tribulation ten days: be thou faithful unto death, and I will give thee a crown of life.	**Suffering is brief** The devil refers to Nero/Rome, but the Devil metaphor here is the Apocalypse Devil, and refers to the Red Dragon/Devil in Rev. 12: 3: *And there appeared another wonder in heaven; and behold a great red dragon, having seven heads and ten horns, and seven crowns upon his heads. This Devil trial of ten days matches the ten horns. (see ?)* *10 days / 10 horns / Time (repeated many times throughout Book of Revelation) Horns- referring to 10 divisions of 12,960 comet years.* Horns (the curved tail of a comet). Mythology begins counting down from TEN horns. Moses is DEPICTED WEARING 2 horns in the colored glass of the Cathedral of Saint John the Divine in NYC. Many mythologies describe show only one horn left. The Unicorn. The one horned Goat. We are in the days of the LAST HORN. THE NEXT HORN of the Apocalypse IS THE LAST. Be not afraid. Overcoming Fear=LIFE
2;11	He that hath an ear, let him hear what the Spirit saith unto the churches; He that overcometh shall not be hurt of the second death.	**Historical:** Whenever you see Jesus use a word such as "perish" or "destroy," the original word is "apollumi." Some evangelical Christian sects have brought fear into the church by teaching that the Greek word "apollumi" meant "to destroy" suggesting that if a person did not believe in Jesus Christ that their soul would be destroyed at death, losing the opportunity to receive eternal life and become immortal. (Using fear to receive Christ's love is manipulative at worst and misguided at best). Beyond "Apollumi" and Perish, overcoming Fear and receiving the Crown of Life, references the

		possibility of the soul's destruction as the "Second death." There has been throughout religious history a disturbing thought concerning the possibility of the annihilation of the soul. In ancient Egypt, the concept of a Second Death was a belief that a second death, or a final annihilation of the soul and its personality could occur. Anglicans, some Lutherans, all Seventh-day Adventists, and others, oppose the idea of eternal suffering but believe that the Second death is an actual second death, meaning that the soul perishes and will be annihilated after the final judgment at the end of time. In Revelation 20:13-15 it says, "And the sea gave up the dead which were in it, and death and Hades gave up the dead which were in them; and they were judged, every one of them according to their deeds. Then death and Hades were thrown into the lake of fire. This is the second death, the lake of fire. And if anyone's name was not found written in the book of life, he was thrown into the lake of fire." There are no testimonies I could find in the Dead Saints Chronicles regarding specific references concerning the Second Death. It is a theology that remains open to debate. Internal: The second death. The first death is physical. The Second Death relates to the destruction of the personality.
2:12	And to the angel of the church in Pergamos write; These things saith he which hath the sharp sword with two edges;	Historical Pergamos: Historical Pergamos: Pergamum had served as the capital of the Roman province of Asia Minor for over 25 years and was an important religious center for a number of pagan cults. It was the first city in Asia to build a temple to Caesar and it became the capital of the cult of Caesar worship. Of Pergamum an ancient writer said it was "given to idolatry more than all Asia." Angel of Church of Pergamos

		Comet Symbol- Sword of two-edges throughout Book of Revelation.
2:13	I know thy works, and where thou dwellest, even where Satan's seat is: and thou holdest fast my name, and hast not denied my faith, even in those days wherein Antipas was my faithful martyr, who was slain among you, where Satan dwelleth.	**Historical Antipas:** A Christian in Pergamum named Antipas is mentioned as a "faithful witness." Church tradition says that Antipas was a physician suspected of secretly propagating Christianity. The Aesculapians (members of the medical guild) accused Antipas of disloyalty to Caesar. Upon being condemned to death, Antipas was placed inside a copper bull, which was then heated over a fire until it was red-hot. Satan's Seat Denying your Faith Antipas the faithful martyr who was slain where Satan dwells Warning be faithful, or get fried!
2:14	But I have a few things against thee, because thou hast there them that hold the doctrine of Balaam, who taught Balac to cast a stumbling block before the children of Israel, to eat things sacrificed unto idols, and to commit fornication	**Historical Balaam** is mentioned in 2 Peter 2:15, and Jude, verse 11, as one who would do wrong for the love of money. In Revelation 2:14, God tells the church in Pergamos that Balaam taught Balak how to ruin the people of Israel, by involving them in sexual sin, and encouraging them to go to idol feasts. Warning against sexual indulgence.
2:15	So hast thou also them that hold the doctrine of the Nicolaitans, which thing I hate	Warning against sexual indulgences
2:16	Repent; or else I will come unto thee quickly, and will fight against them with the sword of my mouth.	Sword out of mouth / Judgment if not repentant.
2:17	He that hath an ear, let him hear what the Spirit saith unto the churches; To him that overcometh will I give to eat of the hidden manna, and will give him a white stone, and in the stone a new name written, which no man knoweth saving he that receiveth it.	Concordance: Zech 3:9: For behold the stone that I have laid before Joshua: upon one stone shall be seven eyes; behold, I will engrave the graving thereof, saith the Lord of hosts, and I will remove the iniquity of that land in one day. Zech 4:10 For who hath despised the day of small things? For they shall rejoice, and shall see the plummet in the hand of Zerubbabel with those seven; they are the eyes of the Lord

		which run to and fro through the whole earth. 7 Eyes on the white stone (pure foundation) A new identity. Born Again. Called by My Name.

2:18	And unto the angel of the church in Thyatira write; These things saith the Son of God, who hath his eyes like unto a flame of fire, and his feet are like fine brass;	**Historical Thyatira:** Eyes like Flame-
		  The Son of God
		  Feet of Burnished Brass
		
		Dead Saint Sybel.
		*I stood there and suddenly I saw Jesus. It was quite an awesome shock to be standing in His presence. I was in such awe, I could not speak. My eyes were fixed upon Him. **His hair was white like wool and hung down to His shoulders. His skin was like brass without one wrinkle. His eyes were like flames of fire and when He spoke, it was with great authority.** Yet when He spoke it was kind and gentle and loving. When He spoke it sounded like thunder rolling across the North Carolina skies but much louder than any I had ever heard.*
		₂₅₀

2:19	I know thy works, and charity, and service, and faith, and thy patience, and thy works; and the last to be more than the first.	
2:20	Notwithstanding I have a few things against thee, because thou sufferest that woman Jezebel, which calleth herself a prophetess, to teach and to seduce my servants to commit fornication, and to eat things sacrificed unto idols.	Historical Jezebel:
2:21	And I gave her space to repent of her fornication; and she repented not.	
2:22	Behold, I will cast her into a bed, and them that commit adultery with her into great tribulation, except they repent of their deeds.	

2:23	And I will kill her children with death; and all the churches shall know that I am he which searcheth the reins and hearts: and I will give unto every one of you according to your works.	
2:24	But unto you I say, and unto the rest in Thyatira, as many as have not this doctrine, and which have not known the depths of Satan, as they speak; I will put upon you none other burden.	
2:25	But that which ye have already hold fast till I come	
2:26	And he that overcometh, and keepeth my works unto the end, to him will I give power over the nations:	
2:27	And he shall rule them with a rod of iron; as the vessels of a potter shall they be broken to shivers: even as I received of my Father.	And I will plead against him with pestilence and with blood; and I will cause to rain upon him and upon his bands and upon the many peoples that are with him an overflowing shower and great hailstones fire and brimstone. (also see Ezekiel 38:22) Those who overcome will be given a rod of iron to rule. Rod of iron, a rare element is often associated with iron meteorites, and its shape describes a falling star or comet. Or a pillar in the temple of God. The city, New Jerusalem, the seat of the throne.
2:28	And I will give him the morning star	13th Sign / Venus / Death & Rebirth/The Morning Star. Christ's Star. Mayan.
2:29	He that hath an ear, let him hear what the Spirit saith unto the churches.	
Rev 3		
3.1	And unto the angel of the church in Sardis write; These things saith he that hath the seven Spirits of God, and the seven stars; I know thy works, that thou hast a name that thou livest, and art dead.	Your works appear to be alive, but in fact, are dead Says the one holding the seven spirits of the seven stars; and the name you have lived, and art dead.
3.2	Be watchful, and strengthen the things which remain, that are ready to die: for	Strengthen yourself, before you lose what you have.

	I have not found thy works perfect before God.	
3.3	Remember therefore how thou hast received and heard, and hold fast, and repent. If therefore thou shalt not watch, I will come on thee as a thief, and thou shalt not know what hour I will come upon thee.	The impacts happen in seconds. It is sudden, like a thief. No notice. No warning. Meteors traveling at 75,000 mph through atmosphere.... (show film)
3.4	Thou hast a few names even in Sardis which have not defiled their garments; and they shall walk with me in white: for they are worthy	Some have not defiled garments:
3.5	He that overcometh, the same shall be clothed in white raiment; and I will not blot out his name out of the book of life, but I will confess his name before my Father, and before his angels.	
3.6	He that hath an ear, let him hear what the Spirit saith unto the churches	
3.7	And to the angel of the church in Philadelphia write; These things saith he that is holy, he that is true, he that hath the key of David, he that openeth, and no man shutteth; and shutteth, and no man openeth;	What is the key of David? Opens what no man can shut, and shuts what no man can open. The answer lies in analyzing the key and door metaphor, which is found in the writings of the prophet Isaiah. He referred to an individual of his time named Shebna who had charge of the palace of the Judean king. Today, we might call him the chief of staff. The prophet Isaiah said the Lord would replace Shebna with a man named Eliakim. The Lord would "place on his shoulder the key to the house of David; what he opens no one can shut, and what he shuts no one can open" (Isaiah 22:22). Thus, Eliakim would be a kind of gatekeeper with power to control entry into the royal kingdom. As the king's steward, he would decide who could or could not have access to the king. We also must understand why God and the people made David their king. He testified concerning him: 'I have found David son of Jesse a man after my own heart; he will do everything I want him to do'" (Acts

13:22). The obvious question is, how could God call David "a man after His heart" when David was such a terrible a sinner, having committed adultery, murder, slaying tens of thousands?

First, David had absolute faith in God. Nowhere in Scripture is this point better illustrated than in 1 Samuel 17 where David as a young shepherd boy fearlessly slew the Philistine, Goliath. Shortly before the duel, we see direct evidence of David's faith in verse 37 where David says, "'The LORD who delivered me from the paw of the lion and from the paw of the bear will deliver me from the hand of this Philistine.' And Saul said to David, 'Go, and the LORD be with you!'" David was fully aware that God was in control of his life, and he had faith that God would deliver him from impending danger. How else would one venture into a potentially fatal situation with such calm and confidence?

Second, God granted David understanding and wisdom through daily meditation. We would do well to think about God throughout the day. "Blessed are they who keep his statutes and seek him with all their heart. They do nothing wrong; they walk in his ways" (Psalm 119:2-3). To put God first in thought and action. To walk Holy before God.

Revelation suggests the Key of David is controlled by he who is HOLY AND TRUE, Jesus Christ, the Messiah (Logos). He not only holds the key to THE DOOR to the Kingdom of Heaven, HE IS BOTH THE KEY AND THE DOOR and has the authority and the power to allow us access to the Kingdom. (John 10:7, 9) In the book of Revelation, John used this Old Testament metaphor to get across a vital message to the church in Philadelphia, and thereby to all those who follow Christ. That is, Christ has the key of David. He opens the door for us, his royal

		household, to come into the presence of God. No one can deprive us of that access.
3.8	I know thy works: behold, I have set before thee an open door, and no man can shut it: for thou hast a little strength, and hast kept my word, and hast not denied my name.	Open Door: No man can shut it, because you are not strong, but has kept my word (love and faith) and not denied His Name.
3.9	Behold, I will make them of the synagogue of Satan, which say they are Jews, and are not, but do lie; behold, I will make them to come and worship before thy feet, and to know that I have loved thee.	Synagogue of Satan: say they are Jews, but are not, worshipping other gods. Hypocrites. Lord will make them worship at your feet – the true worshippers of God
3.10	Because thou hast kept the word of my patience, I also will keep thee from the hour of temptation, which shall come upon all the world, to try them that dwell upon the earth.	If you follow after God's tenants, God will keep you from temptation (requiring good will choice). There is a temptation coming to the entire world. The future prophecy
3.11	Behold, I come quickly: hold that fast which thou hast, that no man take thy crown.	Temptation/lessons/challenges can happen upon you any moment. (personally or globally). Hold on to your faith and your love. Don't let outside events cause fear or loss of faith.
3.12	Him that overcometh will I make a pillar in the temple of my God, and he shall go no more out: and I will write upon him the name of my God, and the name of the city of my God, which is new Jerusalem, which cometh down out of heaven from my God: and I will write upon him my new name.	A Dead Saint's Vision of the Zodiac William describes 12 spokes of a large wheel representing the constellations of the zodiac, revealed to him during his NDE: I would have to say this occurred about halfway through my last NDE. My last NDE was not one single trip but actually seven trips. I would leave my body, not only to escape the tremendous pain I was in, but to also to continue those things I was being instructed in on the other side. It was during the 3rd or 4th trip when I was taken in a room that had a large wheel on the floor, at the end of each spoke of the wheel stood a pillar. The wheel had 12 spokes and subsequently there were 12 pillars at the end of each spoke. Each pillar also contained 12 crystals and 12 symbols. Some of these symbols were representative

of the astrological star constellations of the zodiac.

As William describes above, at the end of each spoke was a pillar. When I refer to the lessons we experience on Earth University, I am referring to some of these 12 pillars. Each of the 12 pillars represents one of 12 major life lessons:

Purity of Heart
Loyalty
Balance
Sexuality
Judgment / Discernment
Money
Power
Alrightness
Discipline
Intuition / Vision
Creativity
Communication

Later, the meaning of the 12 continued in Jesus's choice of 12 apostles, and symbolically memorialized in the Book of Revelation by the Virgin with a crown of 12 stars and the 12 gates, 12 stones, and 12 fruits of the Tree of Life in the celestial New Jerusalem. Like the 12 apostles and the 12 tribes of Israel, each pillar has its strengths, weaknesses, and purpose. The permutations of pillar lessons combined are many, but ultimately originate from these 12. They are separate from the lesson of love and forgiveness described as the 13th path in chapter 24.

Overcoming means: Because you hold the Key of David, you have made Christ the Ruler (authority) of your heart, mind, and body. You then become a Pillar in the Temple of God, New Jerusalem. When this happens the heavenly Temple will descent upon you, and will appear as a golden crown over the head. (The Shekinah Light shines from the Holies of Holies within your brain/mind. Some call this the God

		spot, the Pineal, or 3rd eye, it doesn't matter). "Go no more out" means: rebirth into a physical body on Earth is no longer required. And because you have graduated from Earth University you will be known by a New Name.
3:13	He that hath an ear, let him hear what the Spirit saith unto the churches.	
3:14	And unto the angel of the church of the Laodiceans write; These things saith the Amen, the faithful and true witness, the beginning of the creation of God;	Laodiceans: Amen: The "I AM." The True Witness of Creation. The Creator God.
3:15	I know thy works, that thou art neither cold nor hot: I would thou wert cold or hot.	
3:16	So then because thou art lukewarm, and neither cold nor hot, I will spue thee out of my mouth.	Lukewarm: Be good or be evil, but lazy I will spew you out
3:17	Because thou sayest, I am rich, and increased with goods, and have need of nothing; and knowest not that thou art wretched, and miserable, and poor, and blind, and naked:	
3:18	I counsel thee to buy of me gold tried in the fire, that thou mayest be rich; and white raiment, that thou mayest be clothed, and that the shame of thy nakedness do not appear; and anoint thine eyes with eye salve, that thou mayest see.	Gold tried by fire: The words attributed to the Laodiceans may mark an ironic over-confidence in regard to spiritual wealth; they are unable to recognize their bankruptcy. However the image may also be drawing on the perceived worldly wealth of the city. The city was a place of great finance and banking. In 60 A.D the city was hit by a major earthquake. The city refused help of the Roman empire and rebuilt the city itself.The reference to the "white raiment" may refer to the cloth trade of Laodicea. The city was known for its black wool that was produced in the area.The reference to eye medication is again often thought to reflect the historical situation of Laodicea. According to Strabo (12.8.20) there was a medical school in the city, where a famous ophthalmologist practiced. The city

			also lies within the boundaries of ancient Phrygia, from where an ingredient of eye-lotions, the so-called "Phrygian powder", was supposed to have originated.
	3:19	As many as I love, I rebuke and chasten: be zealous therefore, and repent.	
	3:20	Behold, I stand at the door, and knock: if any man hear my voice, and open the door, I will come in to him, and will sup with him, and he with me.	
	3:21	To him that overcometh will I grant to sit with me in my throne, even as I also overcame, and am set down with my Father in his throne.	
	3:22	He that hath an ear, let him hear what the Spirit saith unto the churches	
4	**Rev**		
	4:1	After this I looked, and, behold, a door was opened in heaven: and the first voice which I heard was as it were of a trumpet talking with me; which said, Come up hither, and I will shew thee things which must be hereafter.	Trumpet / Voice: Door opened in Heaven: Trumpet Voice of God. Being shown things which will come in the future.
	4:2	And immediately I was in the spirit: and, behold, a throne was set in heaven, and one sat on the throne.	Internal: Throne: The temple within you Celestial: Throne: The heavens
	4:3	And he that sat was to look upon like a jasper and a sardine stone: and there was a rainbow round about the throne, in sight like unto an emerald.	Jasper Stone: Bright, clear, pure. Sardine Stone: reddish brown. Emerald: green. Rainbow like emerald Vision of the Four Wheels (Ezekiel 4: 15-21)- 4 wheels define 4 living creatures 1st wheel upon the earth with four faces Color of beryl (green) One Likeness Appearance of wheel within wheel Followed each other wherever they went Rings above them so high and dreadful

Full of eyes (stars)

And when the living creatures went, the wheels went by them: and when the living creatures were lifted up from the earth, the wheels were lifted up.

The appearance of the wheels and their work was like unto the colour of a beryl: (The name beryl is derived (via Latin: beryllus, Old French: beryl, and Middle English: beril) from Greek "beryllos "which referred to a "precious blue-green color-of-sea-water stone. The living creatures were constellations in the "sea" of in sky above, the heavens.

17 When they went, they went upon their four sides: and they turned not when they went. 18As for their rings, they were so high that they were dreadful; and their rings were full of eyes round about them four.

Vision of the Divine Glory (Ezekiel: 22)

And the likeness of the firmament upon the heads of the living creature was as the colour of the terrible crystal, stretched forth over their heads above. 23And under the firmament were their wings straight, the one toward the other: every one had two, which covered on this side, and every one had two, which covered on that side, their bodies. 24And when they went, I heard the noise of their wings, like the noise of great waters, as the voice of the Almighty, the voice of speech, as the noise of an host: when they stood, they let down their wings. 25And there was a voice from the firmament that was over their heads, when they stood, and had let down their wings.

And above the firmament that was over their heads was the likeness of a throne, as the appearance of a sapphire stone (blue sky): and upon the likeness of the throne was the likeness as the appearance of a man above upon it.

And I saw as the colour of amber, (yellow-orange)

		Fire from loins downward/downward Brightness around it. The appearance of bow like the cloud in a day of rain with brightness around it. Brightness like glory of the Lord.
4:4	And round about the throne were four and twenty seats: and upon the seats I saw four and twenty elders sitting, clothed in white raiment; and they had on their heads crowns of gold	Historical: 24 Greek characters. Ursa Minor (little bear) is outlined by 24 stars. Pole Star is called in Arabic Al Ruccaba. Because of the Precession of the Equinoxes, the slow wobbling in the heavens, the north star—in Draco, marked the central point about 4000 to 3000 B.C. In Greece this star was termed the Cynosure. Because of Precession, the Dragon was cast down from its place in the heavens. Ursa Major: The Book of Job the Bear is mentioned under the name of Ash."Canst thou guide Ash and her offspring? The Arabs still call it Al Naish, or Annaish, the assembled together as sheep in a fold. They are called by others Septentriones, which thus became the Latin word for North. The star in the tail is known as Al Cor, the Lamb. (Lamb that is Slain before the foundation of the world.) Relationship to Auriga, the Shepherd. 24 hours of the day. One hour per seat.
4:5	And voices: and there were seven lamps of fire burning before the throne, which are the seven Spirits of God And out of the throne proceeded lightnings and thunderings	Lightning, thundering and voices. Seven burning Lamps of fire The Seven Spirits of God. Seven Stars of the Pleiades.
4:6	And before the throne there was a sea of glass like unto crystal: and in the midst of the throne, and round about the throne, were four beasts full of eyes before and behind.	The throne is in the midst of a sea of glass. Round with four beasts with eyes before and behind. The eyes are the stars of the four living creatures;
4:7	And the first beast was like a lion, and the second beast like a calf, and the third beast had a face as a man, and the fourth beast was like a flying eagle.	The first Lion beast (Leo), the second calf beast (Taurus), the third Beast, fact of a Man, (Aquarius), the fourth Beast, a flying eagle (Scorpio). All beasts referenced by

		(Ezekiel, Daniel, John) as Living Creatures/Angels/Cherebum). They represent the FOUR CARDINAL POINTS OF THE ZODIAC, representing the 25,960 years; 12 zodiac ages of 2,160 years each!
4:8	And the four beasts had each of them six wings about him; and they were full of eyes within: and they rest not day and night, saying, Holy, holy, holy, Lord God Almighty, which was, and is, and is to come.	The four beasts were full of eyes—eyes that refer again to the stars shining amidst the constellations. Each beast has control over six constellations, totalling 24 constellations. Same story told about king author in the once in future king and the 24 knights of who sat at the round table, and king arthur who wielded excalibur, the sword of the christ. **Isaiah** 6:2: cross reference Above [the throne] stood the seraphim's; each one had six wings; with twain he covered his face, and with twain he covered his feet, and with twain he did fly. (6 x 4 =24 wings) These eyes and wings rest not day and night saying, holy, holy, lord god almighty, which was, and is, and is to come.
4:9	And when those beasts give glory and honor and thanks to him that sat on the throne, who lives for ever and ever,	The Beasts give honor to him who sits on the Throne in the skies above, and in the Heavens above the skies.
4:10	The four and twenty elders fall down before him that sat on the throne, and worship him that lives for ever and ever, and cast their crowns before the throne, saying,	24 Elders: The beasts, (The Lion, the Eagle, the Bull, the Man – the four cardinal points of the Zodiac, cast their crowns before the Throne) worshipped him, and cast down there crowns before the throne, saying.
4:11	Thou art worthy, O Lord, to receive glory and honor and power: for thou hast created all things, and for thy pleasure they are and were created.	That the LORD is worthy to receive honor and power forever: because the LORD has created all things, for His pleasure were they created.
Rev 5		
5:1	And I saw in the right hand of him that sat on the throne a book written within and on the backside, sealed with seven seals.	In the right hand of Him who sat on the Throne, who has a book sealed with seven seals *on the backside*.

		The book of seven seals on the backside of the book—is *the end of the prophecy. It describes* how the end of the world—the Apocalypse—will happen.
5:2	And I saw a strong angel proclaiming with a loud voice, Who is worthy to open the book, and to loose the seals thereof?	Who is worthy to open the book, and loose the seals thereof?
5:3	And no man in heaven, nor in earth, neither under the earth, was able to open the book, neither to look thereon	NO mortal man is qualified to open the book. Only one donning the Mind of Christ can open the book.
5:4	And I wept much, because no man was found worthy to open and to read the book, neither to look thereon.	John wept much, because no man was found worthy to open the book or to look thereon – except....
5:5	And one of the elders saith unto me, Weep not: behold, the Lion of the tribe of Juda, the Root of David, hath prevailed to open the book, and to loose the seven seals thereof.	One of the 24 elders said weep not, the lion of the tribe of juda, the root of david, has prevailed to open the book and loose the seals thereof; In general, the passage refers to jesus as a royal lineage, represented by a lion. The lion of judah is the symbol of the jewish tribe of judah. According to the torah, the tribe consists of the descendants of judah, the fourth son of jacob. The association between judah and the lion, most likely the asiatic lion, can first be found in the blessing given by jacob to judah in the book of genesis.
5:6	And I beheld, and, lo, in the midst of the throne and of the four beasts, and in the midst of the elders, stood a Lamb as it had been slain, having seven horns and seven eyes, which are the seven Spirits of God sent forth into all the earth.	In the midst of the throne Four beasts : lion, bull, scorpion, man – the zodiac – in the midst of the 24 elders represented by 6 wind A Lamb as it had been Slain With 7 Horns And 7 Eyes – The 7 Spirits of God sent forth to the Whole Earth
5:7	And he came and took the book out of the right hand of him that sat upon the throne.	
5:8	And when he had taken the book, the four beasts and four and twenty elders fell down before the Lamb, having every one of them harps, and golden vials full of odours, which are the prayers of saints	

5:9	And they sung a new song, saying, Thou art worthy to take the book, and to open the seals thereof: for thou wast slain, and hast redeemed us to God by thy blood out of every kindred, and tongue, and people, and nation;	
5:10	And hast made us unto our God kings and priests: and we shall reign on the earth.	
5:11		From the perspective of the Dead Saints, many thousands if not millions of angels have voices singing in Heaven.
5:12	Saying with a loud voice, Worthy is the Lamb that was slain to receive power, and riches, and wisdom, and strength, and honour, and glory, and blessing.	On the Right Hand of the Almighty a book written with 7 seals - In his Right Hand of the Almighty are seven stars. The Pleiades?Only the Lion of the Tribe of Judah can open the mysteries of the book. (Leo) or Orion According to Daniel 8: 9-11, "the judgment was set, and the books were opened." Daniel sees the fiery river in his vision of the coming of that day, when the true Orion shall come forth in this glory. He says, "I beheld till the thrones were placed, and one that was ancient of days did sit: "His throne was fiery flames, and the wheels thereof burning fire. A fiery stream issued and came forth from before him." This is the River of the Judge. Who can stand before His indignation, when "His fury is poured out like fire? He is seated on the Milky Way: The River of Life. One like the Son of God is a "Lamb as it had been slain" The star Beta, at the tip of the left horn of Taurus has an Arabic name -- El Nath, meaning wounded or slain. It is also the star which comprises the foot of Auriga, the Shepherd - The Lamb slain The Throne is surrounded by 7 stars of the Pleiades, which means a congregation, (a church). Each of the 7 stars have names, but they were given by the Greeks from the names of the seven daughters of Atlas by Pleione. Born on Mount Cyllene in Arcadia: Alcyone, Celaeno, Electra,

		Maia, Merope, Asterope, and Taygeta. *Isaiah 66:14-16King James Version (KJV)* *14 And when ye see this, your heart shall rejoice, and your bones shall flourish like an herb: and the hand of the Lord shall be known toward his servants, and his indignation toward his enemies.* *15 For, behold, the Lord will come with fire, and with his chariots like a whirlwind, to render his anger with fury, and his rebuke with flames of fire.* *16 For by fire and by his sword will the Lord plead with all flesh: and the slain of the Lord shall be many.*
5:13	And every creature which is in heaven, and on the earth, and under the earth, and such as are in the sea, and all that are in them, heard I saying, Blessing, and honour, and glory, and power, be unto him that sitteth upon the throne, and unto the Lamb for ever and ever.	
5:14	And the four beasts said, Amen. And the four and twenty elders fell down and worshipped him that liveth for ever and ever.	

The Temple in Revelation reflects the Temple of Planet Earth and is symbolized throughout by the number 12. In every account of the Holy City in the Bible, the importance of measuring its dimensions is emphasized.

The number 6 is unique. 1+2+3=6 and 1x2x3=6. Ancient astronomers adopted the mile of 5,280 English feet as a unit of measurement to measure the Earth and the Cosmos. (Pg 36).

Diameter of the Sun: =864,000 miles (12 x 12 x 6000)
Diameter of the Moon: =2160 miles (6 x 6 x 60)
Diameter of the Earth: =7920 miles (12 x 660)
Mean circumference of Earth: =24,883.2 miles (12 x 12 x 12 x 12 x 1.2)
Speed of Earth around the Sun: =66,600 miles per hour

Revelation Overview

The Book of Revelation contains an account of visions in symbolic and allegorical language borrowed extensively from the Old Testament, especially Ezekiel, Zechariah, Isaiah, and Daniel. Ultimate salvation and victory are said to take place at the end of the present age when Christ will come (second Coming) in glory at the Parousia and defeat Satan and his cohorts.

St. John had a dramatic mystical-death experience. He describes "he was in Spirit on the Lord's Day" (Out of the body) and then "falls down on his face as though dead" when he sees "One like unto the Son of Man" a brilliant Being of Light with a "countenance like the Sun," hair white as snow" and "feet of burnished brass," and a double-edged sword coming out of his mouth. It is the sword that should get our attention, a symbol fused throughout St. John's vision, a marvelous secret known only to those who have "ears to hear."

Mark of the Beast Note: for though the significance of the number 666 and 2368 would not have been generally understood, St. John's comparison of the idolized body of Jesus Christ with the beast, described Revelation 13 is not hard to recognize. P.213.

Revelation: Verse /Scripture/Interpretation

Rev 6		
[1]	And I saw when the Lamb opened one of the seals, and I heard, as it were the noise of thunder, one of the four beasts saying, Come and see.	Open FIRST SEAL: Noise of THUNDER One of FOUR BEASTS says look
[2]	And I saw, and behold a white horse: and he that sat on him had a bow; and a crown was given unto him: and he went forth conquering, and to conquer.	 **Four Horsemen of the Apocalypse - 1887 painting by Victor Vasnetsov. The Lamb is visible at the top** **WHITE HORSE:** (which beast?) HE THAT SAT ON HIM HA.D. A BOW AND CROWN TO CONQUER WITH A BOW = AN ARCHER TO SHOOT ARROWS FROM HEAVEN...**Look up Sagittarius – the archer...**
[3]	And when he had opened the second seal, I heard the second beast say, Come and see.	Open SECOND SEAL: SECOND BEAST: come and see (which beast?)
[4]	And there went out another horse that was red: and	**RED HORSE:** (which beast?)

	power was given to him that sat thereon to take peace from the earth, and that they should kill one another: and there was given unto him a great sword.	Given to him who sat on the horse, power to take peace from the earth Causing men to kill one another because they are afraid With a great sword – symbol of the comet / cleaving the sky / parting the heavens The rider of the second horse is often taken to represent War. He is often pictured holding a sword upwards as though ready for battle or mass slaughter. His horse's color is red (πυρρός, from πῦρ, fire); and in some translations, the colour is specifically a "fiery" red. The color red, as well as the rider's possession of a great sword, suggests blood that is to be spilled. Reference also to the First Plague of Blood – Exodus and all other comet. Apocalypse battles begin with blood. See Personification of the Myth, chapter xxxx. In revelation, john has the white horse shooting arrows with a bow first.
[5]	And when he had opened the third seal, I heard the third beast say, Come and see. And I beheld, and lo a black horse; and he that sat on him had a pair of balances in his hand.	Opened THIRD SEAL: **BLACK HORSE**: (which beast?) Given to him that sat on him: a pair of balances= day of judgment
[6]	And I heard a voice in the midst of the four beasts say, A measure of wheat for a penny, and three measures of barley for a penny; and see thou hurt not the oil and the wine.	ONE OF FOUR BEASTS SAYS; ONE PART WHEAT, THREE PARTS BARLEY = FOUR PARTS Also for clarity, from the **World English Bible:** And I heard something like a voice in the center of the four living creatures saying, I heard a voice in the midst of the four living creatures saying, "A choenix of wheat for a denarius, and three choenix of barley for a denarius! Don't damage the oil and the wine!" **Notes from Matthew Henry's Concise Commentary:** The choenix appears to have been the food allotted to one man for a day; while the denarius was the pay of a soldier or of a common labourer for one day (Matthew 20:2, "He agreed with the labourers for a penny a day," and Tacitus, 'Ann.,' 1:17, 26, "Ut denarius diurnum stipendium foret."

Cf. Tobit 5:14, where drachma is equivalent to denarius). **The choenix was the eighth part of the modius, and a denarius would usually purchase a modius of wheat**. The price given, therefore, denotes great scarcity, though not an entire absence of food, since a man's wages would barely suffice to obtain him food.

Barley, which was the coarser food, was obtainable at one third of the price, which would allow a man to feed a family, though with difficulty. A season of great scarcity is therefore predicted, though in his wrath God remembers mercy (cf. The judgments threatened in Leviticus 26:23-26, viz. The sword, pestilence, and famine; also the expression, "They shall deliver you your bread again by weight"). And see thou hurt net the oil and the wine.

The corollary to the preceding sentence, with the same signification. It expresses a limit set to the power of the rider on the black horse. These were typical articles of food (cf. Psalm 104:14, 15, "That he may bring forth food out of the earth; and wine that maketh glad the heart of man, and oil to make his face to shine, and bread which strengtheneth man's heart;" and Joel 1:10, "The corn is wasted: the new wine is dried up, the oil languisheth"). Wordsworth interprets, **"The prohibition to the rider, 'Hurt not thou the oil and the wine,' is a restraint on the evil design of the rider, who would injure the spiritual oil and wine, that is, the means of grace, which had been typified under those symbols in ancient prophecy (Psalm 23:4, 5), and also by the words and acts of Christ, the good Samaritan, pouring in oil and wine into the wounds of the traveller, representing human nature, lying in the road."**

'Αδικήσης ἀδικεῖν in the Revelation invariably signifies "to injure," and, except in one case, takes the direct accusative after it (see Revelation 2:11; Revelation 7:2, 3; Revelation 9:4, 10, 19; Revelation

		11:5). Nevertheless, Heinrich and Elliott render, "Do not commit injustice in the matter of the oil and wine." *Rinek renders, "waste not." The vision is a general prophecy of the future for all time (see on ver. 5); but many writers have striven to identify the fulfilment of the vision with some one particular famine.* Grotius and Wetstein refer it to the scarcity in the days of Claudius; Renan, to that in the time of Nero; Bishop Newton, to the end of the second century. Those who interpret the vision as a forewarning of the spread of heresy, especially single out that of Arius. [251]
[7]	And when he had opened the fourth seal, I heard the voice of the fourth beast say, Come and see.	Fourth seal opened Fourth beast: come and see: (which one?)
[8]	And I looked, and behold a pale horse: and his name that sat on him was Death, and Hell followed with him. And power was given unto them over the fourth part of the earth, to kill with sword, and with hunger, and with death, and with the beasts of the earth.	**A pale horse** He who sat on him was death, and hell followed him Power was given to them over the fourth part of the earth to kill with the sword With hunger With death And with the beasts of the earth
[9]	And when he had opened the fifth seal, I saw under the altar the souls of them that were slain for the word of God, and for the testimony which they	Fifth seal opened: The altar of the souls were slain for the word of god Reference to the sons of light (souls of slain) and the last great war…
[10]	And they cried with a loud voice, saying, How long, O Lord, holy and true, dost thou not judge and avenge our blood on them that dwell on the earth?	How long until the judgment? To avenge our blood on them who dwell on the earth?
[11]	And white robes were given unto every one of them; and it was said unto them, that they should rest yet for a little season, until their fellow servants also and their brethren, that should be killed as they were, should be fulfilled.	All given white robes, but must wait a "season" until their fellow servants should be killed- should be fulfilled

251 http://biblehub.com/revelation/6-6.htm

[12]	And I beheld when he had opened the sixth seal, and, lo, there was a great earthquake; and the sun became black as sackcloth of hair, and the moon became as blood;	Sixth seal opened Great earthquake Sun became black as sackcloth of hair Moon became blood red	
[13]	And the stars of heaven fell unto the earth, even as a fig tree casteth her untimely figs, when she is shaken of a mighty wind.	The stars of heaven fell like a fig that casts her untimely figs when shaken by a might wind	
[14]	And the heaven departed as a scroll when it is rolled together; and every mountain and island were moved out of their places	Heaven departed as a scroll when it is rolled together see *when the sky fell.*	
[15]	And the kings of the earth, and the great men, and the rich men, and the chief captains, and the mighty men, and every bondman, and every free man, hid themselves in the dens and in the rocks of the mountains;	The kings of the earth The great men Rich men Chief captains Hid themselves in dens and rocks of the mountains	
[16]	And said to the mountains and rocks, Fall on us, and hide us from the face of him that sitteth on the throne, and from the wrath of the Lamb:	Mountains and rocks fall on us And hide us from the face that sittith on the throne And from the wrath of the lamb The lamb on the throne oversees the four beasts (zodiac) hid themselves when the christ comet hits....	
[17]	For the great day of his wrath is come; and who shall be able to stand?	The great day of his wrath is come Who shall be able to stand?	
Rev .7			
[1]	And after these things I saw four angels standing on the four corners of the earth, holding the four winds of the earth, that the wind should not blow on the earth, nor on the sea, nor on any tree.	Angels standing on the four corners of the earth- see 6th century beth alpha zodiac and the four angels representing the 4 corners of the world; the 12 signs of the zodiac, the chariot of the four horsemen riding the chariot of the sun through the starry heavens, among the sun and the moon, the seven stars.	

<table>
<tr>
<td></td>
<td></td>
<td></td>
<td>
Four angels holding the four winds</td>
</tr>
<tr>
<td>[2]</td>
<td></td>
<td>And I saw another angel ascending from the east, having the seal of the living God: and he cried with a loud voice to the four angels, to whom it was given to hurt the earth and the sea,</td>
<td>Another angel from the east,
Having the seal of the living god (jesus christ)
Angel cried with a loud voice to the four angels
To whom it was given not to hurt the earth or the sea</td>
</tr>
<tr>
<td>[3]</td>
<td></td>
<td>Saying, Hurt not the earth, neither the sea, nor the trees, till we have sealed the servants of our God in their foreheads.:</td>
<td>Hurt note the earth
Neither the sea
Nor the trees
Till we have sealed all the servants of god in their foreheads
Revelation 12:14. Christ set a mark upon them, as was upon the houses of the israelites, when the destroying angel passed through egypt, and destroyed the firstborn in it; and as was upon the foreheads of those that sighed and cried in jerusalem, when orders were given to slay young and old, exodus 12:23. Christ will have a people in the worst of times; he knows who are his, and he will take care of them; he has his chambers of protection to hide them in, till the indignation is over past: the sealers, "we", are either father, son, and spirit, who are all jointly concerned for the welfare of the eject; or christ and his ministering angels that attend him, whom he employs for the good and safety of the heirs of salvation: the seal with which these are sealed is the seal</td>
</tr>
</table>

of the living god, the foreknowledge, love, care, and power of god; and the name of god, even christ's father's name, and their father's name, in their foreheads; the new name of children of god, by and under which they are known and preserved by him: and this is said to be "in their foreheads", in allusion to servants, who used to be marked in their foreheads; hence they are called by apuleius (c) "frontes literati"; and by martial, a servant is called "fronte notatus" (d): but then these were such who had committed faults, and this was done by way of punishment (e); wherefore it can hardly be thought that the servants of god should be sealed, in allusion to them: but rather with reference to the mitre on the high priest's forehead, as some think; or it may be to ezekiel 9:4, and shows, that though these persons were hid and concealed from men, they were well known to god and christ; nor were they ashamed to make a public and open confession of christ before men, as did the true and faithful witnesses of christ, the waldenses and albigenses, in the midst of the greatest darkness of popery, and of danger from men; and who seem to be chiefly intended.252

[4] And I heard the number of them which were sealed: and there were sealed an hundred and forty and four thousand of all the tribes of the children of Israel.

144,000 = Whole Earth Measurement. *"The macrocosmic City of 12,000 furlongs (stadian and furlong are the same length) and square and the microcosmic citadel wall of 144 cubits differ in scale but belong to one geometric figure. When they are brought to commensurable proportions, it is found that a square of 12 furlongs (1 furlong equals 660 English feet) exactly contains a circle of 24,890 feet or 14,400 cubits round.*

The nucleus of St John's New Jerusalem can thus be identified as a cube contained within a sphere which is in fact a model of the earth on the scale 1 foot:1 mile, for the diameter of the sphere is 7920 feet, and the earth's

252 Excerpted from Gill's Exposition of the Entire Bible, (c) Metamorph. l. 9. p. 130. (d) Epigr. l. 3. Ep. 20. (e) Vid. Popma de Operis Servorum, p. 170, & c.; http://biblehub.com/revelation/7-3.htm.

		mean diameter is 7920 miles. City of Revelation, pg 32.
[5]	Of the tribe of Juda were sealed twelve thousand. Of the tribe of Reuben were sealed twelve thousand. Of the tribe of Gad were sealed twelve thousand.	Each tribe sealed 12,000 each Reuben & gad
[6]	Of the tribe of Aser were sealed twelve thousand. Of the tribe of Nepthalim were sealed twelve thousand. Of the tribe of Manasses were sealed twelve thousand.	Each tribe sealed 12,000 each Aser, nepthalim, & manassas
[7]	Of the tribe of Simeon were sealed twelve thousand. Of the tribe of Levi were sealed twelve thousand. Of the tribe of Issachar were sealed twelve thousand.	Each tribe sealed 12,000 each Simeon, levi, Issachar
[8]	Of the tribe of Zabulon were sealed twelve thousand. Of the tribe of Joseph were sealed twelve thousand. Of the tribe of Benjamin were sealed twelve thousand.	Each tribe sealed 12,000 each Zubulon, joseph, and benjamen
[9]	After this I beheld, and, lo, a great multitude, which no man could number, of all nations, and kindreds, and people, and tongues, stood before the throne, and before the Lamb, clothed with white robes, and palms in their hands;	A; great multitude No man could number All kindreds, people, tongues, Stood before the throne and the lamb Clothed with white robes, with palms in their hands (palm sunday ride into Jerusalem.
[10]	And cried with a loud voice, saying, Salvation to our God which sitteth upon the throne, and unto the Lamb.	Salvation comes from our God, who sits on the throne, who is the Lamb
[11]	And all the angels stood round about the throne, and about the elders and the four beasts, and fell before the throne on their faces, and worshipped God,	All the Angels, the (24) Elders, the four beasts (Lion, Eagle/ Scorpion, Bull/calf/ox, Man) fall on their faces and worshipped God
[12]	Saying, Amen: Blessing, and glory, and wisdom, and thanksgiving, and honour, and power, and might, be unto our God for ever and ever. Amen.	Saying Amen (It is finished) Blessing, and glory, and wisdom, and thanksgiving, and honour, and power, and might, be unto our God for ever and ever. Amen (twice Amen. Alpha and Omega.)
[13]	And one of the elders answered, saying unto me, What are these which are	Who are these in white robes?

		arrayed in white robes? And whence came they?	
	[14]	And I said unto him, Sir, thou knowest. And he said to me, These are they which came out of great tribulation, and have washed their robes, and made them white in the blood of the Lamb..	Survivors of the Great Tribulation, their robes made white through the blood of the Lamb: Those in white robes have graduated from the 12 lessons/cycle of birth and death BY MERCY/redemption by the blood of the Lamb-not just from the Age of Pisces, but all time and ALL AGES.
	[15]	Therefore, are they before the throne of God, and serve him day and night in his temple: and he that sitteth on the throne shall dwell among them.	Those in white robes serve him day and night
	[16]	They shall hunger no more, neither thirst anymore; neither shall the sun light on them, nor any heat.	They will hunger and thirst no more. Description of the afterlife and the state of the holy city – no sunlight or heat shall affect them.
	[17]	For the Lamb which is in the midst of the throne shall feed them, and shall lead them unto living fountains of waters: and God shall wipe away all tears from their eyes.	Lamb in the midst of the throne, leads them to living waters, and wipes all tears from their eyes.—the afterlife state...no more tears (or suffering)
Rev .8			
	[1]	And when he had opened the seventh seal, there was silence in heaven about the space of half an hour.	7th seal opened. 30 minutes of silence: $30°$ or one whole sign = 30 years $1°$ or $60'$ = 1 year $30'$ = 6 months $15'$ = 3 months $5'$ = 1 month $1'$ = 6 days If john meant $30°$ of the transition moving from pisces to aquarius, 30 years of a zodiac age of 2160 years.
	[2]	And I saw the seven angels which stood before God; and to them were given seven trumpets.	7 angels were give 7 trumpets (to announce the end of the age.
	[3]	And another angel came and stood at the altar, having a golden censer; and there was given unto him much incense, that he should offer it with the prayers of all saints upon the golden altar which was before the throne.	Golden censer to offer it the prayers of the saints on the golden altar Before the throne

[4]	And the smoke of the incense, which came with the prayers of the saints, ascended up before God out of the angel's hand.	The smoke of the incense came with the prayers of the saints Ascended up before god out of the angel's hand
[5]	And the angel took the censer, and filled it with fire of the altar, and cast it into the earth: and there were voices, and thunderings, and lightnings, and an earthquake.	The angel took the censor Filled it with fire from off the altar And cast it onto the earth There were voices and thunderings Lightnings & an earthquake
[6]	And the seven angels which had the seven trumpets prepared themselves to sound.	
[7]	The first angel sounded, and there followed hail and fire mingled with blood, and they were cast upon the earth: and the third part of trees was burnt up, and all green grass was burnt up.	The *Kalevala* of th Finns, tell of a time when *hailstones of iron* fell from the sky, followed by a period of darkness.253 Here again, we find iron meteors [stones of barad], not ice. Misrashic and Talmudic sources say the stones which fell in Egypt were hot. But this fact by itself does not prove the stones of barad were meteorites. While meteorites can be hot enough to spark a fire, very often when they are recovered immediately after impact, they have been found to be frost covered and icy-cold to the touch. This is because the core of the meteorite, cooled to near absolute zero in space, still retains a temperature far below freezing despite its fiery passage through the atmosphere. Let's research the matter further. The Exodus plague included "thunder and hail" [barad] and the fire ran along upon the ground." 254 The fall of large meteorites or bolides are usually accompanied by crashes or explosion-like noises. In a similar manner, the fall of the stones of barad were accompanied by "loud noises," a description in Hebrew interpreted as "thunderings." It is a translation which is only figurative, and not literally correct because the word for "thunder" is *"raam,"* which is not used here. According to the Exodus narrative, the stones of barad made such a roar that the people in the

253 Kaleva, (transl. J. M. Crawford, 1888), p. xiii., cit. op., Velikovsky, p. 61
254 Exodus 9:23

		palace were terrified as much by the din of the falling stones as by the destruction they caused.255
[8]	And the second angel sounded, and as it were a great mountain burning with fire was cast into the sea: and the third part of the sea became blood;	2nd Angel: Celestial impact phenomena: burning asteroid, bolide impacts the ocean.
[9]	And the third part of the creatures which were in the sea, and had life, died; and the third part of the ships were destroyed.	Celestial impact phenomena: 1/3 of ocean affected, killing sea creatures, and destroying ships
[10]	And the third angel sounded, and there fell a great star from heaven, burning as it were a lamp, and it fell upon the third part of the rivers, and upon the fountains of waters	3rd angel: Celestial impact phenomena: Great Star is a comet, burning like a lamp, part falls on earth's rivers and earth seas.
[11]	And the name of the star is called Wormwood: and the third part of the waters became wormwood; and many men died of the waters, because they were made bitter.	Wormwood means bitter. Comet impact causes intense acid rain making water undrinkable.
[12]	And the fourth angel sounded, and the third part of the sun was smitten, and the third part of the moon, and the third part of the stars; so as the third part of them was darkened, and the day shone not for a third part of it, and the night likewise.	4TH ANGEL: IMPACT EFFECTS: Sun and Moon Darken. Third part of stars darken. Day and night do not shine for 1/3 of the time.
[13]	And I beheld, and heard an angel flying through the midst of heaven, saying with a loud voice, Woe, woe, woe, to the inhabiters of the earth by reason of the other voices of the trumpet of the three angels, which are yet to sound!	Impact effect woes are not over.
Rev .9		
[1]	And the fifth angel sounded, and I saw a star fall from heaven unto the earth: and to him was given the key of the bottomless pit.	5th angel: a star / bright comet wil fall from heaven to earth. Will be monitored by astronomers for five months. Astronomers will determine there will be a great likely hood of comet debris impact in five months. It will be world-wide news. Men will not be able to hide from it.

---

255 Exodus 9: 28

		It will open the key to the bottomless pit; fear
[2]	And he opened the bottomless pit; and there arose a smoke out of the pit, as the smoke of a great furnace; and the sun and the air were darkened by reason of the smoke of the pit.	Smoke, ash, spew from impact site, darkening the skies—creating great fear
[3]	And there came out of the smoke locusts upon the earth: and unto them was given power, as the scorpions of the earth have power.	Out of the smoke came locusts, who had the power of scorpions, to create fear. Greek word john uses; it acts like a drug. (get greek word from gary
[4]	And it was commanded them that they should not hurt the grass of the earth, neither any green thing, neither any tree; but only those men which have not the seal of God in their foreheads.	The scorpions and locusts create fear—fear is our spiritual enemy. But they do not harm those who have the seal of god—the seal of love in their foreheads and hearts. .
[5]	And to them it was given that they should not kill them, but that they should be tormented five months: and their torment was as the torment of a scorpion, when he striketh a man.	Fear will not kill them, but they will torment them for five months, It is like the torment of a scorpion, when it stings a man
[6]	And in those days shall men seek death, and shall not find it; and shall desire to die, and death shall flee from them.	In those days when the comet approaches, men shall seek death, and will not find it, and death shall flee from them.
[7]	And the shapes of the locusts were like unto horses prepared unto battle; and on their heads were as it were crowns like gold, and their faces were as the faces of men.	And the shapes of the locust were like horses prepared for battle – reference to four horsemen On their hads were crowns of gold —conquerers Their faces were the faces of men—reference to the fourth beast—man / aquarius and as to the timing of this apocalyptic event
[8]	And they had hair as the hair of women, and their teeth were as the teeth of lions.	The horses had hair of women – incoming bolides – like the one that exploded over russia in 2013, or any atmosphere penetrating bolode produce bright, long trails of smoke and fire, Teeth of lions having the teeth of lions.,,who roar (make great noise) when they explode.
[9]	And they had breastplates, as it were breastplates of iron; and the sound of their wings	Bolides are made of iron Sound of their wings like many chariots runnning into battle—they

	was as the sound of chariots of many horses running to battle.	are very loud—they will shake the ground. As in russia, the sonic boom of the explosion will be like the pressure wave of a nuclear explosion, accept this will be world-wide and not just one explosion, but many, many thousands of them, with many impacting the earth.
[10]	And they had tails like unto scorpions, and there were stings in their tails: and their power was to hurt men five months.	The power to sting was in their tails to hurt people five months
[11]	And they had a king over them, which is the angel of the bottomless pit, whose name in the Hebrew tongue is Abaddon, but in the Greek tongue hath his name Apollyon.	King over them was the angel of the bottomless pit—fear. All describe the appearance of multiple meteors, "hair of women" – meteor tails, iron bolides, loud explosions in the atmosphere. It doesn't stop for five months. Hebrew —abaddon / greek—apolloyon Means destruction-in much the same manner that the falling star "wormwood" calls attention to the bitter effects of its impact, the result of the fifth trumpet is "destruction." a still further possibility is that the name "apollyon" was intended also to suggest the greek god apollo, who in john's time was widely associated with prophecy.
[12]	One woe is past; and, behold, there come two woes more hereafter.	One woe is past. Tw0 more woes coming
[13]	And the sixth angel sounded, and I heard a voice from the four horns of the golden altar which is before God,	6th angel sounded. Voice from the four horns of the golden altar – four horns of the four corners of the golden altar before god – 4 horns (designate time-2160 years each.
[14]	Saying to the sixth angel which had the trumpet, Loose the four angels which are bound in the great river Euphrates.	
[15]	And the four angels were loosed, which were prepared for an hour, and a day, and a month, and a year, for to slay the third part of men.	The four angels were loosed: they had been prepared for an hour, and a day, and a month, and a yaer to slay the third part of men. John says it was an appointed time. A time discussed by jewish, christian, and islamic prophecies. We have reached the time of the apocalypse.

[16]	And the number of the army of the horsemen were two hundred thousand thousand: and I heard the number of them.	The number of the horsemen — the stones of the apocalypse were 200,000,000. See carolina bay impacts. In 1933, geologists Melton and Schriver of the University of Oklahoma aroused world-wide attention when they announced the bays were depressions left by a "meteoric shower of a colliding comet."256 Current estimates based on actual counts in limited regions conservatively indicate that half a million bays257may have been gouged out by catastrophic meteoric explosions. These detonations, in the air and close to the ground, created shock waves which formed shallow, sand-rimmed depressions. Each exploding fragment produced a trough of unknown depth, circular or oval according to the angle at which the fragment struck the ground. A remarkable feature about these bays is that all of the longitudinal axes are parallel extending from northwest to southeast. The fragments must therefore have been traveling parallel to each other. Around the bays are thrust walls, mounds of earth thrown up from impact or air explosion, and elevated on the southeastern end, indicating the direction of cosmic impact coming from the northwest. Measurements made on more prominent craters show that the large bays average 2200 feet in length, and in several cases exceed six miles in length.
[17]	And thus I saw the horses in the vision, and them that sat on them, having breastplates of fire, and of jacinth, and brimstone: and the heads of the horses were as the heads of lions; and out of their mouths issued fire and smoke and brimstone.	The horses, having breast plates of fire – of jacinth and brimstone — the color of orange and red —the same color a streaking meteor would shine during its 70,000 mph descent through the atmosphere. These would have the heads of lions (roaring loudly as they explode) And out of their mouths issue smoke, fire and brimstone
[18]	By these three was the third part of men killed, by the	Effects. 1/3 of humankind will die.

256 F. A. Melton and W. Schriver, "The Carolina Bays--Are They Meteorite Scars?", Journal of Geology, XLI (1933).

257 W.F. Prouty, The Bulletin of the Geological Society of America , Vol. 63, March, 1952, page 167.

	fire, and by the smoke, and by the brimstone, which issued out of their mouths.		By fire and by smoke, and by the brimstone —the stones of barad, which issued our of there mouths. 2.1 billion people will die.
[19]	For their power is in their mouth, and in their tails: for their tails were like unto serpents, and had heads, and with them they do hurt.		Power of the impacting meteors is in their mouth and in their tails Their meteor heads looked like serpents piercing the atmosphere at 70,000 mph And when they hit the ground, with their meteor heads they do hurt
[20]	And the rest of the men which were not killed by these plagues yet repented not of the works of their hands, that they should not worship devils, and idols of gold, and silver, and brass, and stone, and of wood: which neither can see, nor hear, nor walk:		The rest of men who were not killed by the impact plagues And repented not of the works of their hands That they not worship devils or idols of gold, silver and brass, stone, and wood; Which can nether see, hear or walk All meaning – just as the people of israel worshipped the golden calf when moses came down from the sacred mountain, john is saying here that even during the apocalypse, some still hang on and depend on money, gold and material things - worshipping them instead of turning their thoughts to god.
[21]	Neither repented they of their murders, nor of their sorceries, nor of their fornication, nor of their thefts.		They repent not of their murders or thefts. They hang on to their material life and will kill each other to survive the apocalypse
Rev .10			
[1]	And I saw another mighty angel come down from heaven, clothed with a cloud: and a rainbow was upon his head, and his face was as it were the sun, and his feet as pillars of fire:		Another mighty angel come down from heaven. He is the comet clothed with a cloud and rainbow on his head (michael) – face as the sun, feet as pillars of fire.
[2]	And he had in his hand a little book open: and he set his right foot upon the sea, and his left foot on the earth,		Had a little book open – Right foot on the sea, left foot on the earth
[3]	And cried with a loud voice, as when a lion roareth: and when he had cried, seven thunders uttered their voices.		Lion roars and severn thunders utters their voices: **Mythological concordance**; warning voice in the deluge myths of the world, we find stories of a very great number of warnings given by human, heroic, and divine persons. But even these men, heroes, and gods

		must have derived their knowledge from the observation of the happenings around them. They only put the warning into words, they only interpreted the aspects of the earth and sky. Later, those who escaped the terrible disaster of the great flood gratefully remembered the words of warning and raised the man who had uttered them to divine rank. The best known of all deluge warnings is that recorded in genesis vi. 12-18, when god spoke to noah concerning the impending flood: jewish tradition says that noah spread the news of the impending danger abroad, but was met with incredulity and scorn. Disregard of the warning words of a divinity or hero is the subject of a great number of myths of different peoples. The warning comes as 'a voice from heaven.' (trumpets in revelation) can we take this literally, or is this a cosmic sign, an astronomical warning of some impending doom? In the chaldean report xisuthros got word from chronos (time); utnapishtim, the babylonian, was warned by ea; deukalion was warned by prometheus (one eyed monster who gave fire to the world). In mexican mythology, the god titlacuhuan warned the just man nata and his wife nena. The wichita indians heard a voice from heaven.
[4]	And when the seven thunders had uttered their voices, I was about to write: and I heard a voice from heaven saying unto me, Seal up those things which the seven thunders uttered, and write them not.	Seal up the seven thunders and write them not
[5]	And the angel which I saw stand upon the sea and upon the earth lifted up his hand to heaven,	Angel who stood on the sea and earth
[6]	And sware by him that liveth for ever and ever, who created heaven, and the things that therein are, and the earth, and the things that therein are, and the sea, and the things which are therein, that there should be time no longer:	Angel swears that there shall be time no longer The appointed time of the apocalypse has arrived A new calender has begun

[7]	But in the days of the voice of the seventh angel, when he shall begin to sound, the mystery of God should be finished, as he hath declared to his servants the prophets.	The days of the voice of the seventh angel means the mystery of god is finished by the saying of the prophets, ezekiel, isaiah, daniel, and john.
[8]	And the voice which I heard from heaven spake unto me again, and said, Go and take the little book which is open in the hand of the angel which standeth upon the sea and upon the earth.	John, go and take the little book which is hope in the hand of the angel
[9]	And I went unto the angel, and said unto him, Give me the little book. And he said unto me, Take it, and eat it up; and it shall make thy belly bitter, but it shall be in thy mouth sweet as honey.	The little book is the book of revelation describing prophecy. It is good to know, sweet in the mouth like honey, but when it's fulfillment, when it happens, is bitter in the belly. It will sicken us when it happens.
[10]	And I took the little book out of the angel's hand, and ate it up; and it was in my mouth sweet as honey: and as soon as I had eaten it, my belly was bitter.	John experiences the sweet and the bitterness
[11]	And he said unto me, Thou must prophesy again before many peoples, and nations, and tongues, and kings.	John must prophecy again before many people, nations, tongues and kings. The woes are not over...
Rev .11		
[1]	And there was given me a reed like unto a rod: and the angel stood, saying, Rise, and measure the temple of God, and the altar, and them that worship therein.	Take a reed and measure the temple of god and the altar, and them that worship therein. This is the whole earth
[2]	But the court which is without the temple leave out, and measure it not; for it is given unto the Gentiles: and the holy city shall they tread under foot forty and two months.	Do not measure the temple courtyard of the gentles, they shall tread it underfoot for 42 months-4200 years.
[3]	And I will give power unto my two witnesses, and they shall prophesy a thousand two hundred and threescore days, clothed in sackcloth	2 witness are sodom and egypt. Sodom occurred 2200 bc. Egypt occurred 1500 bc – 4200 years ago – 42 months ago. Their prophecies have been in sackcloth since then. The silent.
[4]	These are the two olive trees and the two candlesticks	These events are the two olive trees – ancient trees that are like candlesticks standing before the god of

	standing before the God of the earth.	the earth.), they were built into solomon's temple as cedars of lebanon, sun and moon, fire and ice.
[5]	And if any man will hurt them fire proceedeth out of their mouth, and devoureth their enemies: and if any man will hurt them, he must in this manner be killed..	No man may destroy their meaning, because they are serpents waiting in the wilderness, waiting in sackcloth, for their appointed time to devour their enemies, for if any man destroy their meaning, they must in like manner be killed.
[6]	These have power to shut heaven, that it rain not in the days of their prophecy: and have power over waters to turn them to blood, and to smite the earth with all plagues, as often as they will.	They have the power to shut heaven that it rain not in the days when they fall from t he skies, to turn waters to blood, to smite the earth with all manner of plagues, as often as they will.
[7]	And when they shall have finished their testimony, the beast that ascendeth out of the bottomless pit shall make war against them, and shall overcome them, and kill them.	And when they have finished their testimony, the beast that ascends out of the bottomless pit will make war against them and kill them.
[8]	And their dead bodies shall lie in the street of the great city, which spiritually is called Sodom and Egypt, where also our Lord was crucified.	There dead bodies lie in the street of the great city – jerusalem / babylon / earth Sodom and gomorrah where fire and brimstone stones of barad fell Egypt where fire and brimstone fell – stones of barad fell Where our lord jesus christ was crucified on which tree of earth? A mythical crucifixion
[9]	And they of the people and kindreds and tongues and nations shall see their dead bodies three days and a half, and shall not suffer their dead bodies to be put in graves.	Their memories of their dead bodies, have never been buried, but have lain on the streets of history for all to see since that time.
[10]	And they that dwell upon the earth shall rejoice over them, and make merry, and shall send gifts one to another; because these two prophets that tormented them that dwelt on the earth.	And they that dwell on the earth have rejoiced that they have not returned, because these prophets (michael and gabriel) that dwelt on the earth (there bodies laying buried in sodom and egypt
[11]	And after three days and an half the Spirit of life from God entered into them, and they stood upon their feet; and great fear fell upon them which saw them.	After 4200 years, the spirit of life entered into them, and they stood on their feet (became comets again) returned after 4200 years and appeared over the earth Great frear fell again to those that saw them.

[12]	And they heard a great voice from heaven saying unto them, Come up hither. And they ascended up to heaven in a cloud; and their enemies beheld them.	A great voice from heaven said, come up hither from sodom and from egypt. They ascended in a cloud – comet cloud – Their enemies beheld them
[13]	And the same hour was there a great earthquake, and the tenth part of the city fell, and in the earthquake were slain of men seven thousand: and the remnant were affrighted, and gave glory to the God of heaven.	A great earthquake Tenth part of the city fell Slain 7,000 (similar to 70,000 islam killed during comet strike/earthquake Remnant were affrighted and gave glory to god in heave.
[14]	The second woe is past; and, behold, the third woe cometh quickly.	Second woe past. Third comes quickly
[15]	And the seventh angel sounded; and there were great voices in heaven, saying, The kingdoms of this world are become the kingdoms of our Lord, and of his Christ; and he shall reign for ever and ever.	7th angel sounded Great voices in heaven The kingdoms of our world have become the kingdoms of our lord As in my dream, "the kings shall become the people."
[16]	And the four and twenty elders, which sat before God on their seats, fell upon their faces, and worshipped God,	24 elders which sat before god on their seats worshipped god
[17]	Saying, We give thee thanks, O Lord God Almighty, which art, and wast, and art to come; because thou hast taken to thee thy great power, and hast reigned.	Thanks given to the lord god almighty because thy great power
[18]	And the nations were angry, and thy wrath is come, and the time of the dead, that they should be judged, and that thou shouldest give reward unto thy servants the prophets, and to the saints, and them that fear thy name, small and great; and shouldest destroy them which destroy the earth.	And the nations were angry that his wrath had come for judgment (at apocalypse) To give reward to the prophets and the saints, And to them that fear thy name small and great **Should destroy them which destroy the earth – mankind is destroying the earth**
[19]	And the temple of God was opened in heaven, and there was seen in his temple the ark of his testament: and there were lightnings, and voices, and thunderings, and an earthquake, and great hail.	*Temple of god opened in heaven.* *The ark of his covenent opened in the seventh day.* *Lightnings and thunderings* *Earthquake and great hail* *The atmosphere has been cleaved.* *The veil ripped asunder.*

			Time has ended. The apocalypse is here again after 12,960 years.
Rev .12			
[1]		And there appeared a great wonder in heaven; a woman clothed with the sun, and the moon under her feet, and upon her head a crown twelve stars:	A great wonder in heaven A woman clothed with the sun and the moon under her feet A crown of twelve stars
[2]		And she being with child cried, trailing in birth, and pained to be delivered.	The woman was with child, ready for birth In pain to be delivered Who is the child? Virgo, the virgin woman Mary the mother of god (see gospel in the stars)
[3]		And there appeared another wonder in heaven; and behold a great red dragon, having seven heads and ten horns, and seven crowns upon his heads.	A great red dragon (comet) ' having 7 heads – 7 periods / 7 day since creation- since the last great apocalypse Ten horns (divisions) Horns represent comet time – horns are comet time divisions. Each horn equals 1290 years each (12,960 / 10).
[4]		And his tail drew the third part of the stars of heaven, and did cast them to the earth: and the dragon stood before the woman which was ready to be delivered, for to devour her child as soon as it was born.	The dragon of the apocalypse returned Drew 1/3 part of the stars from heaven Cast them on the earth Dragon stood before the woman which was ready to bare a child To devour her child before it was born
[5]		And she brought forth a man child, who was to rule all nations with a rod of iron: and her child was caught up unto God, and to his throne.	The woman brought for a man child, aquarius, was to rule with a rod of iron – sword, scepter, the nations,

[6]	And the woman fled into the wilderness, where she hath a place prepared of God, that they should feed her there a thousand two hundred and threescore days.	Woman fled into the wilderness Prepared of god For 1,260 days Same as the 42 months
[7]	And there appeared a great wonder in heaven; a woman clothed with the sun, and the moon under her feet, and upon her head a crown of twelve stars,	*Great wonder in heaven repeats rev. 12:1*
[8]	And she being with child cried, travailing in birth, and pained to be delivered.	Woman in heaven cries in childbirth from her pain
[9]	And the great dragon was cast out, that old serpent, called the Devil, and Satan, which deceiveth the whole world: he was cast out into the earth, and his angels were cast out with him.	The great dragon That old serpent The ancient comet, personified as the devil That deceives the whole world Cast out onto earth And his (serpent angels) with him
[10]	And I heard a loud voice saying in heaven, Now is come salvation, and strength, and the kingdom of our God, and the power of his Christ: for the accuser of our brethren is cast down, which accused them before our God day and night.	God says, now comes salvation and strength And the kingdom of god And the power of christ Because the accuser (the destroyer) has been cast down Who accused them both day and night
[11]	And they overcame him by the blood of the Lamb, and by the word of their testimony; and they loved not their lives unto the death.	And they that overcame him -the devil, by the blood of the lamb And by the word of their testimony That they loved not their lives until death Recalling passover during exodus, when the angel of death passed over any who brushed the blood of the lamb over their doors. Same analogy
[12]	Therefore, rejoice, ye heavens, and ye that dwell in them. Woe to the inhabiters of the earth and of the sea! For the devil is come down unto you, having great wrath, because he knoweth that he hath but a short time.	Those who have confidence in the blood Woe to those in the earth and sea who do not! The angel destroyers coming down to you with wrath, fire,,smoke, destruction This devil knows he has only a short time to make you afraid, and steal your love from god from you.

[13]	And when the dragon saw that he was cast unto the earth, he persecuted the woman which brought forth the man child.	And when the dragon saw he was cast to the earth He persecuted the woman who brought for the the manchild
[14]	And to the woman were given two wings of a great eagle, that she might fly into the wilderness, into her place, where she is nourished for a time, and times, and half a time, from the face of the serpent.	The woman was given 2 wings of a great eagle (scorpio) Fled into the wilderness into her place Where she was nourished for 3.5 times –that is 42 months.
[15]	And the serpent cast out of his mouth water as a flood after the woman, that he might cause her to be carried away of the flood.	Serpent cast a flood after the woman That she might be carried away in the flood Last apocalypse -the great flood
[16]	And the earth helped the woman, and the earth opened her mouth, and swallowed up the flood which the dragon cast out of his mouth.	And the earth helped the woman Swallowed up the flood Which the dragon spewed out of his mouth
[17]	And the dragon was wroth with the woman, and went to make war with the remnant of her seed, which keep the commandments of God, and have the testimony of Jesus Christ.	Dragon was angry at the woman And went to make war with the remnant of her seed Reference to the curse of the serpent in the garden of eden. ouroboros suggestion of a previous impact-god told adam and eve to replenish the earth
Rev .13		
[1]	And I stood upon the sand of the sea, and saw a beast rise up out of the sea, having seven heads and ten horns, and upon his horns ten crowns, and upon his heads the name of blasphemy.	Daniel vision: sawa beast rise up with Seven heads Ten horns On his heads the name of blasphemy
[2]	And the beast which I saw was like unto a leopard, and his feet were as the feet of a bear, and his mouth as the mouth of a lion: and the dragon gave him his power, and his seat, and great authority.	Daniel's beasts Beast like a leopard — **speed** Leopards feet of a bear - **power** Mouth of a lion- aurhority to **kill** Dragon gave leopard beast great power and authority
[3]	And I saw one of his heads as it were wounded to death; and his deadly wound was healed: and all the world wondered after the beast.	42 hundred years ago, one of the dragons heads was wounded to death But his deadly wound was healed And the world wondered after the leopard beast who moved so fast, who was so powerful, who spoke from the heavens with great authority, and

		whose breakup in the heavens, raining down fire, smoke and brimstone, should stop raining down upon the earth. All this happened and recalled from the babylonian captivity by Daniel
[4]	And they worshipped the dragon which gave power unto the beast: and they worshipped the beast, saying, Who is like unto the beast? Who is able to make war with him?	Many worshipped this great sight in heaven Who is like this beast Who can make war with him? In Sodom and in Egypt
[5]	And there was given unto him a mouth speaking great things and blasphemies; and power was given unto him to continue forty and two months.	Power was given to this beast for 42 months to speak great blasphemies This leopard beast, those who witnessed sodom and egypt, that same beast carried authority and power for the last 4200 years
[6]	And he opened his mouth in blasphemy against God, to blaspheme his name, and his tabernacle, and them that dwell in heaven.	And now the leopard beast opened his mouth in blasphemy against god To curse his name, his tabernacle – that is the earth – god's tabernacle And them who dwell in heaven
[7]	And it was given unto him to make war with the saints, and to overcome them: and power was given him over all kindreds, and tongues, and nations.	To make war- a war between the sons of light and the sons of darkness The light – the saints and attempt to overcome them And power was give to the leopard beast over all kindreds and all nations
[8]	And all that dwell upon the earth shall worship him, whose names are not written in the book of life of the Lamb slain from the foundation of the world.	And to tempt all who drell on the earth to worship him Whose names are not written on the book of life Of the lamb slain from the foundation of the world – The stone which the builders rejected The stone not made by human hands
[9]	If any man have an ear, let him hear.	He who has an ear
[10]	He that leadeth into captivity shall go into captivity: he that killeth with the sword must be killed with the sword. Here is the patience and the faith of the saints.	He that becomes captive He that kills with the sword Here is the patience and faith of the saints
[11]	And I beheld another beast coming up out of the earth; and he had two horns like a lamb, and he spake as a dragon.	Another beast came out of the earth He had two horns like a lamb, but spoke as a dragon

[12]	And he exerciseth all the power of the first beast before him, and causeth the earth and them which dwell therein to worship the first beast, whose deadly wound was healed.	Lamb beast(new recreation of the leopard beast) had all the power of the first beast-of course. Same beast, newname.	Whose deadly wound was healed
[13]	And he doeth great wonders, so that he maketh fire come down from heaven on the earth in the sight of men,	Leopard / lamb beast also makes fire come down from heaven in the sight of men. A massive firefall event	
[14]	And deceiveth them that dwell on the earth by the means of those miracles which he had power to do in the sight of the beast; saying to them that dwell on the earth, that they should make an image to the beast, which had the wound by a sword, and did live.	This new lamb beast deceives them that dwell on the earth	By means of those miracles which he had power to do in the sight of the beast Saying to those who dwelt on earth That that should make an image to the beast Which had a wound of a great sword Enter the story of the death of jesus christ, pierced by the centurion's' spear Who received a wound and did live through the resurrection
[15]	And he had power to give life unto the image of the beast, that the image of the beast should both speak, and cause that as many as would not worship the image of the beast should be killed.	Power given to the image of the beast to both speak	And cause that many who did not worship his image should be killed
[16]	And he causeth all, both small and great, rich and poor, free and bond, to receive a mark in their right hand, or in their foreheads:	The beast caused both small, great rich and poor,	Free and bond, to receive a mark in their right hand or their foreheads
[17]	And that no man might buy or sell, save he that had the mark, or the name of the beast, or the number of his name.	That no man buy or sell without this mark, or the mumber of his name	
[18]	Here is wisdom. Let him that hath understanding count the number of the beast: for it is the number of a man; and his number is Six hundred threescore and six.	Number of the beast. Lamb beast with 2 horns. It is the number of a man.	It is six hundred, sixty, and six. It is written in greek gematria with a line across the top of these three letters X chī, 22nd greek letter. Sound ch (chi)

Ξ (xi), 14th greek letter. Sound xi (xylophone)

Σ (sigma) 18th greek letter. Sound s (serpent)

There is no letter j in ancient greek, latin, or hebrew. But jesus' birth name, jesh'ua. When we add up the phonetic values of these 3 greek letters transliterated into english we get ch-z-s =jesus.

The mark-notes from city of revelation

The total value of the phrase, which gives the number of the beast, is there for 2368, the number of jesus christ. There is no question hear of a chance of coincidence of numbers, for the same chapter and john appetizers his meaning by the use of the phrase, the image of the beast, which is repeated three times and verse 15. Now the number of greek word, the image of the beast is 2260 and this numbers already familiar as belonging to the antichrist, whom st. Paul describes and second thessalonians 2.3, using two phases both of obviously traditional significance:

'Let no man to see if you by any means: for that they shall not come except there be a falling i'm away first, and that man of sin be revealed, the son of perdition.'

Amanda sent is greek, what's the value was 2260, same number as the image of the beast. Son of perdition is the number to 385. **The number of 2260** occurs again in revelation 14, the chapter following st. John's account of the beast. Inverse 14 the son of man appears on a white cloud, evidently standing in contrast to the beast from the sea. Yet by the number they are made identical, but the **son of man greek has number 2260**. P. 142-143

Thus:

2260 equals the image of the beast equals a man of sin equals the son of man and **2365 equals the son of perdition the power of christ** the baptism of john and luke 20.4

Whatever may now be thought of the scriptural interpretation by good mantra, there can be no doubt that the early christian scholars the practice it, would have recognized st. John's intention to find jesus christ and the son of man with the image and number of the beast, to draw attention of us fell on the shift to the state affairs as i live five followers who raise the image of his body and temples, observe observing the letter of the law to neglect the spirit. That's awesome statement hey jay good morning just reading into my morning. P. 142-143,

Brought the revelation, the symbolic fingers what you need to stand on the side of the in opposition to each other or reveal that is essentially want to say. I duality which one of you obtain universally is an allusion of this world and exist neither in the world of archetypes knowing their corresponding numbers. It is often been remarked at the greek words, what you reply to the contrasting figures in the apocalyptic fission are strikingly similar inform the bride is great, the whore is great, the beast is great and the lamb create a rare word which scarcely occurs in the new testament outside revelation. The scene of the vision continues greek, so that in one moment the prophecies the splendor corruption of babylon, followed immediately by the holy city, jerusalem, with its fresh springs and walls of sparkling crystal. The woman quote with the sun, the beast with seven heads and 10 horns, the lamb on mount psion and the whore with the scarlet beast are described intern. **Babylon is destroyed, jerusalem is his review. But these are not too different cities. According to the numbers they are identical, for gematria of babylon and the holy city of jerusalem is: babylon 1285 holy city of jerusalem 1285.**

It is been observed at 666 is related to 2368, and that they are both multiples of 37, and that the word greek beast, the cursed 37 times a revelation. This is not without significance or coincidence and cannot nicole literature, or the works of her as such as revelation planned so that

sounds about elements within the various episode should be repeated in the book as a whole, as well in the entire body of sacred texts to which of the longs. That's the first chapter in the book of the bible concerned adamant eve and the tree of knowledge, or the last chapter in the book of revelation ends up with the spirit and the bride in the tree of life.p. 145-146.

St. John's purpose in writing the number of the beast in a phrase with the value 2368, number of jesus christ, and only be understood in the context of the better to steve's throughout the history of the early church between the bishops of rome in the prophetic liters of easton communities. The chief source of the gnostics heresy in the eyes of the church play in the distinction they made between the body of jesus and the spirit of christ and their insistence that that the spirit partook of the divine nature. The further implication was that while the life of jesus was mythical, his spirit was eternal, in reality within the possible expensive all and not through the offices of the church.

The argument is: if jesus suffered on the cross, therefore his body must've been that of an ordinary man, and not the logos, messiah. Wrong. Jesus was born messiah, suffered as a man died and resurrected himself as the christ there is no separation.

The same point, whether it should be the image of a man that is worship or spiritual like to you, that is dispute it in russia today by the eastern orthodox church, was also the chief issue between the profits in the face of the early church. The gnostics we chatted a little literal image of the christian story what's the priest emphasize, and particular representation of the wanted body on the cross. There is well known gnostic graffito of an ass crucified, twice with the risingsun avenue h, but the aspect of his nature that dominated the teaching of the roman church was related above all to the midday sun, the symbol of imperial splendor. The balance between between the dionysian christ and the appolonian christ was upset in the favor of the latter, and this was reflected in the decline of the old prophetic ministry and in the drawing power of the bishops, who were originally concern with the matters of organization and finance. Particularly in the eastern churches there were many portals tennessee on the part of rome to glorify the image of the body spirit. *It was due to their influence that revelation was included in the scriptural canon, and even this was decided by the nearest possible majority, for though the significance of the number 666 and 2368 would not have been generally understood, st. John's comparison of the idolized body of jesus christ with the beast, described revelation 13 is not hard to recognize. P.146-147.*

In revelation john describes the first beast, marked with a deadly wound, rises up out of the sea, this is rises up out of the piscean see to rain in the age of aquarius is a god of a new age and the god of the new age is the second beast. Second base to tears as a lamb but his authority is that of an imperial dragon. And i quote from revelation. And he exercises all the power of the first phase before him, and cause of the earth and them which 12 there and to worship the first beast, who's deathly won't was healed... saying to them that dwell on the earth, that they should make an image to the beast, and which had been wounded by sword and did live. And he had the power to give life into the image of the beast, but the image of the beasts do both speak, i cause that as many as were not worship the image of the fees should be killed. P. 147,

St. Paul, his works were much quoted by the gnostics in their disputes with rome made the same point in the first chapter of a pistol to romans, running at those who change the glory of the uncorrectable god into an image made like to corruptible man and to the birds and to the four for the beasts

			and creeping things... who change the truth of god into a lie and worship and serve the creature more than the creator.p.147. For those who have not yet guess the identity of the beast, is one of the image was set up as an object of compulsory worship on the second base, st. John then give the number of his name in a phrase with value 2368, great. The son of man, 2260, becomes the image of the beast also 2260, and in revelation the great city which is spiritually called sodom and egypt, 1480, where our lord was also crucified. Authors note an amazing statement. Egypt and gematria 1480, where plagues of fire and referenced as christ crucified, referenced as catastrophic event of firefall.
Rev .14			
[1]	And I looked, and, lo, a Lamb stood on the mount Sion, and with him an hundred forty and four thousand, having his Father's name written in their foreheads.	The lamb stood on mount sion With him a 144,000 Having the father's name written on their foreheads	
[2]	And I heard a voice from heaven, as the voice of many waters, and as the voice of a great thunder: and I heard the voice of harpers harping with their harps:	Heard a voice from heaven A voice of many waters A voice of great thunder A voice of harpers with harps	
[3]	And they sung as it were a new song before the throne, and before the four beasts, and the elders: and no man could learn that song but the hundred and forty and four thousand, which were redeemed from the earth.	They sung a new song before the throne Before the four beasts The 24 elders And no man could learn the song expt the 144,000 Which were redeemed from the earth	
[4]	These are they which were not defiled with women; for they are virgins. These are they which follow the Lamb whithersoever he goeth. These were redeemed from among men, being the firstfruits unto God and to the Lamb.	And they were not defiled with woman For they are virgins They follow the lame wherever they go Redeemed among men Being the first fruits of god	
[5]	And in their mouth was found no guile: for they are	In there mouth no guile and no fault	

	without fault before the throne of God.	
[6]	And I saw another angel fly in the midst of heaven, having the everlasting gospel to preach unto them that dwell on the earth, and to every nation, and kindred, and tongue, and people,	Angel flying in the midst of heaven Having the everlast gospel to preach to those that dwell on the earth To every nation, tongue and people
[7]	Saying with a loud voice, Fear God, and give glory to him; for the hour of his judgment is come: and worship him that made heaven, and earth, and the sea, and the fountains of waters.	Saying with a loud voice Fear god – be in awe of god Give glory to him His hour, his appointed time has come And to worship him that made heaven, earth and the sea, and the fountain of waters
[8]	And there followed another angel, saying, Babylon is fallen, is fallen, that great city, because she made all nations drink of the wine of the wrath of her fornication.	Bababylon has fallen – babylon equates / to jerusalem on earth. They are the same.
[9]	And the third angel followed them, saying with a loud voice, If any man worship the beast and his image, and receive his mark in his forehead, or in his hand,	3rd angel, if any man worships the beast and his image, they will receive the mark in his forehead and his hand The beast image, 666 is not the christ, but nearly all the world worships it! The image of the christ is not the true christ
[10]	The same shall drink of the wine of the wrath of God, which is poured out without mixture into the cup of his indignation; and he shall be tormented with fire and brimstone in the presence of the holy angels, and in the presence of the Lamb:	God is indignant that we should worship an image. Those who worship this image shall drink of the wrath of god It is poured out as a mixture into his cup of indignation With fire and brimstone, in the presence of his holy angels, In the presence of the lamb
[11]	And the smoke of their torment ascendeth up for ever and ever: and they have no rest day nor night, who worship the beast and his image, and whosoever receiveth the mark of his name.	The smoke of their torment ascended up for ever and ever Who worship the beast and his image, Who receiveth the make of his name
[12]	Here is the patience of the saints: here are they that keep the commandments of God, and the faith of Jesus.	Here is the patience of the saints that keep the commandments of god, and the faith of jesus
[13]	And I heard a voice from heaven saying unto me, Write,	Voice from heaven heard

	Blessed are the dead which die in the Lord from henceforth: Yea, saith the Spirit, that they may rest from their labours; and their works do follow them.	Write, blessed are the dead which die in the lord from now on; Now they may rest from their labours, and their works do follow them
[14]	And I looked, and behold a white cloud, and upon the cloud one sat like unto the Son of man, having on his head a golden crown, and in his hand a sharp sickle.	A white cloud sat on one like unto the son of man Having on his hand a golden croan, and in his hand a sharp sickle (image of the grim reper —an image of leo, the scythe to reap mankind off the face of the earth
[15]	And another angel came out of the temple, crying with a loud voice to him that sat on the cloud, Thrust in thy sickle, and reap: for the time is come for thee to reap; for the harvest of the earth is ripe.	An angel who came out of the temple, crying with a loud voice To reap for the time has come to reap, for the harvest of the earth is ripe The appointed time to reap Sharp Sickle #1 – 1st Corner of the Earth Reaped
[16]	And he that sat on the cloud thrust in his sickle on the earth; and the earth was reaped.	And he that sat on the cloud thrust his sickle in and the earth was reaped Sharp sickle # 2- 2nd corner of the earth reaped
[17]	And another angel came out of the temple which is in heaven, he also having a sharp sickle.	And another angel came down from the temple in heaven, also having a sharp sickle Sharp Sickle #3 – 3rd Corner of the Earth Reaped
[18]	And another angel came out from the altar, which had power over fire; and cried with a loud cry to him that had the sharp sickle, saying, Thrust in thy sharp sickle, and gather the clusters of the vine of the earth; for her grapes are fully ripe.	Another angel came out of the altar, which had power over fire, and he cried with a loud cry, saying thrust in thy sharp sickly, to gather the cluster of vines in the earth, for her grapes are fully ripe. Sharp Sickle # 4 – Forth Corner of the Earth Reaped
[19]	And the angel thrust in his sickle into the earth, and gathered the vine of the earth, and cast it into the great winepress of the wrath of God.	Sickle is also related to LEO. Winepress is the destruction of humans, plants, animals, cities, world
[20]	And the winepress was trodden without the city, and blood came out of the winepress, even unto the horse bridles, by the space of a thousand and six hundred furlongs.	World-wide destruction. 1600 furlongs is a global measurement of the temple. The horse bridles refer to the 4 horsemen, sitting at the 4 corners of the earth. They have the four sickles of god represented by the four horses of the apocalypse, at the four corners of the earth, each holding the arrows of god, preparing death and destruction on the earth below, prepared for a

			year, a month a week and a day, and an hour to release the seven last plagues The winepress was trodden without the city and the blood came out of the winepress, even unto the horses bridles, by the space of 1600 furlongs – the whole world. 1600 furlongs equals the width of the planet.
Rev **.15**			
[1]		And I saw another sign in heaven, great and marvellous, seven angels having the seven last plagues; for in them is filled up the wrath of God.	And i saw another sign in heaven, great and marvellous, seven angels having the seven last plagues; for in them is filled up with the wrath of god Wrath – cause and effect not an angry god
[2]		And I saw as it were a sea of glass mingled with fire: and them that had gotten the victory over the beast, and over his image, and over his mark, and over the number of his name, stand on the sea of glass, having the harps of God.	John, saw a sea of glass mingled with fire. And them that had gotten the victory over the beast. Victory over his image Victory over his mark Over the number of his name Stand on the sea of glass, Having harps of god
[3]		And they sing the song of Moses the servant of God, and the song of the Lamb, saying, Great and marvellous are thy works, Lord God Almighty; just and true are thy ways, thou King of saints.	And they sing the son of moses the servant of god And the song of the lamb, saying Great and marvellous are thy works, lord god almighty; just and true are thy ways, Thou king of the saints
[4]		Who shall not fear thee, O Lord, and glorify thy name? For thou only art holy: for all nations shall come and worship before thee; for thy judgments are made manifest.	Who shall not fear thee, oh lord, and glorify thy name? For thou only art holy: for all nations shall come and and worship before thee; for thy judgments are made manifest
[5]		And after that I looked, and, behold, the temple of the tabernacle of the testimony in heaven was opened:	After that, i looked The tabernacle of the testimony in heaven was opened
[6]		And the seven angels came out of the temple, having the seven plagues, clothed in pure and white linen, and having their breasts girded with golden girdles.	The seven angels came out of the temple, having seven plagues Clothed in pure white linen Having their breast girded with golden girdles Comet golden girdles
[7]		And one of the four beasts gave unto the seven angels seven golden vials full of the	One of the four beasts gave to the seven angels

		wrath of God, who liveth for ever and ever.	With seven vials full of wrath of god, who lives for ever and ever
	[8]	And the temple was filled with smoke from the glory of God, and from his power; and no man was able to enter into the temple, till the seven plagues of the seven angels were fulfilled.	And the temple was filled with smoke from the glory of god, And from his power; And no man was able to enter the temple Till the seven plagues of the seven angels were fulfilled Preparing again for the impact plagues described in chapters 1-14 to be fulfilled
Rev .16			
	[1]	And I heard a great voice out of the temple saying to the seven angels, Go your ways, and pour out the vials of the wrath of God upon the earth.	A great voice came out of the temple saying to the angels, go your ways, pour out the vials of wrath on the four corners of the earth
	[2]	And the first went and poured out his vial upon the earth; and there fell a noisome and grievous sore upon the men which had the mark of the beast, and upon them which worshipped his image.	The first vial was poured upon the earth. There fell a noisome and grievous sore upon men which had the mark of the beast And upon them which worshipped the image
	[3]	And the second angel poured out his vial upon the sea; and it became as the blood of a dead man: and every living soul died in the sea.	The second angel Poured his vial upon the sea And it became the blood of a dead man And every living soul died in the sea
	[4]	And the third angel poured out his vial upon the rivers and fountains of waters; and they became blood.	The third angel Poured out his vial upon the vial upon the rivers and fountains of waters (sweet waters) And they became as living blood
	[5]	And I heard the angel of the waters say, Thou art righteous, O Lord, which art, and wast, and shalt be, because thou hast judged thus.	And the angel of the waters said Thou art righteous, oh lord, which art, and was, and shall be, because thou has judged thus
	[6]	For they have shed the blood of saints and prophets, and thou hast given them blood to drink; for they are worthy.	For they have shed the blood of saints and prophets, And had have given them blood to drink; for they are worthy
	[7]	And I heard another out of the altar say, Even so, Lord God Almighty, true and righteous are thy judgments.	And i heard another out of the altar say, even so, lord god almighty, true and righteous are the judgment.
	[8]	And the fourth angel poured out his vial upon the	And the fourth angel Poured out his vial like the sun

	sun; and power was given unto him to scorch men with fire.	And power was given to him to scorch men with fire
[9]	And men were scorched with great heat, and blasphemed the name of God, which hath power over these plagues: and they repented not to give him glory.	And men were scorched with great heat Comet blast effects raise atmospheric heat several hundred degrees And from the heat, they cursed the name of god, which has power over these plagues, And repented not to give him glory
[10]	And the fifth angel poured out his vial upon the seat of the beast; and his kingdom was full of darkness; and they gnawed their tongues for pain,	And the fifth angel Poured out his vial upon the seat of the beast And the kingdom was full of darkness And they gnawed on their tongues for pain, From the heat, from the darkness. They were being burned alive
[11]	And blasphemed the God of heaven because of their pains and their sores, and repented not of their deeds.	And they cursed god because of their pain and their sores Not only from heat, but from acid rain
[12]	And the sixth angel poured out his vial upon the great river Euphrates; and the water thereof was dried up, that the way of the kings of the east might be prepared.	The sixth angel Poured out his vial over the river euphrates And the water was dried up So the way of the kings of the east might be prepared
[13]	And I saw three unclean spirits like frogs come out of the mouth of the dragon, and out of the mouth of the beast, and out of the mouth of the false prophet.	And john saw three unclean spirits like frogs (reference exodus frogs)
[14]	For they are the spirits of devils, working miracles, which go forth unto the kings of the earth and of the whole world, to gather them to the battle of that great day of God Almighty.	They are the spirits of devils Working miracles Which go forth unto the kings of the earth and the whole world To gather them to the battle of the great day of god almighty The great war of armageddon The final battle of the sons of light and the sons of darkness
[15]	Behold, I come as a thief. Blessed is he that watcheth, and keepeth his garments, lest he walk naked, and they see his shame.	Behold i come as a their (reference Matthew) Blessed is he who watches, keepeth his garments, lest he walkes nake, and they see his shame

[16]	And he gathered them together into a place called in the Hebrew tongue Armageddon.	And he gathered them together in a place in the hebrew tongue called Armageddon
[17]	And the seventh angel poured out his vial into the air; and there came a great voice out of the temple of heaven, from the throne, saying, It is done.	And the seventh angel sounded And he poured out his vial into the air And there came out a great voice out of the temple of heaven, from the throne, Saying it is done.
[18]	And there were voices, and thunders, and lightnings; and there was a great earthquake, such as was not since men were upon the earth, so mighty an earthquake, and so great.	And there were voices, and thunders, and lightnings And a great earthquake – such as was not since men were upon the earth, so might an earthquake and so great
[19]	And the great city was divided into three parts, and the cities of the nations fell: and great Babylon came in remembrance before God, to give unto her the cup of the wine of the fierceness of his wrath.	And the great city (babylon -the whole earth Was divided into 3 parts, And the nations fell; And babylyon came into remembrance before god, To give unto her the cup the wine the fierceness of her wrath
[20]	And every island fled away, and the mountains were not found.	And every island fled away, and the mountains were not found
[21]	And there fell upon men a great hail out of heaven, every stone about the weight of a talent: and men blasphemed God because of the plague of the hail; for the plague thereof was exceeding great.	And there fell upon men a great hail out of heaven, every stone weight of a talent about 130 pounds...stones of barad, 200,000,000 of them Men cursed god because of the plague of hail, because it was so great
Rev .17		
[1]	And there came one of the seven angels which had the seven vials, and talked with me, saying unto me, Come hither; I will shew unto thee the judgment of the great whore that sitteth upon many waters:	
[2]	With whom the kings of the earth have committed fornication, and the inhabitants of the earth have been made drunk with the wine of her fornication.	
[3]	So he carried me away in the spirit into the wilderness:	

	and I saw a woman sit upon a scarlet coloured beast, full of names of blasphemy, having seven heads and ten horns.	
[4]	And the woman was arrayed in purple and scarlet colour, and decked with gold and precious stones and pearls, having a golden cup in her hand full of abominations and filthiness of her fornication:	
[5]	And upon her forehead was a name written, Mystery, Babylon the Great, the Mother of Harlots and Abominations of the earth.	Judgment of the Great Whore sitting on Many Waters* A woman sat on a scarlet colored beast, with 7 heads and 10 horns, full of the names of blasphemy.*On her forehead: mystery, babylon the great, the mother of harlots, and the abominations of the earth.
[6]	And I saw the woman drunken with the blood of the saints, and with the blood of the martyrs of Jesus: and when I saw her, I wondered with great admiration.	
[7]	And the angel said unto me, Wherefore didst thou marvel? I will tell thee the mystery of the woman, and of the beast that carrieth her, which hath the seven heads and ten horns.	
[8]	The beast that thou sawest was, and is not; and shall ascend out of the bottomless pit, and go into perdition: and they that dwell on the earth shall wonder, whose names were not written in the book of life from the foundation of the world, when they behold the beast that was, and is not, and yet is.	
[9]	And here is the mind which hath wisdom. The seven heads are seven mountains, on which the woman sitteth.	
[10]	And there are seven kings: five are fallen, and one is, and the other is not yet come; and when he cometh, he must continue a short space.	
[11]	And the beast that was, and is not, **_even he is the_**	Egyptian Day of Judgment: 42 Judges

	eighth, and is of the seven, and goeth into perdition.	42 judges oversee the ceremony of the weighing of the heart against a feather. Thoth records the judgment. Judgment day is the 8th day. Hermes conquered the dragon with three tails or heads. He was represented by the Ibis, which could be the sickle of the lion, the hind leg there. Hermes was the recorder of time. The seventh planet is Saturn, gate of chaos, the seventh sphere. 42 is also Michael with a flaming sword. The eighth sphere is allegorically the next "octave" of existence after death. The eighth card of the Tarot is the Lion/ the eighth beast of Revelation (Renewer of the Ages)
[12]	And the ten horns which thou sawest are ten kings, which have received no kingdom as yet; but receive power as kings one hour with the beast.	
[13]	These have one mind, and shall give their power and strength unto the beast.	
[14]	These shall make war with the Lamb, and the Lamb shall overcome them: for he is Lord of lords, and King of kings: and they that are with him are called, and chosen, and faithful.	
[15]	And he saith unto me, The waters which thou sawest, where the whore sitteth, are peoples, and multitudes, and nations, and tongues.	
[16]	And the ten horns which thou sawest upon the beast, these shall hate the whore, and shall make her desolate and naked, and shall eat her flesh, and burn her with fire.	
[17]	For God hath put in their hearts to fulfil his will, and to agree, and give their kingdom unto the beast, until the words of God shall be fulfilled.	
[18]	And the woman which thou sawest is that great city, which reigneth over the kings of the	

Rev .18			
[1]	And after these things I saw another angel come down from heaven, having great power; and the earth was lightened with his glory.	
[2]	And he cried mightily with a strong voice, saying, Babylon the great is fallen, is fallen, and is become the habitation of devils, and the hold of every foul spirit, and a cage of every unclean and hateful bird.	
[3]	For all nations have drunk of the wine of the wrath of her fornication, and the kings of the earth have committed fornication with her, and the merchants of the earth are waxed rich through the abundance of her delicacies.	* All nations have partook of the wrath of her fornification-that you receive not her plagues.
[4]	And I heard another voice from heaven, saying, Come out of her, my people, that ye be not partakers of her sins, and that ye receive not of her plagues.	
[5]	For her sins have reached unto heaven, and God hath remembered her iniquities.	
[6]	Reward her even as she rewarded you, and double unto her double according to her works: in the cup which she hath filled fill to her double.	
[7]	How much she hath glorified herself, and lived deliciously, so much torment and sorrow give her: for she saith in her heart, I sit a queen, and am no widow, and shall see no sorrow.	
[8]	Therefore shall her plagues come in one day, death, and mourning, and famine; and she shall be utterly burned with fire: for strong is the Lord God who judgeth her.	
[9]	And the kings of the earth, who have committed fornication and lived deliciously with her, shall bewail her, and lament for	

	her, when they shall see the smoke of her burning,	
[10]	Standing afar off for the fear of her torment, saying, Alas, alas, that great city Babylon, that mighty city! For in one hour is thy judgment come.	
[11]	And the merchants of the earth shall weep and mourn over her; for no man buyeth their merchandise any more:	
[12]	The merchandise of gold, and silver, and precious stones, and of pearls, and fine linen, and purple, and silk, and scarlet, and all thyine wood, and all manner vessels of ivory, and all manner vessels of most precious wood, and of brass, and iron, and marble,	
[13]	And cinnamon, and odours, and ointments, and frankincense, and wine, and oil, and fine flour, and wheat, and beasts, and sheep, and horses, and chariots, and slaves, and souls of men.	
[14]	And the fruits that thy soul lusted after are departed from thee, and all things which were dainty and goodly are departed from thee, and thou shalt find them no more at all.	
[15]	The merchants of these things, which were made rich by her, shall stand afar off for the fear of her torment, weeping and wailing,	
[16]	And saying, Alas, alas, that great city, that was clothed in fine linen, and purple, and scarlet, and decked with gold, and precious stones, and pearls!	
[17]	For in one hour so great riches is come to nought. And every shipmaster, and all the company in ships, and sailors, and as many as trade by sea, stood afar off,	Within an hour the serpent-firefall-comet-bolide impacts have caused their damage world-wide
[18]	And cried when they saw the smoke of her burning,	

	saying, What city is like unto this great city!	
[19]	And they cast dust on their heads, and cried, weeping nd wailing, saying, Alas, alas, that great city, wherein were made rich all that had ships in the sea by reason of her costliness! For in one hour is she made desolate.	
[20]	Rejoice over her, thou heaven, and ye holy apostles and prophets; for God hath avenged you on her.	
[21]	And a mighty angel took up a stone like a great millstone, and cast it into the sea, saying, Thus with violence shall that great city Babylon be thrown down, and shall be found no more at all.	
[22]	And the voice of harpers, and musicians, and of pipers, and trumpeters, shall be heard no more at all in thee; and no craftsman, of whatsoever craft he be, shall be found any more in thee; and the sound of a millstone shall be heard no more at all in thee;	
[23]	And the light of a candle shall shine no more at all in thee; and the voice of the bridegroom and of the bride shall be heard no more at all in thee: for thy merchants were the great men of the earth; for by thy sorceries were all nations deceived.	
[24]	And in her was found the blood of prophets, and of saints, and of all that were slain upon the earth.	
Rev .19	* Completion of a cycle is equated to the Marriage Supper of the Lamb* Out of heaven a white horse and upon him one who is faithful and True; judge to make war. * Clothed in vesture dipped in blood. Called the Word of God. * The armies in the sky which followed him. A sharp sword issued from his	

	mouth. King of Kings, Lord of Lords. * The beast and the false prophet were taken and cast alive into a burning lake of fire. And the fowls of the earth ate their flesh.
[1]	And after these things I heard a great voice of much people in heaven, saying, Alleluia; Salvation, and glory, and honour, and power, unto the Lord our God:
[2]	For true and righteous are his judgments: for he hath judged the great whore, which did corrupt the earth with her fornication, and hath avenged the blood of his servants at her hand.
[3]	And again they said, Alleluia. And her smoke rose up for ever and ever.
[4]	And the four and twenty elders and the four beasts fell down and worshipped God that sat on the throne, saying, Amen; Alleluia.
[5]	And a voice came out of the throne, saying, Praise our God, all ye his servants, and ye that fear him, both small and great.
[6]	And I heard as it were the voice of a great multitude, and as the voice of many waters, and as the voice of mighty thunderings, saying, Alleluia: for the Lord God omnipotent reigneth.
[7]	Let us be glad and rejoice, and give honour to him: for the marriage of the Lamb is come, and his wife hath made herself ready.
[8]	And to her was granted that she should be arrayed in fine linen, clean and white: for the fine linen is the righteousness of saints.
[9]	And he saith unto me, Write, Blessed are they which are called unto the marriage supper of the Lamb. And he

	saith unto me, These are the true sayings of God.	
[10]	And I fell at his feet to worship him. And he said unto me, See thou do it not: I am thy fellowservant, and of thy brethren that have the testimony of Jesus: worship God: for the testimony of Jesus is the spirit of prophecy.	
[11]	And I saw heaven opened, and behold a white horse; and he that sat upon him was called Faithful and True, and in righteousness he doth judge and make war.	
[12]	His eyes were as a flame of fire, and on his head were many crowns; and he had a name written, that no man knew, but he himself.	
[13]	And he was clothed with a vesture dipped in blood: and his name is called The Word of God.	
[14]	And the armies which were in heaven followed him upon white horses, clothed in fine linen, white and clean.	
[15]	And out of his mouth goeth a sharp sword, that with it he should smite the nations: and he shall rule them with a rod of iron: and he treadeth the winepress of the fierceness and wrath of Almighty God.	
[16]	And he hath on his vesture and on his thigh a name written, KING OF KINGS, AND LORD OF LORDS.	
[17]	And I saw an angel standing in the sun; and he cried with a loud voice, saying to all the fowls that fly in the midst of heaven, Come and gather yourselves together unto the supper of the great God;	
[18]	That ye may eat the flesh of kings, and the flesh of captains, and the flesh of mighty men, and the flesh of horses, and of them that sit on them, and the flesh of all men, both free and bond, both small and great.	

[19]	And I saw the beast, and the kings of the earth, and their armies, gathered together to make war against him that sat on the horse, and against his army.	
[20]	And the beast was taken, and with him the false prophet that wrought miracles before him, with which he deceived them that had received the mark of the beast, and them that worshipped his image. These both were cast alive into a lake of fire burning with brimstone.	
[21]	And the remnant were slain with the sword of him that sat upon the horse, which sword proceeded out of his mouth: and all the fowls were filled with their flesh.	
Rev .20	.	
[1]	And I saw an angel come down from heaven, having the key of the bottomless pit and a great chain in his hand.	
[2]	And he laid hold on the dragon, that old serpent, which is the Devil, and Satan, and bound him a thousand years,	
[3]	And cast him into the bottomless pit, and shut him up, and set a seal upon him, that he should deceive the nations no more, till the thousand years should be fulfilled: and after that he must be loosed a little season.	
[4]	And I saw thrones, and they sat upon them, and judgment was given unto them: and I saw the souls of them that were beheaded for the witness of Jesus, and for the word of God, and which had not worshipped the beast, neither his image, neither had received his mark upon their foreheads, or in their hands; and they lived and reigned with Christ a thousand years.	

[5]	But the rest of the dead lived not again until the thousand years were finished. This is the first resurrection.	
[6]	Blessed and holy is he that hath part in the first resurrection: on such the second death hath no power, but they shall be priests of God and of Christ, and shall reign with him a thousand years.	
[7]	And when the thousand years are expired, Satan shall be loosed out of his prison,	
[8]	And shall go out to deceive the nations which are in the four quarters of the earth, Gog and Magog, to gather them together to battle: the number of whom is as the sand of the sea.	
[9]	And they went up on the breadth of the earth, and compassed the camp of the saints about, and the beloved city: and fire came down from God out of heaven, and devoured them.	
[10]	And the devil that deceived them was cast into the lake of fire and brimstone, where the beast and the false prophet are, and shall be tormented day and night for ever and ever.	
[11]	And I saw a great white throne, and him that sat on it, from whose face the earth and the heaven fled away; and there was found no place for them.	
[12]	And I saw the dead, small and great, stand before God; and the books were opened: and another book was opened, which is the book of life: and the dead were judged out of those things which were written in the books, according to their works.	
[13]	And the sea gave up the dead which were in it; and death and hell delivered up the dead which were in them: and they were judged every man according to their works.	

[14]	And death and hell were cast into the lake of fire. This is the second death. Concordance: The theme of fire being used to consume is also used by Jesus Himself. "I am come to send fire on the earth; and what will I, if it be already kindled?" (Luke 12:49). This "fire" that Jesus was speaking of was not to be literal fire, but the fire of His wrath and justice to take place with the change of religious systems (the phasing out of the Old Covenant and the bringing in of the New Covenant)	
[15]	And whosoever was not found written in the book of life was cast into the lake of fire	
Rev .21		
[1]	And I saw a new heaven and a new earth: for the first heaven and the first earth were passed away; and there was no more sea.	* New Heaven and a New Earth. The Holy city coming down out of heaven, prepared as a bride adorned for her husband. * No more pain, no death, no tears. * All things are made new. All who seek are given the water of life. * He that overcomes will be the Son and daughters of God. * Anything other than God will be cast into a lake of fire. * New Jerusalem: 12 gates, 12 foundations. * Reed to measure the new city. 12,000 furlongs. Wall is 144,000 cubits. - measure of a man* 12 stones.-- City had no need for sun or moon for the Lamb is the Light (Light casts no shadow)

New Heaven & New Earth "Look again..... What do you see NOW?" Suddenly I saw what the voice saw - "I see our planet and there are no borders dividing countries... The borders are gone!!"... He said "This is why you're going back. You have a mission."... And that's how I came back. I became more conscious of the environment, of clean air, clean water, world hunger, wars and poverty." |
| [2] | And I John saw the holy city, new Jerusalem, coming down from God out of heaven, prepared as a bride adorned for | |

	her husband. Concordance: The Holy City, the New Jerusalem, was no earthly city, but a city "whose builder and maker is God." Heb. 11:10-16	
[3]	And I heard a great voice out of heaven saying, Behold, the tabernacle of God is with men, and he will dwell with them, and they shall be his people, and God himself shall be with them, and be their God.	
[4]	And God shall wipe away all tears from their eyes; and there shall be no more death, neither sorrow, nor crying, neither shall there be any more pain: for the former things are passed away.	
[5]	And he that sat upon the throne said, Behold, I make all things new. And he said unto me, Write: for these words are true and faithful.	
[6]	And he said unto me, It is done. I am Alpha and Omega, the beginning and the end. I will give unto him that is athirst of the fountain of the water of life freely.	
[7]	He that overcometh shall inherit all things; and I will be his God, and he shall be my son.	
[8]	But the fearful, and unbelieving, and the abominable, and murderers, and whoremongers, and sorcerers, and idolaters, and all liars, shall have their part in the lake which burneth with fire and brimstone: which is the second death.	
[9]	And there came unto me one of the seven angels which had the seven vials full of the seven last plagues, and talked with me, saying, Come hither, I will shew thee the bride, the Lamb's wife.	
[10]	And he carried me away in the spirit to a great and high mountain, and shewed me that great city, the holy Jerusalem,	

	descending out of heaven from God,	
[11]	Having the glory of God: and her light was like unto a stone most precious, even like a jasper stone, clear as crystal;	
[12]	And had a wall great and high, and had twelve gates, and at the gates twelve angels, and names written thereon, which are the names of the twelve tribes of the children of Israel:	
[13]	On the east three gates; on the north three gates; on the south three gates; and on the west three gates.	
[14]	And the wall of the city had twelve foundations, and in them the names of the twelve apostles of the Lamb.	
[15]	And he that talked with me had a golden reed to measure the city, and the gates thereof, and the wall thereof.	
[16]	And the city lieth foursquare, and the length is as large as the breadth: and he measured the city with the reed, twelve thousand furlongs. The length and the breadth and the height of it are equal.	
[17]	And he measured the wall thereof, an hundred and forty and four cubits, according to the measure of a man, that is, of the angel.	
[18]	And the building of the wall of it was of jasper: and the city was pure gold, like unto clear glass.	
[19]	And the foundations of the wall of the city were garnished with all manner of precious stones. The first foundation was jasper; the second, sapphire; the third, a chalcedony; the fourth, an emerald;	
[20]	The fifth, sardonyx; the sixth, sardius; the seventh, chrysolite; the eighth, beryl; the ninth, a topaz; the tenth, a chrysoprasus; the eleventh, a	

	jacinth; the twelfth, an amethyst.	
[21]	And the twelve gates were twelve pearls; every several gate was of one pearl: and the street of the city was pure gold, as it were transparent glass.	
[22]	And I saw no temple therein: for the Lord God Almighty and the Lamb are the temple of it.	
[23]	And the city had no need of the sun, neither of the moon, to shine in it: for the glory of God did lighten it, and the Lamb is the light thereof.	
[24]	And the nations of them which are saved shall walk in the light of it: and the kings of the earth do bring their glory and honour into it.	
[25]	And the gates of it shall not be shut at all by day: for there shall be no night there.	
[26]	And they shall bring the glory and honour of the nations into it.	
[27]	And there shall in no wise enter into it any thing that defileth, neither whatsoever worketh abomination, or maketh a lie: but they which are written in the Lamb's book of life.	
Rev .22		
[1]	And he shewed me a pure river of water of life, clear as crystal, proceeding out of the throne of God and of the Lamb.	
[2]	In the midst of the street of it, and on either side of the river, was there the tree of life, which bare twelve manner of fruits, and yielded her fruit every month: and the leaves of the tree were for the healing of the nations.	
[3]	And there shall be no more curse: but the throne of God and of the Lamb shall be in it; and his servants shall serve him:	

[4]	And they shall see his face; and his name shall be in their foreheads.	
[5]	And there shall be no night there; and they need no candle, neither light of the sun; for the Lord God giveth them light: and they shall reign for ever and ever.	
[6]	And he said unto me, These sayings are faithful and true: and the Lord God of the holy prophets sent his angel to shew unto his servants the things which must shortly be done.	
[7]	Behold, I come quickly: blessed is he that keepeth the sayings of the prophecy of this book.	
[8]	And I John saw these things, and heard them. And when I had heard and seen, I fell down to worship before the feet of the angel which shewed me these things.	
[9]	Then saith he unto me, See thou do it not: for I am thy fellowservant, and of thy brethren the prophets, and of them which keep the sayings of this book: worship God.	
[10]	And he saith unto me, Seal not the sayings of the prophecy of this book: for the time is at hand.	
[11]	He that is unjust, let him be unjust still: and he which is filthy, let him be filthy still: and he that is righteous, let him be righteous still: and he that is holy, let him be holy still.	
[12]	And, behold, I come quickly; and my reward is with me, to give every man according as his work shall be.	
[13]	I am Alpha and Omega, the beginning and the end, the first and the last.	
[14]	Blessed are they that do his commandments, that they may have right to the tree of life, and may enter in through the gates into the city.	

[15]	For without are dogs, and sorcerers, and whoremongers, and murderers, and idolaters, and whosoever loveth and maketh a lie.	
[16]	I Jesus have sent mine angel to testify unto you these things in the churches. I am the root and the offspring of David, and the bright and morning star.	
[17]	And the Spirit and the bride say, Come. And let him that heareth say, Come. And let him that is athirst come. And whosoever will, let him take the water of life freely.	
[18]	For I testify unto every man that heareth the words of the prophecy of this book, If any man shall add unto these things, God shall add unto him the plagues that are written in this book:	John threatens the plagues like a curse -not a jot or tittle to be changed. It was common for copiests to make changes to serve their own interpretation of the Gospels.
[19]	And if any man shall take away from the words of the book of this prophecy, God shall take away his part out of the book of life, and out of the holy city, and from the things which are written in this book.	John warns of the severe consequences to not take away the "words of this prophecy," or add to the prophecy or God will take away his part out of the Book of Life, and out of the Holy City, and out of the things written in this book – don't change or add anything. Same threat John threatens a curse: not a jot or tittle to be changed. It was common for copiests to make changes to serve their own interpretation of the Gospels.
[20]	He which testifieth these things saith, Surely I come quickly. Amen. Even so, come, Lord Jesus.	
[21]	The grace of our Lord Jesus Christ be with you all. Amen. The End.	

TABLE 1 | **3 8 3**

Table 1 - Impact Chronology

Biblical & Islamic synoptic prophecies - Biblical - Islamic Sources

According to Costa, the fact that the Madhi is not a human being is alluded to in many Muslim texts, "My brothers' second error: they ascribe a mortal identity that is prone to decay... Since this time is one of a spiritual entity, such great and eternal truths cannot be built on mortal, helpless identities that are liable to error.' (Turkish scholar Bediuzzaman Said Nursi (1878-1960) in his 1946 work, The Ratifying Stamp of the Unseen, p. 10) pg.27, Adam to Apophis.

Is the 12th Hazrat Madhi human?

"And a hail of fire mingled with blood cast upon the earth...And a greats star fell from heaven burning as if it were a lamp.....(Rev 9:10) Galileo Code.

-Signs of the Islamic End times before Hazrat Madhi appears:

"The last of this strife will be the killing of innocent people and then Hazrat Madhi will appear, to the approval of all (Al-muttaqi al-Hindi (1567A.D.), Al-Burnhan fi Alamat al-Mahdi Akhir az-Zaman, p. 38)

"Hazrat Madhi will not appear until the innocent people are slaughtered. He will appear when those on Earth and in the skies no longer can bear the killing..." (Ibn Hajar al-Haytam (1566 A.D.) Al-Qawl al-Mukhtasar fi Alamat al-Madhi al-Muntadhar, p 37)

"They will kill mothers, fathers, daughters, men, everyone, and inflict great suffering on the community." ((Al-muttaqi al-Hindi (1567A.D.), Al-Burnhan fi Alamat al-Mahdi Akhir az-Zaman, p. 36)

-An Al-Qaeda leader, Abu Mus'ab Al-Suri was reported to have said, "It is our destiny to fulfill the prophecies."(cited by Yamin Ramzi, a former jihadist cleric in an interview with Marissa Allison, November 22, 2009. -I cannot imagine a greater evil than this. It appears that ISIS and Al-Qaeda are attempting to force the Hazrat Madhi to appear by slaughtering innocents. They completely misunderstand their own prophecies. Hazrat Madhi is not a man. He (it) is a celestial event.

- People will be mass exterminated; their deaths will be like passing souls through 'a sieve' Only one in ten will survive. Note: Other prophetic sources describe similar mass extinction rates.

- "I swear that a flame will engulf you. That flame is presently in an extinguished state...That flame swallows up people with terrible pain inside it, burns down and destroys people and property, and spreads all over the world by flying like a cloud with the assistance of winds. Its heat at night is much higher than its daytime temperature. By coing as deep as the center o the earth from above the heads of the people that flame becomes a terrible noise, just like the lightning between the ground and the sky...." (Ash sha'rani (d 1565A.D.), Mukhtasar Tazkirah al-Qurtubi, p.461)

The Madhi, also known as the "Guided One," according to Islamic hadiths and the Holy Quran is destined to lead mankind into righteousness.

Page 27. Excerpts from Adam to Apophis. The belief that Hazrat Madhi would return began in 1979, a date equated with great significance, as it was the first day of I Muharram), year 1400 of the Islamic Calendar. It was not surprising that a Muslim, Muhammad bin abd Allah al-Qahatani, was proclaimed by his brother-in-law, to be Madhi in that same year. He and 300 followers seized the Grand Mosque in Mecca, in November 1979. After a two-week siege, and hundreds of deaths, the uprising was put down.

Jewish, Christian, Islam literature – Why is the Messiah, Mahdi kept a secret?

It was kept a secret because who wants to know when the end of the world will occur?

Yet in the Holy Quran, it says, "But none except Allah and those who are deeply rooted in knowledge will know its final meaning..." (Surah Al Imran, 7)

Imam in the 10th century stated, "Never make the name of Hazrat Mahdi (AS) well-known..." (Sheik Muhammad ibn Ibrahim Numani, al-Ghaybah al-Numani, p. 174.

According to Costa, the fact that the Madhi is not a human being is alluded to in many Muslim texts, "My brothers' second error: they ascribe a mortal identity that is prone to decay... Since this time is one of a spiritual entity, such great and eternal truths cannot be built on mortal, helpless identities that are liable to error.' (Turkish scholar Bediuzzaman Said Nursi (1878-1960) in his 1946 work, The Ratifying Stamp of the Unseen, p. 10) pg.27, Adam to Apophis.

"Only the father knoweth the Day and Hour," Islam – "They ask you, [O Muhammad], about the Hour: when is its arrival? Say, "Its knowledge is only with my Lord. None will reveal its time except Him. It lays heavily upon the heavens and the earth. It will not come upon you unexpectedly." They ask you as if you are familiar with it. Say, "Its knowledge is only with Allah, but most people do not know." (Surah: 7:187)

"A comet will signal his appearance. A great fire in the sky will illuminate the night. The Madhi's face will shine like a star in the heavens and will shine on the surface of the moon."

"Behold, he cometh with clouds; and every eye shall see him....and all kindreds of the earth shall wail because of him. I am Alpha and

TABLE 1 | 385

Omega, the beginning and ending, saith the Lord, which is, and which was, and which is to come, the Almighty." Rev. 1: 7-

"When he appears, he will call out and the whole world will hear and see him wherever they are."

And hand will extend from the sky and a voice will cry out and the people will look and see it. He will be so powerful that when he stretches out his hand to the largest tree in the world, he will uproot it and cast it away and if he shouts among the mountains hard rock will turn into powder."

"He will be accompanied by companions who have hearts of iron, and by thousands of angels who will strike the faces and the backs of those who oppose him. (Rev 10:7-10: and the shapes of the locusts were like horses prepared for battle...they had hair as the hair of women, teeth as the teeth of lions,..and had breastplates of iron; and the sound of their wings was the sound of chariots of many horses running into battle... and they had tails like scorpions).

"Stones will rain down from the sky and there will be earthquakes and great cities will be flattened."

"And a hail of fire mingled with blood cast upon the earth....And a greats star fell from heaven burning as if it were a lamp. (Rev 9:10) on prophecy. This following is an interesting Islamic prophecy of a personal and planetary Judgment Day from Quran 82: al-Infitar. The Cleaving Asunder

1. When the heaven is cleft asunder.

2. When the planets are dispersed.

3. When the seas are poured forth,

4.And the sepulchers are overturned.

5. A soul will know what it hath sent before (it) and what is left behind

6. O man! What hath made thee careless concerning thy Lord, the bountiful,

7. Who created thee, then fashioned, then proportioned thee?

8. Into whatsoever form he will, He casteth thee

9. Nay, but ye deny the Judgment

10. Lo! There are above you guardians

11. Generous and recording

12. Who know (all) that ye do

13. Lo! The righteous verily will be in delight.

14. And Lo! The wicked verily will be in hell;

15. They will burn therein in the Day of Judgment

16.And will not be absent thence.

17. Ah, what will convey unto thee what the Day of Judgment is?

18. Again, what will convey unto thee what the Day of Judgment is?

19.A day on which no soul has power at all for any (other) soul, the (absolute) command on that day is Allah's.

Number of dead after impact: Zechariah 13:8 forecasting that two of every three people will die. "And it shall come to pass that in all the land saith the LORD two parts therein shall be cut off and die; but the third shall be left therein."

Number of dead after impact:

BP Calendar   Impact Phoenix Cycle   Comet/meteor Storm   Ice Core Cooling Evidence Tree Ring Data Extinction Data/global Contemporary History Biblical History

14,800 BP   c. 12,800 B.C.   1st Impact Significant impact event Yes. higher CO2   Adam/Eve – Serpent enters garden

12,900 BP   c. 10,900 B.C.   1,900 years Significant   Impact   event Younger Dryas Extinction Level Event /World Flood   Banishment from Garden of Eden

12,450 BP   c. 10,450 B.C.   Cold Period Sphinx/Osiron/early Pyramids built Enoch – Atlantis Period

11,700 BP   c. 9,700 B.C.   1,200 years Ice Age ends within 40 years

11,500 BP   c. 9,500 B.C.   Gobekli-Tepi Monument built

9,700 BP   c. 7,640 B.C.   2,060 years Significant   Impact   event None

Yes. Tollman Data Jericho Built

8,200 BP   c. 6,200 B.C.   1,200 years Significant Impact event Cooling

6,019 BP   c. 4,004 B.C.   Samarian Culture begins Old standard. World created

5,776 BP   c. 3, 776 B.C.   Significant   Impact   Event: Baranger Crater legend (less than 5,000 years old (science places it at 49,000 years None   Jewish calendar year begins. May 17. Halloween. November Impact

5,114 BP   c. 3,114 B.C.   Mayan Calendar Begins

4,920 BP   c. 2,920 B.C.   Legends of Flood/ Deucalion Flood (Bible Flood: 2,750 B.C. Traditional Flood dates) Saqqara Pyramid built

4,254 BP   c. 2,574 B.C.   Great Pyramid construction started

4,470 BP   c. 2,470 B.C.   Cheops/Khufu (buried)

4,200 BP   c. 2,200B.C.   Significant   Meteor   storm   event. Volcano-Santorini explodes Famine, evidence of change in weather, changing Sahara to desert. Egyptian   Dynasty   change   (500   years). Abraham- Sodom and Gemorrah

3,623 BP   c. 1,623 B.C.   Significant Comet Yes.   Santorini Explosion Moses/Joshua - 645 years / Sun/Moon Stand Still -31 kingdoms destroyed by falling stones from the sky

TABLE 1 | **387**

BP Calendar   Impact Phoenix Cycle   Comet/meteor Storm   Ice Core Cooling Evidence Tree Ring Data Extinction Data/global

Contemporary History Biblical History

2,003 BP   c. 12 B.C.   Halley Birth of Jesus of Nazareth

1,949 BP   c. 66 A.D.   Halley

1,945 BP   c. 70 A.D.   Fall of Jerusalem/Destruction of Temple

c. 530 A.D. Halley

c. 540-41 A.D. Comet of Gaul evidence   evidence   cooling

c. 571 A.D. Leonid Meteor Storm evidence   evidence   Dark Ages, Famine, plague Mohammed Born (some say 569, or 570 A.D.) – Year of the Elephant   Kaaba erected with Meteorites

c. 1347 A.D.   Meteor Storm Dark Ages- Famine, plague evidence

c. 1833 A.D.   Leonid Meteor storm   Earthquakes, fall of stones noted   Lincoln observed – thought end of the world was near

c. 1866 A.D.   Leonid Meteor Storm   Earthquakes, fall of stones noted   107 BP c. 1908 A.D. – June 28 Enke Impact Tunguska, Russia- 10 MT

105 BP c. 1910 Halley's Comet

49 BP   c. 1966 A.D.   Leonid Meteor storm

29 BP   c. 1986 Halley's Comet

16 BP   c. 1999 A.D.   Leonid Meteor Storm

10 BP   c. 2004 Designated 2007 PA8   1100 foot asteroid Apophis discovered (324 cycle/7 year

2 BP   c. 2013 A.D. Feb 15 Chelyabinsk explosion, bright as the sun, within clouds and thunder – estimated at

DA14 is approaching Earth from the south, so any fragment of that rock would also appear to move south-to-north. So again, I think this is unrelated to 2012 DA14. But wow, what a huge coincidence! The fireball was incredibly bright, rivaling the Sun! There was a pretty big sonic boom from the fireball, which set off car alarms and shattered windows. I'm seeing some reports of many people injured (by shattered glass blown out by the shock wave). I'm also seeing reports that some pieces have fallen to the ground, but again as I write this those are unconfirmed.

1   c. 2014 A.D.- Sept 7   Near miss- 2014RC

c.2021 A.D. Apophis flyby

c.2028 A.D. Xf11

c.2029 A.D. April 13   Apophis   close   approach-19,000 Kilometers 12,000 miles-will0 flyby within geosynchro orbit of satellites

c.2032 A.D. Nov 19 Leonid meteor storm

c. 2036 A.D. April 13 Passover. Apophis flyby. Right now, it stands at 1-in-48,000. One more time: Whew. Just for reference, there is a 1-in-354,319 chance that you'll be killed in an airplane accident. NASA estimates the energy from this particular asteroid to be roughly the same as if 65,000 nuclear bombs were dropped on us. What happens next depends on where it hits. It would certainly trash the immediate area. So, after astronomers had determined that the April 2029 encounter wasn't going to impact the Earth, they ran some simulations and found that the orbit of the asteroid will bend about 28 degrees, altering its course. The flyby will make the orbit a bit bigger and Apophis will travel a bit slower. How much the orbit changes depends on how close it gets to us. If it flies through a specific 610-meter wide region of space as it goes past us in 2029, then Apophis' and the Earth will be in the exact same spot 7 years later on Sunday, April 13th, 2036. The most well known being the asteroid 1950DA, which is estimated to have a 1 in 300 chance of impacting the Earth in 2880.

     c. 2061 A.D.   Halley's
     c. 2065 A.D.   Nov 230   Leonid Meteor Storm
     c.2880 A.D.March 16, 1950 DA

I found this in a separate document David had been working on dated only a few weeks before he died.

The Secret Place of the Lion
The Lion / Serpent number is 33.33
10 Kings ruled 432,000 Years before the Deluge:

Divide numbers by 33.333333
Divide then by 12 or 24 for a cycle

<u>828 Year theory / 10,800 units</u>

There was a belief that the world would be shaken up in the 1656 year of its creation. If this is so, then a Flood was expected in 3113 B.C. . (Corresponds with Mayan Calendar)

According to Censourinus, the great year of Heraclitus comprised of a period of 10,800 "years." (The chaldaic world month also consisted of 10,800 years.) "Amont other ancient calendars... the most frequently used cycle was 10,800 years, common to the Hindus, the Sumerians, and the Babylonians. Forty of these cycles made the great cycle of the Hindus and the great year of Berossus, high priest of Babylon.

"The figure 10,800 also repeats itself in many other places. Multiples or fractions of this number can be found in sacred texts from all around the world. The Rig-Veda, the most important sacred book of the Hindus, has 10,800 verses and the altar of the Vedic god Agnni was built of exactly 10,800 bricks... Another example of this number is to be found in

TABLE 1 | **3 8 9**

Cambodia in the temple of Angkor Wat. This place is much older than most people think, and it is decorated with 540 statues along 5 avenues. Each statue represents 20 years of time. They are erected exactly like stone pillars, or stelae of the Mayas which also represent 20 years each, and when you add them all, they represent 10,800 years. Finally, the German legend of the Nibelungs speaks of the 10,800 souls of dead warriors who enter the gates of Walhalla -- the abode of immortal heroes " (Chatelain, 1978: 50-51)

*Note: If each avenue represent 1 age, or 2160 years, 5 would be 10,800 years. 540 statues of 20 years each would be 10,800 years. 21,600 years would be a "day" and a "night", or the 6 days of creation, within the precession of the equinoxes. 10 Avenues are 10 Kings????? 600 x 60 x 6 = 216,000 Divided by 10 kings = 21,600

The Most important great period of Mayan Chronology is the Baktun: 144,000 days, about 395 years. 144 plays an important role in world circle calculations. The 3 fold of the 144,000 is used most of the time (3,600 x 3) = 10,800

Therefore, , Berossus writes, based on ancient traditions, that before the Deluge ten kings reigned in Old Babylon during 432,000 years. This is also a main time-circle. The Chinese main time circle is identical (10 wings)

(Egyptians began a new epoc after Deluge) Their ancient king and hero of the time of the deluge is identified with the historic king Menes. Also, Eratoshtenes maints that Menes' reign falls in the time of the Deluge. Narmer-Menes and Nimrod-Menrot, also Hungarian traditions maintain that their furst forefaterh Nimrod settled with his two sons Hunor and Magyar in Persia, just after the Deluge. He, too, had the famous tower built.

In our opinion, the cycle of Heraclitus mean St:

* Brahma world year= 8,640,000 = 25,920 x 1
* Hindu world Year = 4,320,000 = 12,960 x 2
* Babylonian Year = 2,160,000 = 6,480 x 4
* Yuga = 432,000 = 1,296 x 20
* Babylonian Month = 10,800 = 32.4 x 800

The 12,000 Brihati verses consists of 432,000 narasamsasvaras, or complete syllables of the Rigveda (Shamasastry, 1908:143).

216,000	216,000
(18 x 12,000)	(18 x 12,000)
108,000	108,000
(9 x 12,000)	(9 x 12,000)

432,000 year Yuga = 12,960 years
864,000 year Yuga = 25,920 years

*	Babylonian Day	=	360
*	3 periods of 144,000	=	432,000
*	Chinese Great Age	=	432,000
*	Hindu Yuga	=	432,000
*	Babylonian Age	=	432,000
*	Iranian Age	=	432,000
*	Edda 540 gates x 800 Einherier	=	432,000 Walhalla Heroes
*	Brihati - 12,000 Verses	=	432,000 Naramsasvaras
*	10,800 periods of 4 weeks each	=	43,200
*	The Edda 5,400 of 8 weeks each	=	43,200
*	1 week of 43,200	=	43,200
*	7 days of 43,200	=	302,400 periods

Divided tropical Year (365.242199) = 827.94378 years (close to 828), 40 cycles of 10,800 = 432,000

The inundation of the Nile was 4 months = 150 days (5 months)

The Secret Place of the Lion
The Lion / Serpent number is 33.33
10 kings ruled 432,000 Years before the Deluge:

1 King	43,200 years	1,296 years	1296	
2 King	43,200 years	1,296 years	2592	
3 King	43,200 years	1,296 years	3888	
4 King	43,200 years	1,296 years	5184	4
5 King	43,200 years	1,296 years	6480	5
6 King	43,200 years	1,296 years	7776	6
7 King	43,200 years	1,296 years	9072	7
8 King	43,200 years	1,296 years	10,368	8
9 King	43,200 years	1,296 years	11,664	9
10 King	43,200 years	1,296 years	12,960	10

432,000 units   12,960 years   =   1/2 Precession = 12 hours

11 King	43,200 years	1,296 years
12 King	43,200 years	1,296 years
13 King	43,200 years	1,296 years
14 King	43,200 years	1,296 years
15 King	43,200 years	1,296 years
16 King	43,200 years	1,296 years
17 King	43,200 years	1,296 years

TABLE 1 | 391

```
18  King    43,200 years    1,296 years
19  King    43,200 years    1,296 years
20  King    43,200 years    1,296 years
864,000 units   25,920 years    = 1 Precession = 24 hours
```

In Anchor Watt: 540 statues in 5 columns x 20 years each = 10,800 (period between deluges)

Thus, there are 108 statues in each column. x 20 = 2160:

In 3,600 years, there are 108 cycles of the Leonid storms (33.33) = 1 Saros

2160 x 5 = 3,600 x 3=   In 10,800 years, there are 324 cycles of the Leonid storms

2160 x 10= 3,600 x6=   In 21,600 years there are 648 cycles of the Leonid storms

1st way =   108 statues X 20 years = 2160 1st God / Sun   108   x 33.33   =3600

2nd way=   108 statues X 20 years = 2160 2nd God / Sun 108   x 33.33   =3600

3rd way=   108 statues X 20 years = 2160 3rd God / Sun 108   x 33.33   =3600

4th way=   108 statues X 20 years = 2160 4th God / Sun 108   x 33.33   =3600

5th way=   108 statues X 20 years = 2160 5th God / Sun 108   x 33.33   =3600

540 = 10,800   18,000

3,600 (108) x 5 = 18,000

540 = Judgment (greek) years

1080 = 36,000

40 X 10,800 = 432,000/25,920 years = 33.3333333 (Great Cycle of the Hindus, and the Great Year of Berossus.)

40 = the Trial Period

25,920/ 40 = 648 cycles of 40 or 40 x 648 divided by 4 = 10 x 648

25,920 /4 = 6,480

3,600 divided by 33 1/3 = 108

108 x 8 = 864

108 x 800 = 86,400

# of seconds in a day   = 86,400 = 24 hours

# of units in Precessional year = 864,000 = 1 day

25,920 divided by 33 1/3   = 777.6

600 x 60 x 6 = 216,000

1260 = 7.2

9072/42 = 216

9072/ 3.5 = 2592

Time, Time, Time, + 1/2 time = 3.5 times rest between serpent ages

70 weeks of years = 9,072
70 x 7 = 490
9072/490= 18.51 (Saros Number?)

The Paul Solomon Story

The son of a Southern Baptist minister, Paul (born William Bilo Dove) grew up into "a fourth generation of Southern Baptist Ministers from the buckle of the Bible Belt, where the ministry is the assumed career. Paul's great-grandfather was one of the old "circuit-rider" preachers during the expansion of the West. Paul described him as "a preacher with a Bible in one hand, Colt Peacemaker in the other, for snakes...of all kinds."[258]

Growing up, he watched his parents seek guidance daily and noted they based decisions on the silent replies they received. God was accessible; the Holy Spirit was an active participant in their everyday lives.

Growing up in a fundamentalist Christian home, a child is taught and believes God is always available for guidance and takes comfort in that belief. But for the young Paul Solomon there was a hitch; a very real dilemma. He exhibited uncommon abilities which did not jibe with a good Baptist life. From an early age, he saw colors around people. He called them "good lights" and "bad lights," depending on how he perceived the individual. As a young child, he would point at someone and sound the alarm, "He's bad!" His embarrassed parents would scold him for being rude. Eventually he learned to keep what he saw to himself. He knew what people were thinking and learned early that most people say the opposite. He also knew about events beforehand; he could predict the future, which frustrated and undermined his attempt to be "like everyone else." While these "gifts of knowledge," these psychic experiences, came naturally to Paul, they were obviously not "normal."

At 14, he was ordained. He began his pastoral tasks by ministering to the incarcerated and their families as the youngest Baptist Minister in the Arkansas Prison system. Soon after High School, Paul enlisted in the Army, married and had a daughter. After the Army, he enrolled in a Baptist Seminary. During this period, he preached in a Spanish speaking Mexican Baptist church in the Panhandle of Texas.

He was well on this way to a successful career when his wife filed for divorce, leaving him for another man. Back in 1967, there was no such thing as a divorced Southern Baptist Minister. So in one day he lost his family and his career.

The traumatic experience left him very, very angry ... at God. He had given over his entire life to serving the Lord, and God had pulled the rug out from under him. And he rebelled; first smoking, then drinking.

---

258 Description of Paul's great-grandfather excerpted from: Michael McCarthy 2012. On Death and Living: Memoir of a Soul. Virginia Beach, VA: Paul Solomon Institute Publications. p. 154.

Southern Baptists don't even use wine for communion. It was an entirely new experience. He spent the next five years working, partying, trying not to think about God, and surrounding himself with people who were doing the same.

One day, watching his pet hamsters running endlessly on their treadmill, the lesson struck home: "That's me." he thought. "I am running as hard as I can just to keep myself from thinking about my life." With that sudden realization, and little further ado, he kicked out the roommates he'd been partying with, and went into seclusion. He closed the blinds, locked his door, and stopped answering the phone.

He had an evening job as a night club manager and still reported to work, but his life had gone from frenetic activity to avoiding everybody. Days were spent in the dark in his apartment, and nights locked in his night club office. For about thirty days, the night club ran itself, which the waitresses loved.

He began to worry about his sanity. Maybe someone could tell him "he was alright" and help him out of an unrelenting depression. At his Seminary, he had majored in psychology, but he did not think a doctor could give him useful answers. He didn't want to talk a Minister, because he knew what they would say. Then he remembered his work in the Army at the Brooke Army Medical Center. Under the supervision of doctors, he had been part of a pilot program that used hypnosis in battlefield situations when pain killers were not available. He had also witnessed hypnosis help soldiers through a variety of personal problems, including depression. And given the success of that program, he thought that a good hypnotherapist would be helpful to him. He found a hypnosis clinic in the yellow pages, called and made an appointment. The therapist seemed confident that Paul's problem was a deep-seated guilt complex brought on by his strict religious upbringing and was sure that successful treatment was possible. However, when Paul was told the exorbitant price of the twice-weekly sessions, he left feeling more miserable than ever. Even when he actively sought help, it remained out of reach. His life was at a crisis point.

Hypnosis Experiment Triggers a Near-Death Experience

Paul went to work one night angry and bitter, but at least he had finally talked about his problem. He revealed his predicament to Harry Snipes, a regular at the nightclub. "How in the world am I supposed to afford hypnosis at those prices?" he complained.

"I hypnotized someone once," Harry offered, his voice trailing off as Paul ranted on about the unfairness of life. Incredulous, Paul stared at him.

"You hypnotized someone once?" he echoed.

There was a dubious confidence in Harry's expression, and Paul supposed "once" could happen again. "Let's go, then!" And off they went to Paul's apartment where he laid down, closed his eyes, and submitted, with no little apprehension, to Harry's unprofessional instructions. Paul's last conscious thought was, "This'll never work. This guy has no idea what he's doing. If anything happens, it'll be because he bored me to sleep."

The night was February 14, 1972. In response to Harry's bland attempts at hypnotism, Paul appeared to fall asleep. Without warning, his body jerked erect as if his soul had been shot from a cannon out of his body. He began speaking in an intense, authoritative voice, apparently unconscious on the couch. Paul awoke from the hypnosis session doubled up with pain from cramps in his stomach and Harry jumping up and down with excitement insisting they had talked to "spirits."

His only recollection of the hypnosis session upon awaking was a dream, which would later be recognized as some of the "classic" core death elements of the near-death experience:

I am approaching the entrance to a tunnel. It is similar to the mouth of a cornucopia because it seems to spiral inward and upward. I can see two figures at the opening. They seem to be waiting for me. One is Merle who was my girlfriend in high school until she died in our senior year. The other is my young friend, Jaida, who remained my secret companion for years following his death when we were seven. They each take me by the hand and lead me through the tunnel. We come out the other side onto a grassy hill. As we begin to climb, I see that we are approaching a temple at the top of the hill.... The two figures are waiting to take me through the tunnel. As I come out the other side, I find myself in a meadow of wild flowers where a soft breeze blows, and I can hear the sound of a brook. Ahead of me is a mountain. As I climb, I pass through seven terraced gardens of glorious color. At the top, I enter a temple. The air is rich with music, though I see no one. I see rows and rows of books with names on the bindings. I am in an enormous library.[259]

As Paul recovered from the violent post-session stomach cramps, Harry recalled all he could of "the voice" —which to Paul, sounded too "religious" to be true.

He said, "Harry, what do you expect to get when you hypnotize an ex- Southern Baptist Preacher? He didn't want to hear more and told Harry to leave. But Harry returned the next day insisting they try the hypnosis experiment again. This daily process lasted about a week, until Harry made a remark that intrigued him.

---

259 Transcribed Paul Solomon Reading Excerpt, February 15, 1972.

He said, "I know that wasn't you because you're not that smart."

Harry brought a tape recorder and together they formulated a test question; a question that neither knew the answer to but that was provable. Paul wanted to make sure that the answer was not coming from his subconscious mind.

Paul's great grandfather had been murdered before Paul was born and died without disclosing the whereabouts of a large sum of money he had been saving but hiding. In spite of a thorough search, the family had never found the money.

The question: "Where was Great Grandfather's money?"

They set up another hypnosis session again to clarify "who" was talking and this time recorded it. Again, the authoritative, booming voice spoke:

You have not attained sufficient growth or spiritual awareness to understand contact with these records...That which you perform is a foolish experiment, for you attempt to harness powers you do not understand and to contact sources, records and intelligences you are not familiar with. How will you try the spirits should you attain that you seek? Would you recognize Him whom you do not know, have not been familiar with?[260]

During hypnosis the "conductor" asked Paul, "Who is speaking?" Paul stated emphatically:

This is not a spirit. It is not some other personality. You are talking with the Source of his Mind, which gave birth to his mind. It pre-existed his physical body and created his physical body.[261]

Harry asked the voice what they should do to prepare spiritually for further hypnosis sessions. They were given a list of instructions, which included a specific diet, exercise, prayer, meditation, and a recommendation to read the Sacred Scriptures in the Bible, and study the Search for God books from the ARE." Harry asked, "What's the ARE?" The voice explained it was Edgar Cayce's organization, the Association for Research and Enlightenment in Virginia Beach, Virginia.

This time when Paul awoke, he was able to listen to the voice on tape. He heard a stronger, more authoritative version of his own voice, providing detailed information on a number of subjects, including a thorough description of his great grandfather's death and the whereabouts of the hidden money: in the chimney of the old family home (The money was later retrieved by Reverend Dove. Attempts to identify the amount of Grampa's stash were unsuccessful, primarily because he kept it a private family matter).

---

260 Transcribed Paul Solomon Reading Excerpt, February 21, 1972

261 Transcribed Paul Solomon Reading Excerpt, February 21, 1972

The voice also explained the stomach cramps Paul experienced upon returning to normal consciousness:

Your consciousness is disengaging itself from the physical body, and when the physical body feels the consciousness leaving, it associates that with death, and will do anything in its power to hold consciousness in the body. And for that reason, those muscles are cramping in an attempt to sustain life.[262]

At the conclusion of the session, as a kind of finale, the voice described a man who addressed a comment to Reverend Dove—Paul's father; "You will know who this is and you'll remember our agreement, and that this is the conclusion of our agreement." Paul realized he had never encountered anything like this during his Southern Baptist Ministry. He knew this wasn't just religion emanating from his own subconscious. He was both excited and disturbed by it.

But he needed to know: was this of God?

So with the cassette tape in hand, he went to South Carolina to consult the man whom he most trusted to know the answer, the Southern Baptist Preacher who would know: his own father, the Reverend Dove.

After playing the tape, he asked his father, "So, is it of God or is it of the Devil?" Looking his son straight in the eye, his father didn't hesitate and said, "It's of God."

Paul, playing Devil's advocate, asked, "Dad, how can you say that? Southern Baptists don't believe in this stuff."

The Reverend replied, "There are two reasons I know it's of God. You know, this is not the first time in history that something like this has occurred. The prophet Daniel found himself in a situation where his life depended on receiving word from God regarding the interpretation of King Nebuchadnezzar's disturbing dream—information that was not available to his conscious mind. Under threat of death, he went into a room with his three friends, and according to the Bible story, he 'prayed all night.' I believe after their prayers, Daniel laid down and went to sleep and he awoke in the morning with the dream's true meaning, which he passed on to the king, saving their lives. It seems to me that Daniel did something like what you just did."

His father continued, "And the second reason is this. Ever since you left the ministry five years ago, your mother and I have prayed every day for God to find a way to bring you back to a right relationship with Him. I knew it had to be something dramatic. I know it's of God because, "I asked God for bread, and I know God did not give us a scorpion or a stone." [263]

---

262 Transcribed Paul Solomon Reading Excerpt, February 21, 1972

263 Matt 7: 9-10. "Or what man is there of you, whom if his son ask bread, will he give him a stone? Or if he asks for a fish, will he give him a serpent?"

398 | ARMAGEDDON STONES

Reverend Dove advised Paul go back to Atlanta and carefully pursue his ministry. He expressed his belief that every minister, when he stands before a congregation, represents himself as communicating with God and should have that ability.

Finally, he shared with Paul the meaning of the little reference at the end of tape. Twenty-five years earlier, Reverend Dove had a discussion with a friend about whether or not there was awareness after death. His father's response at the time was that he didn't know, but they both were intrigued by the possibility of life after death. The two of them made an agreement that whichever one passed on first, if it were possible, would attempt to make contact and communicate. So the message on the tape was the completion of their agreement that had been made over two decades before.

The encounter with "the Source" as he would later call it, was for Paul Solomon, a life-transforming awakening. His "Source readings" and later his lectures and classes became the catalyst for a world-wide ministry.

INDEX

www.ingramcontent.com/pod-product-compliance
Lightning Source LLC
Chambersburg PA
CBHW051411090426
42737CB00014B/2613